To Jean and Pat

CONTENTS

AUTHORS' PREFACE

This book contains nineteen programmes incorporating material which will be found in most undergraduate and HND syllabuses in Strength of Materials and Structural Analysis at first- and second-year level. The book is intended for use by students of Engineering and Building as a main text in Structural Mechanics as taught in most College and University departments of Engineering and Construction.

The aim of the text is to provide a sound understanding of fundamental principles of Structural Mechanics by presenting the principles and concepts involved in a distinctive programmed learning format whereby the student can work and learn at his/her own pace and test his/her understanding by answering a series of carefully constructed questions and graded practical problems.

Although it is assumed that students using the book will have some grasp of elementary mechanics, the first chapter provides useful revision of the basic principles of statics which provide the foundation material for the rest of the book. Subsequent chapters deal with concepts of structural form and types of loading. Techniques for analysing statically determinate pin jointed frame structures and simply supported and cantilevered beams are introduced. The drawing of shear force and bending moment diagrams for beams subjected to different types of loading is given comprehensive coverage.

A number of programmes cover stress analysis including analysis and design of structures subjected to direct, bending, shear or torsional stress. Problems involving combined bending and axial loading are also considered. The analysis of complex stress situations using Mohr's Circle of Stress forms the basis of one complete programme, as does the analysis of composite sections under different stress situations including thermal loading.

Later programmes show how the deflection of statically determinate beams may be calculated using integration techniques, and Strain Energy and Virtual Work methods are introduced as a foundation for further more advanced study.

The techniques of moment distribution and of slope-deflection are presented as important manual methods for the analysis of statically indeterminate structures; the latter being the basis of modern computer methods which are likely to be studied at a more advanced level. The effect of moving loads acting on structures is introduced through the use and application of influence lines and, in the penultimate programme, the theory of elastic buckling of slender members which are loaded in axial compression is explained and considered.

The final programme looks at the application of plastic methods to the analysis and design of steel framed structures and acts as an introduction to analysis techniques that consider the response of structures strained beyond the elastic range of material behaviour.

The book is suitable for general class use and for individual study supported by seminar and/or tutorial work. It is appreciated that the coverage of the text may be a little greater than that required for some courses and that the last few programmes will, in many courses, form part of second-year work. This is intentional as the earlier programmes are suitable for use by HND students or those undergraduates on Building Courses where the depth of study of Structural Mechanics may not be so great. Engineering undergraduates will find that most of the text will be required in both their first-year and second-year studies.

Throughout the book the emphasis is on student centred learning with the intention of providing the student with a sound grasp of the fundamental principles of Structural Mechanics. Although nowadays computers are frequently used to carry out the more routine and/or complex calculations associated with Structural Mechanics, at this level of study it is important that fundamental principles and concepts of structural behaviour are studied and understood. It is only with a sound grasp of basic theory that the student will be able to progress satisfactorily to the study of more advanced concepts of structural behaviour and competently to apply the theories learnt to the design of structures and structural elements.

We are indebted to those who have assisted us in writing this book including the Building and Civil Engineering students at Coventry University who co-operated with us in the testing of the material that forms the basis of this text. With their help we were able to refine the original draft to its present form. Thanks also to the final year Civil Engineering M.Eng. students at Coventry University for reading and commenting on parts of the text, and to various reviewers who made useful and constructive comments.

The first edition of the book has been widely adopted, over the last ten years, as a standard first-year text. This second edition has been extended by the addition of five additional programmes which, together with the original text, should now cover the major part of the Structural Mechanics/Analysis syllabuses for the first two years of most Civil Engineering degree courses. The opportunity has also been taken to make some very minor modifications to improve the original text. It is hoped that this newly presented text will continue to meet the needs of another generation of students.

RAY HULSE
JACK CAIN

HOW TO USE THIS BOOK

This book contains nineteen 'programmes'. Each programme is a self-contained unit of study which deals with one specific topic. However, as the later programmes build on material studied in earlier parts of the book, you are advised to work systematically through the text studying the programmes in the given sequence.

You *must* start at the beginning of and work sequentially through each programme. Every programme is subdivided into a number of short 'frames', each of which contains a limited quantity of information. The frames are designed to enable you to learn at your own pace, and most of them end with a short question or problems for you to tackle. These enable you to test your understanding of the material that you have just studied. The correct answer to each problem is given at the top of the next frame.

To use the book most effectively you should use a piece of paper or card to conceal the next frame until you have answered the given question. You can use the paper for rough working if necessary. Only when you have made a response should you look at the answer. If you have given an incorrect answer you should not proceed until you have found out why you made a mistake. Usually a worked solution or some further explanation will be given immediately below each answer so that you should be able to find where and why you went wrong. If you still cannot understand how to get the correct answer, you should make a note to discuss it with another student or with your tutor.

At the end of and at intermediate stages throughout each programme you will find sets of problems for you to tackle. You must attempt as many of these problems as possible. They are graded in difficulty and will give practice in applying the techniques that you have learnt.

The most important thing to remember is that you should work systematically, sequentially and carefully through the book at whatever pace of learning you find comfortable. Do not miss out any part of any programme, don't 'cheat' by looking at the answers to questions before offering your own solution and do not proceed at any stage until you are satisfied that you have grasped the information in the frame that you have just studied.

Programme 1

REVISION OF THE
FUNDAMENTALS
OF STATICS

On completion of this programme you will be able to:

- Recognise and explain the nature of force systems
- State and apply the conditions for equilibrium
- Combine force systems into force resultants and equilibrants using graphical methods
- Resolve forces into components
- Determine the resultant of force systems using analytical methods
- Calculate unknown forces in systems of forces in equilibrium
- Recognise and explain the nature of the moment of a force
- Calculate the moment of a force system about a point
- Determine the centroid of regular and irregular geometrical shapes

1

In this programme we will review the basic principles of Statics which will be needed in following programmes. You may have learnt these principles in your previous studies, in which case this will be useful revision. If you have not studied Statics before, do not worry; by working conscientiously through the following frames and exercises you will be able to master the subsequent programmes. You will need a scale rule and a protractor.

2

Statics is the study of bodies at rest. A body is a general word used to describe, for example, a building, a bridge or a component of a structure such as a beam, wall or foundation. All bodies are subject to the action of forces, such as their own self-weight, which are usually expressed in units of kiloNewtons (kN). It may be difficult to visualise a kN but it may be helpful to think of a small car as weighing approximately 10 kN. Self-weight forces act vertically downward but other forces, such as the effect of wind blowing on the side of a building may act in a horizontal direction.

If a wind force of 10 kN is acting on a body is it sufficient to say 'the wind force is 10 kN'?

3

> No—force is a *vector* quantity and must be specified
> both in magnitude and direction

We need to say, for example, that the wind force is 10 kN in a horizontal direction. We can indicate forces in various ways, such as:

(a) by the use of arrows on 'free body' diagrams (diagrams showing the geometrical arrangement of the structure and load system) such as shown in figure (a) below;
(b) by the use of *vectors* (see figure (b) below).

Vectors will be used in *vector diagrams* later in this programme and in Programme 3.

(a)   (b)

In figure (b) the vector **AB**, drawn parallel to the force denoted F_1 and to some convenient scale measuring 10 kN in length, represents the force F_1. Similarly vector **CD** drawn parallel to the force F_2 and to the same scale measuring 20 kN, represents the force F_2. Note that the direction in which the force acts is shown by the arrow head and by the way the vector is lettered. Thus **AB** implies the force acts in the direction A to B. If the force acted in the opposite direction, we would refer to the vector as vector **BA**.

It should be apparent to you that the combined effect of the two forces acting on the body shown in figure (a) will be to move the body downwards and to the right (if no other forces are acting on the body).

If there are only two forces acting on a body and the body does not move, can you suggest what conditions must be satisfied by those forces?

4

> (1) They must be equal in magnitude.
> (2) They must act in opposite directions.
> (3) They must be collinear (that is, act along the same straight line).

The forces F_1 and F_2 in figure (a) below satisfy these conditions. The body does not move and is said to be in *equilibrium* under the action of forces F_1 and F_2.

If these same forces were not collinear but acted as in figure (b) then the body on which they act would tend to rotate (in this case in a clockwise direction) and the body is not in equilibrium.

(a)   (b)

Now look at the diagram below showing a body acted upon by two forces.

Is the body in equilibrium?

5

$$\boxed{\text{NO}}$$

In fact, the body would move in the direction indicated in figure (a) below. It is as if the body was acted upon by a single force R as shown in figure (b). The force R which has the same effect upon the body as the original forces is described as the *resultant* of the original two forces. If we wished to prevent the body moving under the action of the forces F_1 and F_2, we would have to apply to the body a force equal in magnitude but opposite in direction to the resultant R as shown in figure (c). The force E, which when acting on the body in conjunction with F_1 and F_2 maintains the body in a state of equilibrium, is described as the *equilibrant* of the original two forces (F_1 and F_2). Note that in figures (a) and (c) all the forces pass through a single point.

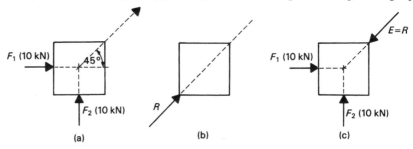

RESULTANTS OF FORCES

In our future work it will be necessary for us to determine the resultant (or equilibrant) of any number of forces acting at a single point. Such a point might be the centre of gravity of a solid body or a joint in a structural framework. We will now consider *graphical methods* of determining resultants and in later frames study *analytical methods*.

 The following diagrams illustrate how two, three and four forces may be combined graphically to give their respective resultants.

In each case, and preferably drawn on graph paper, line AB is drawn parallel to F_1 and to a convenient scale to represent the magnitude of force F_1. Line CD similarly represents F_2 and so on. *The vectors must be drawn such that the arrows showing the direction of force progress sequentially* (that is, they follow one another round the diagram). Thus if you put a pencil at A in vector diagram (b) and trace along the lines ABCDEF, your pencil moves in the direction of each force as it traces along the line representing that force. It is not however necessary to draw the vectors in any particular order. Thus in figure (b) the sequence is F_1, F_2, F_3, but in figure (c) it is F_1, F_3, F_2 then F_4.

In each case we have drawn an open-sided polygon. The length and direction of the open side gives us the magnitude and direction of the resultant of the force system. *The direction of action of the resultant is always from the start point to the end point of the polygon.* Thus, measuring from the scaled diagrams:

for figure (a) Resultant $=$ AD $= 14.1$ kN acting from A to D, $\theta = +315.0°$

for figure (b) Resultant $=$ AF $= 20.6$ kN acting from A to F, $\theta = \;\;\;+1.7°$

for figure (c) Resultant $=$ AH $= 17.5$ kN acting from A to H, $\theta = \;\;+13.6°$

Note that the angles θ have been measured from the X-axis assuming that they are positive if drawn in a clockwise sense.

Now try drawing the polygon of forces for figure (c) taking the forces in the order F_1, F_4, F_2, F_3. (Hint: use a scale of 10 mm $= 1$ kN)

6

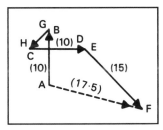

Now consider a number of forces acting at a point, P, as shown in figure (a) below. If we draw the force polygon in figure (b) we can determine the resultant by measuring line AF and, if on the free body diagram we draw a line through P parallel to AF, we have located the resultant in space. If the resultant, R, acted alone on the point P as in figure (c), it would have the same effect on P as the original forces.

What is the effect of the three forces acting on the point P?

7

> P would move in the direction of action of the resultant R.

EQUILIBRANTS OF FORCES

If we wished P to be in equilibrium, we would need to apply another force acting on P and equal in magnitude but opposite in direction to the resultant, R. Such a force would be the *equilibrant* of the original system of three forces. The equilibrant could be determined by measuring the length of line FA on the vector diagram. Figure (a) below shows the equilibrant plotted on the diagram. In this case the equilibrant and the original three forces have no net effect on P and the system is in *equilibrium*. When a system of forces acting at a point is in equilibrium, as in figure (a), then the force polygon for those forces is a closed polygon. The vectors representing the forces follow each other in sequence from the start point, around the polygon and back to the start point as in figure (b).

(a)

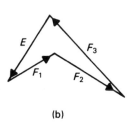

(b)

You should now be able to construct polygons of forces for any number of forces acting at a point and use the polygons to determine the resultant or equilibrant. Practise this by solving the following problems.

8

PROBLEMS

1. The following figures show four systems of forces. Determine the *resultant* of the systems in figures (a), (b) and (c) and determine the *equilibrant* for the system in figure (d). An A4 size sheet of paper is large enough and a scale of 1 cm to 5 kN big enough to give acceptable answers. You may find it helpful to use squared paper. The direction of the forces should be specified by quoting the value of the angle θ measured from the XX axis, clockwise being taken as positive. You will remember that this convention was used in Frame 5.

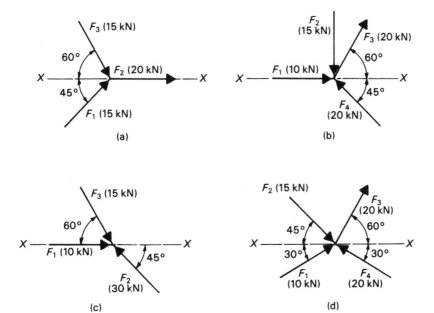

Ans. ((a) 38.2 kN, +3.6° (b) 17.5 kN, +289.6° (c) 9.0 kN, +245.7° (d) 24.8 kN, +118.8°). The angles are given as the angle measured clockwise from the X axis.

9

Now that you are familiar with the technique for constructing force polygons, we will simplify the lettering of the diagrams in the following way:

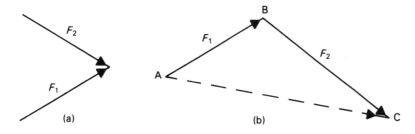

Figure (a) shows two forces acting at a point. Figure (b) is the force polygon (in this case a triangle) in which vector **AB** represents force F_1, vector **BC** represents force F_2 and the vector **AC** represents the resultant. You will see that we now have only one letter at each corner of the force polygon. This method of lettering is simpler and enables us to develop a very useful reference system when analysing the forces in the members of frameworks. You will learn about this system in Programme 3.

10

In Frame 7 you learnt that if a number of forces acting at a point are in equilibrium then they have no resultant and the vector diagram of the force system will consequently be a closed polygon. In the case of three forces, this fact is traditionally quoted as the principle of the *triangle of forces* which can be stated as: 'three coplanar forces in equilibrium acting at a point may be represented in magnitude and direction by the three sides of a triangle'. The term *coplanar* reminds us that we are considering forces acting in a single plane.

An equally useful related fact is that if three non-parallel forces are in equilibrium they must be *concurrent*; that is, they must all pass through a common point.

Let us now see how these two facts may be used to solve a problem. Figure (a) shows a ladder of weight W standing on a rough floor and resting against a smooth wall. Since the wall is smooth there is no friction and the force (R_1) exerted by the wall on the ladder is at right angles to the wall. The weight force W is, of course, vertical but we do not immediately know the direction of the force (R_2) exerted by the floor on the ladder.

The forces exerted by the wall and the floor are *reactions* and we shall be learning more about such forces in Programme 2.

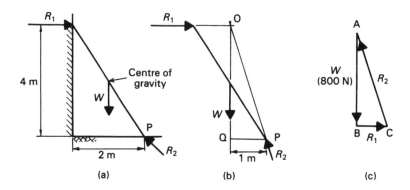

(a) (b) (c)

Figure (b) shows how we can use the fact that three forces in equilibrium must be concurrent to locate the line of action of the force at the foot of the ladder. We extend the lines of action of the forces R_1 and W to intersect at point O. The reaction at the foot of the ladder (P) must also pass through O and hence by joining P and O we obtain the line of action of R_2. It is then possible to construct the triangle of forces in figure (c).

Draw the triangle of forces and assuming that the weight of the ladder is 800 N determine:

(i) the magnitude of reaction R_1
(ii) the magnitude and direction of reaction R_2.

Assume that the centre of gravity of the ladder is half way along its length.

$$R_1 = 200 \text{ N}$$
$$R_2 = 825 \text{ N at } 14° \text{ to the vertical}$$

In the problem you have just done the triangle of forces is a right-angled triangle and you may have noticed that it is similar to the triangle OPQ in figure (b). The ratio $PQ/OQ = \frac{1}{4}$. It follows that the ratio R_1/W in the triangle of forces is also $\frac{1}{4}$ thus:

$$R_1 = \frac{1}{4} \times W$$
$$= \frac{1}{4} \times 800$$
$$= 200 \text{ N}$$

and making use of Pythagoras' theorem $R_2 = \left(800^2 + 200^2\right)^{1/2}$
$$= 824.62 \simeq 825 \text{ N}$$

You should always look for such geometrical relationships to aid calculations.

NON-CONCURRENT FORCES

You can now cope with any number of forces acting at a point but you may occasionally come across a system in which not all forces are concurrent. Such a system of forces would have to be dealt with as follows:

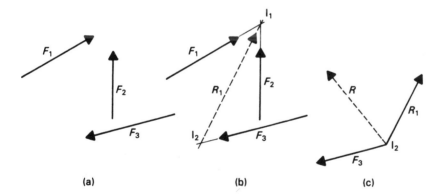

(a)   (b)   (c)

Figure (a) shows three non-concurrent forces. In figure (b) we have extended the line of action of forces F_1 and F_2 to locate their point of intersection at I_1. We can now combine F_1 and F_2 to give their resultant R_1 which will, of course, act through I_1.

In figure (c) we have replaced F_1 and F_2 by their resultant R_1. The problem is now reduced to a two force problem, F_3 and R_1, whose lines of action intersect at I_2. It is now a straightforward matter to determine the resultant R of R_1 and F_3. R is then the resultant of all the original forces. Complex systems may be solved in this way.

12

COMPONENTS OF FORCES

Since any number of forces can be combined into a single force, it conversely follows that a single force can be replaced by any number of forces whose resultant effect is the same as that of the original single force. We say that a single force may be *resolved* into *components*. In practice, it is sufficent for most analytical purposes to resolve a force into two components at right angles to each other. The directions taken for these components are normally, but not necessarily, vertical and horizontal.

Figure (a) shows a single force, F (= 20 kN), acting on a body. In figure (b) the vector AB represents F and lines AC and BC are drawn vertically downwards and horizontally to the left to complete a triangle of forces. If AC (length = 17.3 kN) is the vector for a force V, and CB (length = 10.0 kN) is the vector for a force H we see from figure (b) that F is the resultant of V and H, or conversely V and H are the vertical and horizontal components respectively of F. We can then replace force F by V and H as in figure (c). V and H acting together will have the same effect on the body as the original single force.

(a)　　　　　　　(b)　　　　　　　(c)

You will see from figure (b) above that by choosing to replace F by two components at right angles (V and H) we have drawn a right-angled vector diagram ABC from which $AC = AB\sin 60°$ ($V = F\sin 60°$) and $BC = AB\cos 60°$ ($H = F\cos 60°$). We can express this in more general terms as follows:

Given a force F acting at an angle θ to the horizontal;
Vertical component　　　　$V = F\sin\theta$
Horizontal component　　$H = F\cos\theta$
You should relate these equations to figure (d) below.

When resolving a force into components you must be careful to identify the direction in which each component acts. It is common practice to take forces acting to the right as positive and forces acting upwards as positive. Thus in the above example of the 20 kN force, the components are $V = -17.3$ kN and $H = -10$ kN.

Now determine the horizontal and vertical components of the forces shown below in figures (e), (f) and (g).

(d)　　　　　　(e)　　　　　　(f)　　　　　　(g)

> (e) $V = +6.84$ kN: $H = +18.79$ kN
> (f) $V = +21.65$ kN: $H = -12.50$ kN
> (g) $V = -21.21$ kN: $H = +21.21$ kN

ANALYTICAL DETERMINATION OF THE RESULTANT OF A FORCE SYSTEM

Since the effect of a resultant is the same as the total effect of the system of forces of which it is the resultant, it follows that the vertical component of the resultant must be equal to the sum of the vertical components of all the original forces in the system. Similarly for the horizontal component. This fact provides an *analytical method* for determining the resultant of any system of forces. The calculations are best done in tabular form as in the following example:

Consider the system of forces shown in figure (a). If forces acting to the right are taken as positive and forces acting upwards as positive then:

For F_1 $V = 10\sin 30° = +5.00$ kN: $H = 10\cos 30° = +8.66$ kN

These components are tabulated below together with the corresponding components of F_2, F_3 and F_4:

Force	Value	θ	Components V	H
F_1	10	30	+5.00	+8.66
F_2	15	45	−10.61	+10.61
F_3	20	60	+17.32	+10.00
F_4	20	30	+10.00	−17.32
totals			+21.71	+11.95

Thus: the vertical component of the resultant $= +21.71$ kN
the horizontal component of the resultant $= +11.95$ kN

Then as shown in figure (b) the components can be combined to give the resultant

$$(R) = (21.71^2 + 11.95^2)^{1/2} = \underline{24.78 \text{ kN}}$$

and the direction of the resultant is given by θ where:

$$\tan\theta = 21.71/11.95 \quad \text{or } \underline{\theta = 61.2°} \text{ (see figure (b))}$$

(a)   (b)   (c)

Now calculate the magnitude and direction of the resultant of the forces shown in figure (c). Present your calculations in tabular form as shown above.

14

> Resultant $= 2.84$ kN
> Direction $\theta = +260.3°$

As in previous examples, θ has been quoted as the angle measured from the X axis in a positive clockwise direction. Did you sketch a triangle of forces, as in figure (b) in Frame 13, to help you determine the correct direction of the resultant force?

CALCULATION OF UNKNOWN FORCES IN A SYSTEM OF FORCES IN EQUILIBRIUM

Now suppose that we had to analyse the problems shown in figures (a) and (c) to determine the unknown forces and angle (θ). The two systems of forces are shown acting on a point P and both systems are in equilibrium. The solutions could be obtained graphically by constructing triangles of forces, and figure (b) shows how this is done for problem (a). Vector AB is drawn representing force F_3. A line AC_1 is then drawn from A parallel to F_1 and a line BC_2 from B parallel to F_2. The intersection of AC_1 and BC_2 determines the position of C and completes the triangle of forces. Vector CA then gives the magnitude and direction of force F_1 and vector BC gives the corresponding values of F_2. Note that the arrows follow in sequence around the diagram.

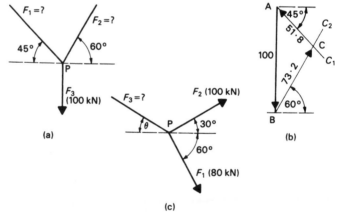

Construct the triangle of forces for the problem shown in figure (c).

15

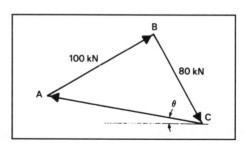

In figure (c) we are given the magnitudes and directions of two forces so the construction of the triangle of forces is straightforward. You should have drawn vector **AB** representing 100 kN in the direction of, and parallel to, the line of action of force F_2. Then from point B you should have drawn **BC** representing 80 kN and parallel to F_1. Joining C to A completes the diagram and vector **CA** then gives the magnitude and direction of the third force.

*What is the difference between vector **CA** and vector **AC**?*

16

> The two vectors represent forces in different directions.
> Vector **AC** represents the resultant of F_1 and F_2.
> **CA** is the equilibrant.

This type of problem can also be solved analytically using the method of resolution into components since **the algebraic sum of all components in any direction must be zero if the original forces are in equilibrium**. Hence the algebraic sum of all the vertical components must be zero otherwise the point P would move vertically. Similarly the algebraic sum of the horizontal components must be zero.

Why must the horizontal components sum to zero?

17

> Otherwise P would move horizontally

Let us apply the method of resolution to figure (a) in Frame 14. Note that the signs used for the components of F_1 and F_2 are based on the assumption that these forces act in a direction away from point P. If we obtain a negative answer for either F_1 or F_2 we will know that that force acts in the opposite direction towards P.

Force	Vertical component	Horizontal component
F_1	$+ F_1 \times \sin 45° = +0.707F_1$	$- F_1 \times \cos 45° = -0.707F_1$
F_2	$+ F_2 \times \sin 60° = +0.866F_2$	$+ F_2 \times \cos 60° = +0.500F_2$
F_3	$= -100$	$= 0.0$

sum of vertical components $\quad 0.707F_1 + 0.866F_2 - 100 = 0 \qquad$ (1.1)

sum of horizontal components $\quad -0.707F_1 + 0.500F_2 \quad\quad = 0 \qquad$ (1.2)

Now solve these simultaneous equations for F_1 and F_2.

18

$$F_1 = 51.77 \text{ kN}$$
$$F_2 = 73.21 \text{ kN}$$

If you had difficulty solving the equations then proceed as follows:

Equation (1.1) was	$0.707F_1 + 0.866F_2 - 100 = 0$
(1.2) was	$-0.707F_1 + 0.500F_2 \quad\quad = 0$
adding these equations we get	$1.366F_2 - 100 = 0$
from which	$F_2 = 100/1.366$
	$= \underline{73.21 \text{ kN}}$

then substituting this value in equation (1.2):

$$-0.707F_1 + 0.5 \times 73.2 = 0$$
$$F_1 = 36.6/0.707$$
$$= \underline{51.77 \text{ kN}}$$

Use the method of resolution to determine the values of the unknown force and angle (θ) in the following problem. This is the same problem as given in figure (c) in Frame 14. Start by assuming that force F_3 acts away from P.

19

$$F_3 = 128.53 \text{ kN}: \theta = 8.64°$$

If you did not get this answer check your solution against the following:

Force	Vertical component	Horizontal component
F_1	$-80 \sin 60° = -69.28$	$+80 \cos 60° = +40.00$
F_2	$+100 \sin 30° = +50.00$	$+100 \cos 30° = +86.60$
F_3	$= +F_3 \sin \theta$	$= -F_3 \cos \theta$

Sum of vertical components $\quad -69.28 + 50.00 + F_3 \sin \theta = 0$

$\therefore \quad F_3 \sin \theta = +19.28$ (1.3)

Sum of horizontal components $\quad +40.00 + 86.60 - F_3 \cos \theta = 0$

$\therefore \quad F_3 \cos \theta = +126.60$ (1.4)

Then dividing equation(1.3) by (1.4) we get $\tan\theta = +0.152$

$$\therefore \quad \theta = \underline{8.64^\circ}$$

Note that this angle is as shown in the figure and is not measured from the positive direction of the X-axis.

Substituting this value of θ in equation (1.3): $\quad F_3 \sin 8.64^\circ = +19.28$

$$\therefore \quad F_3 \times 0.15 = +19.28$$
$$\therefore \quad F_3 = \underline{128.53 \text{ kN}}$$

20

MOMENTS OF FORCES

In Frame 4 you saw how a force could rotate or tend to rotate a body. Consider the figure where a vertical force F is shown acting on one end of a bar with a hole at the other end loosely fitting over a shaft or hinge at O. The bar is thus free to rotate about O.

Force F would, if not resisted, push the whole bar including the hinge O downwards, but as O is fixed in space, being supported on a framework sitting on the floor, then the support would provide an upward resisting reactive force (R). O would thus not move vertically. You should appreciate however that although R is preventing O from moving vertically, there is nothing to prevent the bar rotating clockwise about the hinge. The arrangement shown is in fact a *mechanism* and not a *structure*.

We see that this force, F, tends to turn the body on which it acts and we measure the turning effect by saying that the force has a turning moment or *moment* equal to $F \times x$ where x is the horizontal distance from the hinge to the line of action of the force F.

This simple equation indicates that the turning effect is increased if the magnitude of the force increases and/or if the distance x increases.

Now to the next frame:

21

In the previous frame the concept of moment of force was explained. A more general definition would be:

> The *moment* of a force about *any point* equals the value of the force multiplied by the *perpendicular* distance between that point and the *line of action* of the force

Calculate the moment of the force about O in each of the following figures. Qualify each answer by stating the direction in which the force is tending to rotate the body – that is, clockwise or anticlockwise.

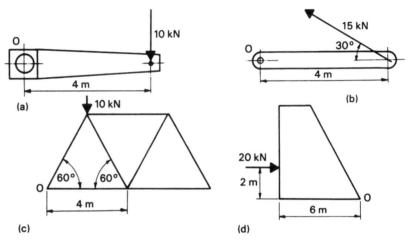

22

(a)	$10 \times 4 = 40$ kN m clockwise
(b)	$15 \times 2 = 30$ kN m anticlockwise
(c)	$10 \times 2 = 20$ kN m clockwise
(d)	$20 \times 2 = 40$ kN m clockwise

In (b) did you calculate the perpendicular distance from O to the line of action of the force?

RESULTANT OF A SERIES OF PARALLEL FORCES

The concept of a moment of force may be used to locate the line of action of the resultant of a number of parallel forces. The three forces F_1, F_2 and F_3 acting on the beam shown in the following figure could be replaced by a single force R of magnitude $F_1 + F_2 + F_3$ but we do not immediately know its line of action. We cannot use vector diagrams as in previous problems because the forces are not concurrent. What we do know is that the turning effect of the resultant must be the same as the total turning effect of the original forces. We can use this fact to calculate the line of action of the resultant by taking (calculating) moments about an appropriate point.

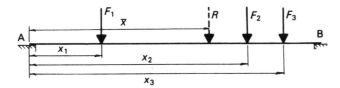

Let's take moments about A:

If the resultant R acts at distance $\bar{x}$ from A as shown then:

The moment of R about A = sum of moments of original forces about A

$$R \times \bar{x} = F_1 \times x_1 + F_2 \times x_2 + F_3 \times x_3$$

from which we can calculate $\bar{x}$.

Note that it is not essential to take moments about A. You can in fact choose any convenient point. You will discover in future work that we often choose to take moments about a point on the line of action of one of the forces in a force system because by so doing the moment produced by that force is zero and it disappears from the calculation.

Calculate the resultant of the system of parallel forces shown in the figure below. Determine the line of action of the resultant by taking moments (a) about A and (b) about C (both methods should give the same answer).

23

> Resultant = 37 kN
>
> Line of action is 6.89m to the right of A

You should have calculated R equal to $10 + 15 + 12 = 37$ kN then:

(a) by moments about A: $37 \times \bar{x} = (10 \times 3) + (15 \times 7) + (12 \times 10)$ where $\bar{x}$ is measured from A

thus
$$\bar{x} = 6.89 \text{ m}$$

(b) by moments about C: $37 \times \bar{x}_1 = -(10 \times 4) + (15 \times 0) + (12 \times 3) = -4$ kN m where $\bar{x}_1$ is measured from C

thus
$$\bar{x}_1 = -0.11 \text{ m (that is, 0.11m to the left of C)}.$$

Note the use of signs. Moments tending to rotate the system clockwise about A in case (a) and about C in case (b) are taken as positive. Anticlockwise moments are taken as negative. Thus in case (b) the moment of the 10 kN force is anticlockwise about C. The total moment about C is also negative thus the resultant must act to the left of C. Also note that although the 15 kN force has zero moment about C, it still makes a contribution of 15 kN towards the total magnitude of the resultant force.

24

CENTROIDS OF SHAPES

A body or structure usually consists of separate individual parts each subject to its own weight force. These forces are vertical and parallel to each other and, as in the previous frames, may be replaced by a single vertical resultant of magnitude equal to the total weight of the body and acting at the *centre of gravity* of the body. The position of the centre of gravity of a body, such as the machine part shown below, may consequently be determined by taking moments.

Side view   End view

Disc A has a weight of 10 N acting through its own centre of gravity, G_A. Shaft B has a weight of 25 N acting at G_B. If we take moments about the end O of the shaft then the distance $\bar{x}$ to the centre of gravity of the whole body from O will be given by:

$$\text{Total weight} \times \bar{x} = \text{the sum of the moments of each part about O}$$
$$= (10 \times 12.5) + 25 \times (250 + 25)$$

thus
$$\bar{x} = (125 + 6875)/(10 + 25)$$
$$= \underline{200 \text{ mm}}$$

Suppose we now wish to locate the centre of gravity (G) of an irregular shaped plate of material of uniform thickness and uniform weight w N/mm^2 as shown in the figure below. Consider the plate laying horizontally. The plan view is as shown with the weight forces acting vertically downwards at right angles to the page.

X and Y are axes in the horizontal plane and intersect at a convenient origin O. The small element shown has an area δA and is located at a distance x from the Y-axis and distance y from the X-axis.

What is the moment of the weight force of the small element about the Y-axis?

$$w\delta A \times x \text{ (or } wx\delta A)$$

The distance from the Y-axis to G will be given by:

Total weight $\times \bar{x} = \Sigma wx\delta A$

where Σ is the usual mathematical symbol to denote 'the sum of'

but total weight = sum of the weight of the small elements $= \Sigma w\delta A$

thus $\qquad \Sigma w\delta A \times \bar{x} = \Sigma wx\delta A$

or $\qquad \Sigma \delta A \times \bar{x} = \Sigma x\delta A$ (since w is uniform)

but $\Sigma \delta A = A$ (the total surface area)

thus $\qquad \underline{A\bar{x} = \Sigma x\delta A}$

In the case of a plate of *uniform thickness*, then the position of G is independent of the weight per unit area and is purely a function of the shape of the plate. The expression $A\bar{x} = \Sigma x\delta A$ may be solved for any shape and defines the position of the centre of area or *centroid* of that shape with respect to the Y-axis. Similarly, solving the expression $A\bar{y} = \Sigma y\delta A$ will locate the centroid with respect to the X-axis.

The concept of the centroid is important and will feature in future programmes, so let's consider a few common shapes. It should be obvious that for a square, rectangle and circle, the centroid will be central as shown below:

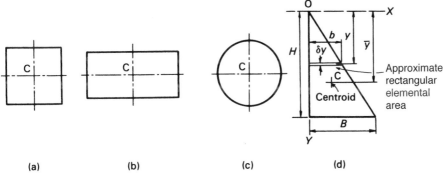

(a)   (b)   (c)   (d)

Now look at the triangle shown in figure (d) above. Taking moments of area of the approximately rectangular shaped elements about an X-axis passing through the apex, then

$$A \times \bar{y} = \Sigma y\delta A$$

that is $\qquad \frac{1}{2}BH \times \bar{y} = \Sigma by\delta y \qquad$ since $\delta A = b\delta y$

$$= \Sigma (By/H)y\delta y \quad \text{since } b/B = y/H$$

$$= \frac{B}{H}\int_0^H y^2 \mathrm{d}y = \frac{B}{H}\left[\frac{y^3}{3}\right]_0^H = \frac{BH^2}{3}$$

thus $\qquad \underline{\bar{y} = \tfrac{2}{3}H \text{ from apex (or } \tfrac{1}{3}H \text{ above base)}}$

What is the distance of the centroid, C, from the Y-axis?

26

$$\boxed{\tfrac{1}{3}B}$$

To locate the centroid for a complex shape, you will need to divide it into simple basic shapes for which you know the position of the centroid. You can then apply the principle that *the total area multiplied by the distance from the centroid to a convenient axis is equal to the algebraic sum of the moments of area of each part about the same axis.*

Thus for the complex shape shown below, divide the area into two rectangles (A and C) and a triangle (B) as indicated. Then taking the corner (O) of the area as the origin and the X and Y axes as shown:

for part A $\qquad\qquad\qquad\qquad\qquad$ area $= 100 \times 50 \qquad = 5000$ mm^2

$\qquad$ distance y of centroid C_A of A from the X-axis $= 50 + 100/2 \quad = 100$ mm

$$\therefore \quad \text{area} \times y = 500\,000 \text{ mm}^3$$

for part B $\qquad\qquad\qquad\qquad\qquad$ area $= \tfrac{1}{2} \times 100 \times 50 = 2500$ mm^2

$\qquad$ distance y of centroid C_B of B from the X-axis $\quad = 50 + 100/3 \quad = 83.33$ mm

$$\therefore \quad \text{area} \times y = 208\,333 \text{ mm}^3$$

for part C $\qquad\qquad\qquad\qquad\qquad$ area $= 100 \times 50 \qquad = 5000$ mm^2

$\qquad$ distance y of centroid C_C of C from the X-axis $= \tfrac{1}{2} \times 50 \qquad = 25$mm

$$\therefore \quad \text{area} \times y = 125\,000 \text{ mm}^3$$

Total moment of area of all parts about the X-axis:

$$= 500\,000 + 208\,333 + 125\,000$$
$$= 833\,333 \text{ mm}^3$$

Total area $= 5000 + 2500 + 5000 = 12\,500$ mm^2

thus $\qquad\qquad\qquad \bar{y} = 833\,333/12\,500 = \underline{66.67 \text{ mm}}$

Now calculate $\bar{x}$ by taking moments about the Y-axis.

$$\boxed{43.33 \text{ mm}}$$

The calculation of $\bar{x}$ together with the calculation for $\bar{y}$ is tabulated below. Tabular methods provide a concise way of setting out this form of calculation and you should adopt them wherever possible.

Part	Area (A)	x	Ax	y	Ay
A	5 000	25.00	125 000	100.00	500 000
B	2 500	66.67	166 675	83.33	208 333
C	5 000	50.00	250 000	25.00	125 000
total	12 500		541 675		833 333

$$\bar{x} = \frac{541\,675}{12\,500} \qquad\qquad \bar{y} = \frac{833\,333}{12\,500}$$
$$= \underline{43.33 \text{ mm}} \qquad\qquad\quad = \underline{66.67 \text{ mm}}$$

28

In the previous two frames we had to calculate both $\bar{x}$ and $\bar{y}$ in order to locate the centroid because the shape of the figure was not symmetrical. In many instances, however, when you are determining the positions of centroids of cross-sectional areas of structural members you will find that the area has at least one axis of symmetry. If there is an axis of symmetry, the centroid will lie on the axis of symmetry and it will only be necessary to locate the position of the centroid on that axis.

Let's do this for the T section shown. The centroid will lie somewhere along the Y-axis since this is an axis of symmetry. Then taking the X-axis through the base:

Part	Area (A)	y	Ay
A	500	55	27 500
B	500	25	12 500
total	1000		40 000

thus $\bar{y} = 40\,000/1000 = 40$ mm

Now to the next frame.

STRUCTURAL MECHANICS

29

To complete this programme we will see how to locate the centroid of a shape with one boundary formed by a parabola, since such shapes will be encountered in later work.

The curve in the figure has an origin at O and the dimensions x and y are measured from axes passing through O. The curve has an equation of the form:

$$y = ax^2$$

Consider area S and an elemental strip of that area of height y and width δx

then the area of the strip $= y\delta x$

and the total area

$$A = \int_0^B y \, dx$$

but as $y = ax^2$ then:

$$A = \int_0^B ax^2 \, dx$$

$$= a\left[\frac{x^3}{3}\right]_0^B = \tfrac{1}{3}aB^3$$

$$= \tfrac{1}{3} \times B \times H \text{ since } H = aB^2$$

$$= \tfrac{1}{3} \times \text{base} \times \text{height}$$

What is the area of the upper part (T) of the diagram?

30

> $$\tfrac{2}{3}B \times H$$
>
> since area S + area T = area of enclosing rectangle = $B \times H$.

To locate the centroid of S relative to the Y=axis we can take moments of area about the Y-axis:

moment of the area of the strip about the Y-axis $= y\delta x \times x$

moment of total area of S about the Y-axis $= \int_0^B yx \, dx$

$$= \int_0^B ax^3 \, dx$$

$$= \left[\frac{ax^4}{4}\right]_0^B = \tfrac{1}{4}aB^4$$

thus

$$\text{area } A \times \bar{x} = \tfrac{1}{4}aB^4$$

$$\tfrac{1}{3}B \times H \times \bar{x} = \tfrac{1}{4}aB^2 \times B^2 = \tfrac{1}{4}H \times B^2$$

thus

$$\bar{x} = \tfrac{3}{4}B$$

Calculate the value of $\bar{x}$ for the upper part (T) of the diagram.

22

$$\boxed{\bar{x} = \tfrac{3}{8}B}$$

Check your working against the following.

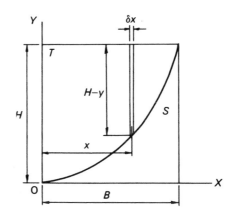

$$\tfrac{2}{3}(B \times H) \times \bar{x} = \int_0^B (H - y)x\,\mathrm{d}x$$

$$= \int_0^B Hx\,\mathrm{d}x - \int_0^B yx\,\mathrm{d}x$$

$$= \int_0^B Hx\,\mathrm{d}x - \int_0^B ax^3\,\mathrm{d}x$$

$$= \left[\frac{Hx^2}{2} - \frac{ax^4}{4}\right]_0^B$$

$$= \tfrac{1}{2}H \times B^2 - \tfrac{1}{4}aB^2 \times B^2$$

$$= \tfrac{1}{2}H \times B^2 - \tfrac{1}{4}H \times B^2$$

$$= \tfrac{1}{4}H \times B^2$$

thus $\qquad \bar{x} = \tfrac{3}{8}B$

How could you check the accuracy of these last two results?

32

> $A\bar{x}$ for the whole rectangle (of area $= B \times H$) must equal the sum of the $A\bar{x}$ values for the two parts (S and T).

For the whole rectangle, area $= B \times H$ and $\bar{x} = \tfrac{1}{2}B$

thus $\qquad A\bar{x} = \tfrac{1}{2}B^2 \times H$

for part area S $\qquad A\bar{x} = \left(\tfrac{1}{3}B \times H\right)\left(\tfrac{3}{4}B\right) = \tfrac{1}{4}B^2 \times H$

for part area T $\qquad A\bar{x} = \left(\tfrac{2}{3}B \times H\right)\left(\tfrac{3}{8}B\right) = \tfrac{1}{4}B^2 \times H$

for whole area $(S + T)$ $\qquad A\bar{x} = \tfrac{1}{4}B^2 \times H + \tfrac{1}{4}B^2 \times H$

$$= \tfrac{1}{2}B^2 \times H$$

The two results agree, so the calculated values of $\bar{x}$ are correct.

STRUCTURAL MECHANICS

33

| TO REMEMBER |

Force is a vector quantity: magnitude and direction must be quoted (units = N or kN).

The resultant and equilibrant of any system of coplanar concurrent forces may be determined by drawing a polygon of forces or by resolving the forces into components and combining the components algebraically.

If three coplanar concurrent forces are in equilibrium, they may be represented in magnitude and direction by the three sides of a triangle.

The moment of a force about any point equals the magnitude of that force multiplied by the perpendicular distance between the line of action of that force and the point (units = kN m).

It is important to distinguish between clockwise and anticlockwise moments.

The resultant and equilibrant of a system of parallel forces may be determined by taking moments about *any* convenient point.

Centroids of areas may be located using $A\bar{x} = \Sigma x\delta A$ and/or $A\bar{y} = \Sigma y\delta A$.

The centroid of a symmetrical shape lies on the axis (axes) of symmetry.

The centroid of a triangular shape is $\frac{1}{3}$ of the height above the base.

34

| FURTHER PROBLEMS |

1. Determine the magnitude and line of action of the resultants for the systems of forces shown in figures Q1(a), Q1(b) and Q1(c).

Ans. ((a) 20.86 kN, $\theta = +90.3°$ (b) 148.36 kN, $\theta = +289.9°$ (c) 71.78 kN, $\theta = +16.0°$: positive values of θ are measured in a clockwise direction from the XX axis)

2. Figures Q2(a), Q2(b) and Q2(c) show force systems in equilibrium. Determine the magnitude and direction of the unknown forces.

Ans. ((a) $F_1 = 77.65$ kN, $F_2 = 109.81$ kN (b) 50.00 kN, $\theta = 120°$ (c) -111.80 kN, $\theta = 26.6°$)

Problems Q1(a) to Q2(c) should be done both graphically and analytically.

Q2 (a)　　　　Q2 (b)　　　　Q2 (c)

3. Calculate the magnitude and locate the line of action of the resultants of the systems of forces shown in figures Q3(a) and Q3(b).

Ans. ((a) 45 kN, 5.09 m to the right of A (b) 85 kN, 7.29 m above O)

Q3 (a)　　　　Q3 (b)

4. Determine the position of the centroids (C) of areas of the figures Q4(a) and Q4(b).

Ans. ((a) C is on the Y-axis and 78.00 mm above the base (b) $\bar{x} = 28.59$ mm, $\bar{y} = 75.41$ mm from the bottom left-hand corner)

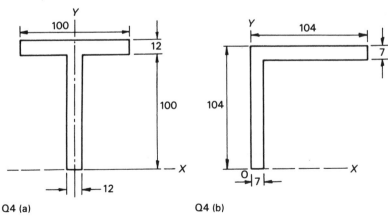

Q4 (a)　　　　Q4 (b)

5. Determine the position of the centroids (*C*) of areas of the figures Q5(a) and Q5(b).

Ans. ((a) C is on the Y-axis and 32.81 mm above the base (b) C is on the Y-axis and 150.00 mm above the base)

Q5 (b)

Q5 (a)

Programme 2

SIMPLE STRUCTURES
(LOADS AND REACTIONS)

On completion of this programme you will be able to:

- Recognise different forms of structural systems
- Identify different forms of structural loading
- Identify different forms of support conditions
- Calculate the magnitude and direction of components of reaction
- Draw free-body diagrams for different forms of structures subject to different loading and support conditions
- Recognise and apply the conditions of equilibrium
- Identify the conditions for external statical determinacy
- Calculate the reactive forces in structures containing internal pins
- Determine the support reactions for three-pinned arch structures
- Determine factors of safety for overturning and sliding for mass structures
- Calculate the components of reaction in three-dimensional space structures

1

In this programme we will look at a number of simple structures. A structure can be defined as an assembly of structural components so arranged that loads can be supported without failure. Consequently we must study the various loads to which a structure may be subjected and the way in which the structure reacts to those loads. We must also understand what is meant by failure. Failure may imply a complete collapse. Collapse may be due to incorrect geometrical arrangement of the components comprising the structure or to a failure of one of the components because of an overloading of that component.

Excessive movement may also be considered to be a failure. You should appreciate that all structures will move or deflect to a certain extent and maximum acceptable limits for deflection are normally specified for a structure at the design stage.

It may be possible, however, to load a structure to such an extent that it moves as a whole, the movement continuing until the loading is changed. Such a situation indicates a lack of stability—the structure is not in equilibrium. This last aspect of failure will be studied in this programme. In subsequent programmes the behaviour of the components of a structure and the estimation of deflections will be studied.

2

STRUCTURAL FORM AND IDEALISATION

Look at the structural systems shown in the following diagrams. You will see a variety of types ranging from a simple beam to three-dimensional space frames. Note how some of the simple types occur as parts of the more complex structures.

(a) Column   (b) Beam   (c) Slab

Two facts emerge from a study of the diagrams:

(1) A part of a complex structure can be considered as a structure in its own right and analysed as such. These parts are referred to as structural elements. Thus you can see columns, beams and slabs as structural elements of the space frame in figure (g).

(2) Although all structures are strictly three-dimensional bodies, many may be satisfactorily analysed as a two-dimensional problem. The following diagrams show the simple beam (figure (b)) and the crane structure (figure (f)) simplified for solution as two-dimensional problems.

Can you draw diagrams to show how the space frames (d) and (g) might be analysed as two-dimensional problems?

3

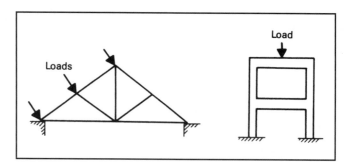

The roof structure could initially be analysed as a series of roof trusses. These are typical examples of *plane frames*, so termed because they are drawn in a two-dimensional plane. Similarly, structure (g) could be analysed as a series of plane frames.

STRUCTURAL LOADING

Now let's think about the forces which might act on our structures. The beam may be supporting the weight of brickwork in a wall above a door opening and the roof truss will be supporting the weight of the roof covering. These forces will be vertical and always present. They do not move and are consequently called *dead loads*. There will also be loads not always present such as those due to wind and snow deposits acting on a roof and the load caused by maintenance gangs of men working on the roof. Since these loads are not always present but vary in intensity they are known as *live loads* or *imposed loads*.

What would be the main source of live load on the deck of a bridge carrying a railway over a river?

4

The weight of
the trains

Wind loading can be a very important factor in design. It is, however, sufficient at the moment for you to understand that wind forces are normally assumed to act at right angles to the surface on which the wind is blowing. You should also appreciate that wind forces vary in direction; they may be positive (pressing down on a roof) or negative (tending to lift the covering off a roof).

Can you think of any loads acting on a structure which we have not yet mentioned?

5

┌─────────────────────────────┐
│   The self-weight (dead load) │
│   of the structure   │
└─────────────────────────────┘

This is one of several possible answers. Other loading could be due to, for example, earth or hydrostatic pressures. There is also a form of loading induced by rotating machinery and which varies continuously and regularly in intensity and may lead to unacceptable vibrations of the structure. Such loading, as in the case of wind loading which may also set up vibration, is beyond the scope of this book.

In the following frames we will consider structures with stated loads. You should realise, however, that in practice structures are subject to a number of different load conditions. For example, a roof truss may be subject at different times to dead loads only, dead load plus wind load only or dead load plus wind load and snow load. Consequently several calculations may have to be made when determining the overall effect of loads on the structure.

The loading on a structure will be assessed at the design stage by the Engineer who may refer to various British Standards or Codes of Practice.

Now let's consider the way in which loads may be applied.

6

CONCENTRATED LOADS

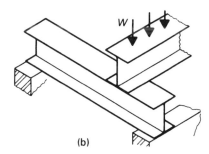

(a)   (b)

Loads are applied over a finite area but if that area is small compared with the dimensions of the structure, it is normally sufficiently accurate to consider the load as being concentrated at a point. For example, if a load is applied to a beam by a wheel (figure (a)) the full force W is transmitted through the contact surface which is a line equal in length to the width of the wheel. In elevation the line of contact appears as a single point and the load is concentrated at that point.

Figure (b) shows one steel beam bearing on top of another. Can you suggest where the load carried by the upper beam should be considered to be transmitted to the lower beam?

7

> Through the intersection in plan of
> the centre lines of the two beams

The contact surface between the two beams is rectangular in plan as shown in figure (b) below. For calculation purposes, however, this load would normally be treated as a single concentrated load acting at the intersection of the centre lines of the two beams. The elevation of the lower beam (figure (c)) shows this single concentrated load.

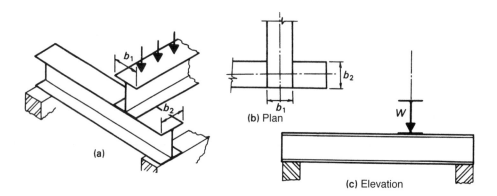

(b) Plan

(a)

(c) Elevation

When a structure is acted on by a number of concentrated loads it is often convenient for calculation purposes to assume them replaced by a single resultant.

The roof truss shown below is subject to the vertical dead loads and inclined wind loads as indicated.

Determine
(i) the resultant of the wind loads and locate the line of action of the resultant by quoting its perpendicular distance from A;
(ii) the resultant of the dead loads and locate the line of action by quoting its perpendicular distance from A.

(i)	80 kN	2.5 m from A
(ii)	160 kN	6.5 m from A

These results are obtained as follows:

(i) For the wind loads (remember that these forces act at right angles to the surface of the roof covering and are thus at right angles to the rafter AB):

Magnitude of resultant $= 20 + 40 + 20 = 80$ kN

Then taking moments about A

$80 \times p_W =$ the algebraic sum of the moments about A of all the wind forces

where p_W is the perpendicular distance from A to the line of action of the resultant

$\therefore \quad 80 \times p_W = (20 \times 0) + (40 \times 2.5) + (20 \times 5)$

$\therefore \qquad p_W = 200/80 = 2.5$ m

Since the wind loading is symmetrical, you should be able to arrive at this result by a simple inspection of the diagram and without calculation.

(ii) For the dead loads (these are weight forces and are thus vertical):

Magnitude of resultant $= 10 + 20 + 40 + 60 + 30 = 160$ kN

Then by moments about A

$160 \times p_D = (10 \times 0) + (20 \times 2) + (40 \times 4) + (60 \times 8) + (30 \times 12)$

$\qquad p_D = 1040/160$

$\qquad\quad = 6.5$ m

In subsequent analysis involving the entire structure, the truss may be considered as being loaded as in the figure below where the actual loads have been replaced by their resultants. You will appreciate that this will simplify the calculations. Remember, however, that these structures only represent the resultant effect of the actual loads on the structure and are not themselves actual loads. Thus there is no actual load at X although X is on the line of action of the resultant.

In the next frame we will look at a different type of loading.

9

DISTRIBUTED LOADS

Figure (a) shows a typical arrangement of a timber floor of planks supported by timber beams at 0.5 m centres. It is clear that beam A carries the weight of an area of flooring equal to 3 m × 0.5 m = 1.5 m².

(a)

(b)

(c)

If the timber weighs 4000 N/m³, what is the total load on beam A?

10

$$\boxed{345 \text{ N}}$$

$$
\begin{aligned}
\text{Weight of flooring} &= & (3 \times 0.5 \times 40/1000) \times 4000 &= 240 \text{ N} \\
\text{Self-weight of beam} &= (3 \times 50/1000 \times 175/1000) \times 4000 &= \underline{105 \text{ N}} \\
& & \text{total} & \quad 345 \text{ N}
\end{aligned}
$$

Since this floor is of uniform thickness, the load is *Uniformly Distributed* (UD) along the entire length of the beam. The intensity of load will be constant along the beam and equal to the total load divided by the span of the beam. Thus the intensity of load= 345/3 = 115 N/m.

This is illustrated in figure (b) by the series of arrows, while alternatively figure (c) is a load distribution diagram in which the height of the diagram represents 115 N/m.

For some calculation purposes, the uniformly distributed load may be considered replaced by its resultant (total load *W*) acting through the centroid of the load distribution diagram. Thus, in this case the centroid is half way along the diagram and the moment about any point of a load *W* acting at mid span is the same as the moment of the original loading about that point.

What is the moment of the original UD loading about point O?

$$\boxed{518 \text{ N m}}$$

$$345 \times \tfrac{3}{2} = 518 \text{ N m}$$

We may have a distributed load which varies in intensity. For example, on beam AB we see a load varying from $w =$ zero at A to $w = 100$ N/m at B. Note that we are using the symbol w for the value of load per unit length (N/m or kN/m), and W for the value of a total load (N or kN).

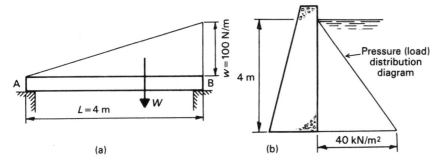

(a)   (b)

The triangular diagram above the beam is the load distribution diagram.

Total load W on the beam = average intensity of load × length of beam

$$= (w/2) \times L$$
$$= (100/2) \times 4 = 200 \text{ N}$$

Note that the value of total load is also given by the area of the load distribution diagram.

Thus for beam AB $W = \tfrac{1}{2} w \times L$

$$= \tfrac{1}{2} 100 \times 4 = 200 \text{ N}$$

The line of action of W will be through the centroid of the load distribution diagram at a distance of $\left(\tfrac{2}{3} \times 4\right)$ m from A.

The total moment (M) of the load about A is then given by:

$$M = 200 \times \left(\tfrac{2}{3} \times 4\right)$$
$$= 533 \text{ N m} = 0.53 \text{ kN m}$$

This type of loading is always present on the face of dams or walls retaining water or soils. Figure (b) shows a vertical wall retaining water to a depth of 4 m. The horizontal loading on the wall varies linearly from zero at the top to 40 kN/m^2 at the base.

Considering a one metre length of the wall, calculate the total load on the wall and the moment of that load about the base.

(Note that this wall may be of considerable length. It is normal practice to analyse such structures by considering a unit length.)

12

$$\boxed{\begin{array}{c} 80 \text{ kN} \\ \hline 106.67 \text{ kN m} \end{array}}$$

Pressure distribution
diagram

Check your working.

Average load $= 40/2 = 20$ kN/m^2 acting on an area of 4×1 m^2

thus load $= 20 \times 4 = 80$ kN.

Or

load = area of load distribution diagram $\times$ length of wall

$= \left(\frac{1}{2} \times 40 \times 4\right) \times 1$

$= 80$ kN

Distance of centroid of load distribution diagram from base$= \frac{1}{3} \times 4$ m (remember that the centroid of a triangle is at a distance $\frac{1}{3}$ of the height above the base)

$\therefore$ Moment about base $= 80 \times \left(\frac{1}{3} \times 4\right)$

$= 106.67$ kN m

13

COUPLES

We will now consider the special case of a load applied to a structure, in this case a beam, in such a way that an additional moment is induced at the point of application of the load. In figure (a) a load W is applied directly to a beam at a point C. In figure (b) the same load is applied via a rigid crank.

To understand the effect of the load in figure (b) we adopt the technique of imagining equal but opposite forces W applied at the point D (see figure (c)). This does not affect the overall loading on the beam since the two forces W at D cancel in any calculations involving the whole system. The three forces may however be considered as a single force W acting downwards at D with its line of action passing through C, together with a *couple* comprising two equal but opposite forces W distance s apart. If moments are taken about C, this couple is seen to be exerting a moment of $W \times s$ about that point.

The effect of the crank can be summarised as in figure (d) where you see a moment (Ws) applied to the beam at C in addition to the vertical load, W. Suppose now that we wished to calculate the total moment of the load system in figure (d) about A. You know that the moment of W about A is $WL/2$, but what about the moment Ws? Look at figure (e) where the two forces W forming the couple are shown in relation to A. If we take moments of these two forces about A we have:

$$\text{clockwise moment about A} = W \times (L/2 + s) - W \times L/2$$
$$= Ws$$

Thus the moment of the couple about A is independent of the distance of the couple from A. *It follows that the moment of a couple is the same about any point irrespective of the position of that point.*

If we now go back to our example we can evaluate the total moment (M) about A as:

$$M = WL/2 + Ws$$

It is interesting to note that you could obtain the same result from figure (b) by simply taking moments of the original force W about A as $M = W(L/2 + s)$.

14

STRUCTURAL SUPPORTS AND REACTIONS

Structures are maintained in equilibrium by reactive forces (reactions) provided by supports of various kinds. A beam may be supported on brick walls, a roof truss by concrete columns, a mass retaining wall by the ground, and so on. Let's start by considering the nature of a reaction.

In figure (a) the weight W is tending to pull the rope down. The ceiling has to provide an upward acting reaction R to prevent the rope being pulled down. For equilibrium, it is clear that R must equal W. The rope itself is being stretched. It is in *tension*. A structural member in tension is called a *tie* and has internal reactive forces developed within it.

By taking a section through the rope and considering the free body diagram of the weight (figure (b)), can you identify the direction of the force exerted by the rope on the weight?

(a)   (b)

15

You can see this by considering the *free-body diagram* of the weight as shown in figure (a) below. The weight is acted upon by an upward force T and the load W. These must obviously be of equal magnitude if the weight is not going to move up or down. Similarly at the top B of the rope $T = R = W$. The tensile force remains constant at value T throughout the length of the rope, because no external loads act on the rope other than at the ends. In figure (b) below, we see a column acted upon by a vertical downward load W. In this case the load is tending to compress the column. By considering the free body diagram of the column we can see that the ground must exert an upward acting reaction $R = W$ for equilibrium and a compressive force $C = W$ will develop internally in the column.

Note that in the diagrams showing the internal forces two arrow heads are used, one each end, and facing in different directions. It is important to remember to do this when marking structural members since the internal forces act in opposite directions. A structural member forming part of a frame and which is in compression is called a *strut*, or a *column*.

(a)   (b)   (c)

If an engineer weighing 700 N stands on the ground, what is the reaction of the ground upon the engineer?

16

Remember that any reaction is only there as long as the external load is applied. Remove the load and the reaction is no longer developed in the support. Thus in the above cases of tie and column, R is not tending to move the members but is only resisting the tendency of load W to cause movement.

Now we will consider some simple problems to illustrate the techniques of calculating forces and reactions.

In figure (a) below, the student weighs 700 N. This force of 700 N acts downwards through her centre of gravity. She is holding a balloon exerting an upward pull of 150 N. What is the effect on the ground? If you look first at the balloon you can see that a tensile force equal to 150 N must develop in the string in order to prevent the balloon rising. Putting the two arrows on the string to indicate the internal forces, you can then see that at the bottom of the string an upward force of 150 N acts on the student. She is now subject to two forces, 700 N down and 150 N up. There is an out-of-balance force of $700 - 150 = 550$ N. Thus an upward reaction of 550 N must develop between the ground and the student if she is not to move.

In figure (b) a rope passing over a frictionless pulley carries a load of 10 kN at each end. What is the reaction at the ceiling? The important thing to realise is that when a rope passes over a frictionless pulley, the internal force in the rope does not change. If you look at the left-hand load first, you will realise that there will be a tensile force of 10 kN developed in the rope. If there is a tensile force of 10 kN in the rope to the left of the pulley, there will also be a tensile force of 10 kN in the rope to the right of the pulley. There are thus two forces acting on the right-hand load, 10 kN up and a weight force of 10 kN down. The system is thus in equilibrium. To determine the reaction in the ceiling, consider the equilibrium of the pulley. There are two downward forces of 10 kN in the rope, and the reaction, all acting on the pulley. For equilibrium therefore, the reaction R must be 20 kN upwards.

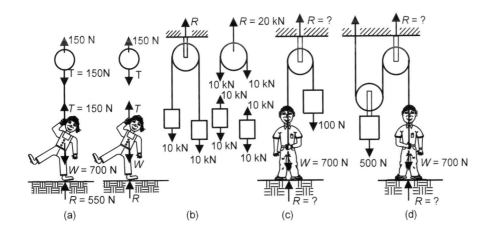

(a)   (b)   (c)   (d)

Now determine the magnitude and direction of the reactions below the engineer and securing the pulley to the ceiling in figures (c) and (d).

Draw free-body diagrams of each part of each problem and consider the equilibrium of each free body in turn.

(Hint in (d), start by considering the equilibrium of the left-hand pulley.)

18

> (c) 600 N up and 200 N up
> (d) 450 N up and 500 N up

The following diagram shows a more complex situation where one rope is no longer vertical.

(a)   (b)   (c)

If we start at the load we see that there must be a tensile force of 200 N in the rope. Then considering the forces acting on the pulley (figure (b)) we can resolve vertically and horizontally to determine the reaction in the ceiling (this will be the equilibrant of the other two forces):

resolving vertically

$$+ R\sin\theta - 200 - 200\sin 45° = 0$$

resolving horizontally

$$+ R\cos\theta - 200\cos 45° = 0$$

from the first equation

$$R\sin\theta = 341.42$$

from the second

$$R\cos\theta = 141.42$$

squaring and adding gives

$$R^2 \sin^2\theta + R^2 \cos^2\theta = 341.4^2 + 141.4^2$$

$$\therefore \quad R^2(\sin^2\theta + \cos^2\theta) = 136\,568$$

$$\therefore \quad R^2 = 136\,568$$

and

$$R = 369.55 \text{ N}$$

then from the second equation

$$\theta = 67.5°$$

You could alternatively determine R graphically using a triangle of forces as in figure (c).

Considering the vertical equilibrium of the student, we note that she is subject to the tension in the rope and her own weight. If we resolve these forces we obtain a total vertically downward force on the ground of value $700 - 200\sin 45° = 559$ N. This will be resisted by an upward reaction of 559 N.

Resolving the forces acting on the student horizontally, we see that there is also a horizontal out-of-balance force of $200\cos 45° = 141$ N.

What effect will this force tend to have on her and what reaction will be required at ground level to maintain her equilibrium?

19

It will tend to pull her to the right, causing her to slide along the ground. A frictional resistance is required between her feet and the ground.

Frictional forces will develop to resist sliding between any two surfaces in contact. The value of such a force depends upon the nature of the materials in contact, the state of the surfaces (rough or smooth) and on the value of the normal force pressing the surfaces together. For any two surfaces there is an upper limit to the value of frictional force which can be developed in response to a given value of normal force, and this relationship is quantified by the *coefficient of friction* (μ) defined as the ratio (maximum frictional force)/(normal force between the surfaces).

In the above example there is a normal force of 559 N between the student and the ground. We need a horizontal frictional force of 141 N to prevent her sliding. This is only possible if the coefficient of friction is at least $141/559 = 0.25$, otherwise she will slide along the ground and will not be in a state of equilibrium.

In this type of problem it may be convenient to quote the reaction in terms of the vertical component (V) and the horizontal component (H), or it may be more useful to quote the total reaction R (as we did at the pulley).

20

PROBLEMS

1. A rope of a small suspension bridge passes over a frictionless pulley at the top of a pier 12 m high as in figure Q1. If the tension in the rope is 50 kN, calculate by considering the components of each force (i) the total vertical load acting downwards on the pier and (ii) the effective horizontal load acting at the top of the pier.

Ans. ((i) 75.44 kN downward, (ii) 13.30 kN to the right)

2. Calculate the reactions at: (i) A, (ii) B, (iii) C, and (iv) the value of load at D in figure Q2 (*Hint*: start by looking at the free-body diagram of the load W_1).

Ans. ((i) 14.64 N upward 30° to the left of vertical, (ii) 25.36 N vertically upwards, (iii) 10.35 N upward 45° to the right of vertical, (iv) load at D = 25.36 N)

21

SUPPORT TYPES

Simple supports

You should now begin to appreciate the nature of the reactions required to keep a structure in equilibrium. It is appropriate therefore now to look in more detail at the type of supports available to provide those reactions in more realistic types of structural systems. The same types of supports can be used both for beams and frames.

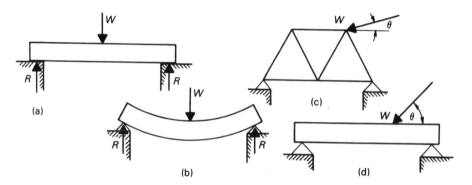

The simplest way to support a beam or a frame would be to place the structure directly on top of a support as in figure (a). In practice this means that the supports are assumed to act like knife edges providing vertical reactions only and permitting a limited amount of rotation at the support, thus allowing a beam to deflect as in figure (b). This is not too unrealistic an assumption if you bear in mind that any rotation would in reality be very small and that the material of both structure and support have elastic properties and are able to change shape under load.

Suppose the loads on a structure are not vertical but are as shown in figures (c) and (d).

Are the structures in figure (c) and (d) in equilibrium?

22

No: they would move to the left

If you have difficulty seeing this, resolve the loads into vertical and horizontal components. The vertical component ($W \sin \theta$) can be resisted by vertical reactions at the supports, but there is no possibility of a horizontal reaction developing to resist the horizontal component ($W \cos \theta$) unless (a) friction exists between the structure and support, (b) the geometry of the support provides horizontal restraint or (c) the structure is *pinned* to the support.

These three possible methods of providing horizontal restraint are illustrated below in figures (a), (b) and (c) respectively. Method (c) is a common solution which provides positive restraint both vertically (up and down) and horizontally (left and right), but at the same time permitting the rotation mentioned previously (note, in figure (c), the way in which we indicate a pin). The 'pins' used for this purpose would act like hinges and would typically be cylindrical bars many centimetres in diameter inserted through holes in fixed supports and the structural member. This hole in the structural member would have sufficient clearance to permit the member to rotate freely about the 'pin'. Figure (d) shows a typical pinned joint at the end of a bridge beam.

(a) (b) (c)

(d)

Would it be necessary to provide a pin at both ends?

23

No

You should have realised that one pin is sufficient to maintain horizontal equilibrium. The other end must in fact be free to move horizontally to allow for the bending and other movement under load and also to allow for movement as a result of temperature variation. Structures will expand in hot weather and contract in cold weather, thus giving rise to longitudinal movement. The type of support must be able to accommodate this movement.

Can you suggest a form of support which would permit unrestricted movement horizontally but provide complete vertical restraint?

24

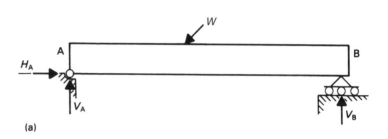

A frictionless roller

(a)

The sketch (a) shows a beam with a pin at one end providing vertical reaction V_A and horizontal reaction H_A, while at end B a roller provides vertical reaction V_B but permits horizontal movement. Both supports permit rotation. A structure with this support arrangement is often referred to as being *simply supported*. The pin at the left-hand end provides two restraints (H_A, V_A) and the roller at the other end provides a single restraint (V_B).

The detail of the roller support shown above and elsewhere is of course only diagrammatic and not indicative of actual construction. In practice, actual cylindrical bar rollers may be used, or the structure may be supported on composite steel and rubber bearings as shown in figure (c) and which permit sufficient horizontal movement to be considered as frictionless rollers.

(b)

(c)

Note that the reaction V_B as shown above can only be upward. You will learn in later frames, however, that some loading conditions may tend to lift a structure from its support.

Would a simple roller support of the types shown above be suitable in a situation where the structure tends to lift from the support?

25

No

The simple roller support will not prevent the structure lifting. Consequently it may be necessary to consider providing a roller support of the type illustrated diagrammatically in figure (a) below.

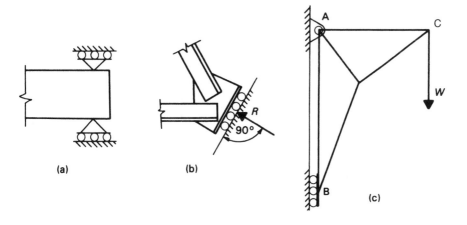

(a)　　　　(b)　　　　(c)

The reaction at a roller is always at right angles to the surface on which the roller runs. Thus in figure (b) the reaction is as indicated.

With reference to the cantilevered crane structure shown in figure (c):

(i) what is the direction of action of the reaction at B?
(ii) would the roller indicated at B be suitable if the downward load at C was replaced by an upward force?

26

(i) Horizontal: acting to the right
(ii) No

As shown in figure (c) above, there is nothing to prevent joint B moving away from the wall if the load acts in an upward direction. Joint B would have to be pinned to this wall in a similar way to joint A.

27

Fixed supports

Some supports, in addition to preventing linear movement, also prevent any rotation of the structure at the support. Such supports are termed *fixed* (or *encastré*) *supports* and provide three restraints: two mutually perpendicular components of a reaction and a fixing moment.

Figure (a) shows a beam with a fixed support at A and carrying a load W acting at an angle θ to the horizontal. Three restraints are indicated at A and you will see that for equilibrium: the horizontal component (H_A) of the reaction at A must equal $W\cos\theta$ and the vertical component (V_A) must equal $W\sin\theta$. In addition, the load W causes a clockwise moment about A of value $W\sin\theta \times S$. To prevent rotation and to ensure equilibrium, therefore, the support must be capable of developing an anticlockwise moment also of value $W\sin\theta \times S$. this reactive moment (fixing moment) is indicated as M_A.

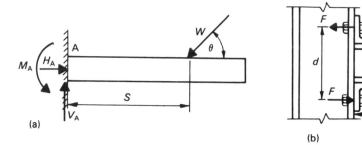

(a)　　　　　　　　　　　　　(b)

To understand how such fixing moments would develop it is useful to visualise how the necessary restraint could be provided. The beam might be bolted to a vertical column as in figure (b). Then when the load was applied, a tensile force of magnitude F would develop in the upper bolts. An equal and opposite compressive force would develop between the lower cleat and the column. These forces distance d apart would form a couple with anticlockwise moment of magnitude $F \times d$. $F \times d$ would have to equal $W\sin\theta \times S$ for equilibrium.

A structure such as the beam shown above which is completely fixed at one end with no support at the other end is called a *cantilevered* structure.

28

CONDITIONS FOR EQUILIBRIUM

If a structure has a net resultant force acting on it, then that resultant force can be replaced by components acting in the horizontal and vertical directions. If the structure is in equilibrium however, the magnitude of the resultant force and hence the magnitude of its components must be zero. Hence for any structure we can say that it is in equilibrium if the algebraic sum of all the vertical components of force acting on the structure is zero ($\Sigma V = 0$) and if the algebraic sum of all the horizontal components is also zero ($\Sigma H = 0$). Although it is usual to consider horizontal and vertical directions, these criteria can be applied to any other pair of mutually perpendicular directions.

Bearing in mind that for a structure to be in equilibrium, $\Sigma V = 0$ and $\Sigma H = 0$, is the plane frame shown below in equilibrium?

29

No

$\Sigma V = 0$ and $\Sigma H = 0$, thus the frame will not move vertically or horizontally, but if moments are taken about A there is an anticlockwise moment of value 10 kN m $\{[(10 \times 1) - (5 \times 2) - (2.5 \times 4)] = -10 \text{ kN m}\}$. The frame will consequently rotate anticlockwise. We realise then that for complete equilibrium we need to ensure that rotation cannot take place. The algebraic sum of all moments about any point must therefore also be zero ($\Sigma M = 0$).

Summarising the above conditions for static equilibrium, we can state that for any structure to be in equilibrium three equations must be satisfied. That is:

$\Sigma V = 0$ **(the algebraic sum of all vertical forces must be zero)**
$\Sigma H = 0$ **(the algebraic sum of all horizontal forces must be zero)**
$\Sigma M = 0$ **(the sum of the moments of all forces about *any point* must be zero)**

We can also state that for any plane structure to be in a state of stable equilibrium, a minimum of three restraints are essential. This statement is generally true, but in some cases the value of one of the restraints may be zero. For example, if a horizontal beam is subjected only to vertical loads, the horizontal component of reaction at the supports will be zero.

For any structure subject to known loading and maintained in equilibrium by support restraints we can write down the three equations quoted above. It follows that if a structure has no more than three restraints of unknown magnitude, their values can be determined by solving the three equations.

Such a structure is said to be *externally statically determinate*.

If a plane structure has more than three unknown restraints it cannot be solved by applying the three conditions for static equilibrium and is said to be *externally statically indeterminate* or *externally redundant*. The number of redundancies in any particular case is the number of unknown restraints less three (because we can write three equations). The solution of redundant structures is considered in later programmes.

30

In the previous frame you learnt that for a plane frame to be externally statically determinate it must not have more than three restraints of unknown magnitude. If a frame does have more than three unknown restraints, it is externally redundant and cannot be solved by applying the three equations of statical equilibrium. Now look at the figures shown below.

Which of the structures shown are redundant? (Hint: a roller provides one restraint, a pin two, and a fixed support three.)

31

b, c and d

Figure	Unknowns		Determinate or redundant
a	$V_A, H_A, \quad , V_B$	3	determinate
b	$V_A, H_A, \quad , V_B, V_C$	4	redundant
c	$V_A, H_A, \quad , V_B, H_B$	4	redundant
d	V_A, H_A, M_A, V_B	4	redundant

You should now be able to recognise whether a structure is statically determinate or not, and if it is to be able to calculate the unknown reactions. Let's do some examples.

32

Suppose we have a beam AD loaded with a combination of concentrated and uniformly distributed loads as shown in the free-body diagram on the next page. There is a pinned support at B and a roller support at C.

First we indicate on the diagram the reactions whose values will have to be determined. These are vertical and horizontal reactions at B and a vertical reaction only at C. Note that we use arrows to show the direction we are assuming for these reactions. If our answers turn out to be negative it simply implies that our initial choice of direction was wrong.

There are only three unknown support reactions, thus the beam is statically determinate and we can proceed.

Bearing in mind that for equilibrium $\Sigma H = 0$ and noting that all loads are entirely vertical, you can see that $H_B = 0$.

To determine the other reactions it is convenient to start by taking moments since, if we choose the point about which we take moments correctly, we can eliminate one of the unknowns and simplify the calculations. In this case we will take moments about B and thus eliminate V_B from the calculation. To take moments of the UD loads, we will consider them replaced by their resultants.

What are the values of those resultants and where do they act?

33

A to B	4 kN acting 1 m left of B
B to C	16 kN acting 2 m right of B

Then the moment about B:

$$\text{of the UD load on AB} = -4 \times 1 = -4 \text{ kN m}$$
$$\text{of the UD load on BC} = +16 \times 2 = +32 \text{ kN m}$$
$$\text{of the concentrated load on BC} = +10 \times 3 = +30 \text{ kN m}$$
$$\text{of the concentrated load at D} = +5 \times 6 = +30 \text{ kN m}$$

and the reaction at C will have a moment about $B = -V_C \times 4$

Then using $\Sigma M_B = 0$, we have:

$$-4 + 32 + 30 + 30 - 4V_C = 0$$

from which $$V_C = +22 \text{ kN}$$

The positive sign implies that the reaction at C is upwards as initially assumed.
We can determine the value V_B by using $\Sigma V = 0$.
The total load is

$$(2 \times 2) + (4 \times 4) + 10 + 5 = 35 \text{ kN}$$
$$\therefore \quad V_B + V_C - 35 = 0$$
$$V_B + 22 - 35 = 0$$

from which $$V_B = +13 \text{ kN}$$

This value for V_B may be checked by taking moments about C and demonstrating that the sum of the moments about C, using the calculated reactions, also equates to zero.

34

We will now calculate the support reactions for a frame subject to both vertical and inclined loads. Consider the roof truss shown below. Dead loads are represented by the vertical forces acting at the joints of the frame, and wind forces are acting at right angles to the rafter. The frame has a pin support at A and a roller support at B.

(a)

(b) Equivalent load system

There are only three unknown reactions (H_A, V_A and V_B) so the frame is statically determinate and we can proceed.

We use the three equations for statical equilibrium: $\Sigma H = 0$, $\Sigma V = 0$ and $\Sigma M = 0$. First we will take moments about A. This is the most convenient point to choose because the two reactions at A have no moment about A and hence do not appear in the equation. You will remember that *you may in fact take moments about any point* and equate the algebraic sum to zero for equilibrium.

There are several loads involved and we may simplify the working by using resultants.

What is the resultant dead load and where does it act?

35

> 16 kN vertically downwards, 3 m to the right of A

If you did not obtain this value, check that you included the 2 kN load at A. This load has to be taken into account even though it acts at A and hence has no moment about A. Why? Because we need the resultant of *all* the dead loads. Similarly we can determine the resultant of the wind loads to be 20 kN acting as shown in figure (b). Then by moments about A:

$$\Sigma M_A = 0: \qquad 16 \times 3 - 20 \times 1.803 - V_B \times 6 = 0$$

from which
$$V_B = +1.99 \text{ kN}$$

The positive sign indicates that the reaction is indeed upwards as initially assumed. We can alternatively take moments about B to give the vertical reaction at A eliminating V_B and H_A from the calculation as these reactions do not have a moment about B.

The perpendicular distance from B to the wind force can be calculated from the geometry of the figure and is shown as $p = 3.19$ m in figure (b). Then by taking moments about B:

$\Sigma M_B = 0$: $\qquad V_A \times 6 + 20 \times 3.19 - 16 \times 3 = 0$

from which $\qquad\qquad\qquad\qquad V_A = -2.63$ kN

The minus sign indicates that the reaction is in the opposite direction to that assumed. It therefore acts downwards.

We can now check our two answers by considering overall vertical equilibrium of the structure using the necessary condition that $\Sigma V = 0$.

What is the vertical component of the total wind load?

36

> 16.64 kN

Note that it is not necessary to calculate the value of the angle θ. From the geometry of the figure, the length of the rafter is $(2^2 + 3^2)^{\frac{1}{2}} = 3.606$ metres, so that $\cos\theta$ is $3/3.606$ and the vertical component of the total wind loading is therefore $20 \times \cos\theta = 20 \times 3/3.606 = 16.64$ kN.

Now to check for vertical equilibrium:

$$\text{Total vertical forces} = V_A + V_B + \text{component of wind force} - \text{dead load}$$
$$= -2.63 + 1.99 + 16.64 - 16$$
$$= 0 \text{ which is correct.}$$

Finally using $\Sigma H = 0$ we obtain $H_A = $ horizontal component of the total wind loading $= 20 \times \sin\theta = 20 \times 2/3.606$

$$= \underline{11.09 \text{ kN}} \text{ (i.e. acting to the right)}$$

Would any special precautions have to be taken when designing the support at A?

37

> Yes. The vertical reaction at A is downward.
> The supports would have to be designed to resist uplift.

We know that there are three conditions defining whether a structure is statically determinate. These conditions were summarised in frame 29 by the equations $\Sigma H = 0$, $\Sigma V = 0$ and $\Sigma M = 0$. You may also recall that in the previous example we actually wrote down two moment equations, one about each support, and consequently wrote a total of four equations $\Sigma H = 0$, $\Sigma V = 0$, $\Sigma M_A = 0$ and ΣM_B. You might think that this would enable us to solve for four unknowns. In fact we are still limited to three unknowns.

Why doesn't the additional moment equation enable another unknown to be determined?

38

If you do not understand this consider figure (a):

$$\Sigma M_A = W_1 x_1 + W_2 x_2 - V_B L$$

and
$$\Sigma M_B = V_A L - W_1 (L - x_1) - W_2 (L - x_2)$$
$$= (W_1 + W_2 - V_B)L - W_1 L + W_1 x_1 - W_2 L + W_2 x_2$$
$$= W_1 L + W_2 L - V_B L - W_1 L + W_1 x_1 - W_2 L + W_2 x_2$$
$$= W_1 x_1 + W_2 x_2 - V_B L$$
$$= \Sigma M_A$$

(a)   (b)

The two equations are thus simply different ways of writing the same equation. We are putting in mathematical terms only one condition: that is, that the whole frame does not rotate in the XY plane.

INTERNAL PINS

If, however, we divide a structure into two independent parts connected by means of an internal pinned connection, as in figure (b) above, then we have a different situation. We can now write an additional equation $\Sigma M_P = 0$ which *is* unique. Why is it? Because it differs from the first moment equation in that it specifies a further condition. The new condition is that either the left-hand or the right-hand part of the structure is free to rotate about the pin independently of the rotation of the other part. A pin acts like the hinge of a door; the structure rotates about it freely *with no resisting moment*.

It is hence important to remember that if there is a pinned connection joining two parts of a structure then the sum of the moments about the pin of all forces (loads and reactions) acting on the part to the right of the pin must be zero. Alternatively, the sum of the moments of all forces to the left of the pin must be zero. Note that the two equations generated by the above statements are again only two versions of the same equation reflecting one condition for equilibrium.

We see that if a structure consists of parts complete in themselves and joined together by internal pins then our rules for judging whether a structure is externally statically determinate must be revised. For a basic structure, three restraints are required for equilibrium and three equations may be formulated enabling three unknowns (reactions or fixing moments) to be determined. For every internal pin in a structure, one additional equation may be written thus one additional unknown may be determined. A structure is statically determinate provided the number of unknown restraints is not greater than the number of unique equations which can be written.

Which of the following structures are statically determinate?

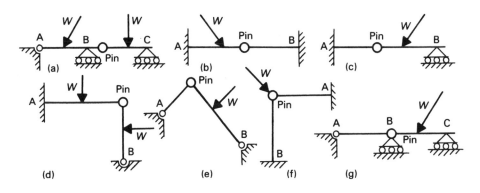

a, c, e and g

In all cases there is one internal pin, thus four unknown support reactions or moments can be determined. No more than four unknowns can be determined by the rules for static equilibrium.

Figure	Unknowns		Determinate or indeterminate	Degree of redundancy
a	V_A, H_A, V_B, V_C	4	determinate	0
b	$V_A, H_A, M_A, V_B, H_B, M_B$	6	indeterminate	2
c	V_A, H_A, M_A, V_B	4	determinate	0
d	V_A, H_A, M_A, V_B, H_B	5	indeterminate	1
e	V_A, H_A, V_B, H_B	4	determinate	0
f	$V_A, H_A, M_A, V_B, H_B, M_B$	6	indeterminate	2
g	V_A, H_A, V_B, V_C	4	determinate	0

In the next frame we will work through an example.

41

Figure (a) shows a beam rigidly fixed at A and on a roller at support B. A pinned connection at P connects the two parts of the beam together.

(a)

There are four unknowns (V_A, H_A, M_A and V_B) the positive directions of which are taken as shown in figures (b) and (c). Because of the internal pinned connection the structure is statically determinate and we can analyse it using the normal principles of statics.

(b)   (c)

Figures (b) and (c) show the free-body diagrams of both parts of the beam to either side of the pinned connection. The pin cannot transmit a moment but can transmit an equal and opposite pair of reactive forces between the two parts of the structure. These reactive forces will have horizontal and vertical components, shown as H_P and V_P in the figure. Effectively the right-hand part of the structure, PC, is a simply supported beam, being supported at P by the left-hand part, AP. The left-hand part, AP, is effectively a cantilevered beam supporting the uniformly distributed load of 1 kN/m together with point loads at its end arising from the effect of PC bearing down on it.

Now let's consider the part of the structure from P to C as shown in figure (c). This part of the structure can be immediately analysed as there are only three unknown reactions (V_B, H_P and V_P). Note that the inclined 20 kN load has been replaced by its horizontal and vertical components as shown.

Can you take moments about P to calculate the value of the reaction V_B?

42

$$V_B = 25.76 \text{ kN}$$

Did you set out your calculations as follows?

$$\Sigma M_P = 0: \quad (4 \times 2) + (20 \sin 60° \times 4) - (V_B \times 3) = 0$$
$$\therefore \quad 8 + 69.28 - 3V_B = 0$$
$$\therefore \quad V_B = 25.32/3 = 25.76 \text{ kN (upwards)}$$

Did you note that the horizontal component of the 20 kN load does not have a moment about P as its line of action passes through P? Also, when calculating the moment of the uniformly distributed load did you (i) only take that part of the load which is acting on PC (i.e. a total load of $1 \times 4 = 4$ kN) and (ii) consider the resultant of this load acting through its centroid at $4/2 = 2$ m from P?
Now take moments about B to calculate the value of the reactive force V_P.

43

$$\boxed{-4.44 \text{ kN}}$$

Check your calculations:

$$\Sigma M_B = 0: \quad -(4 \times 1) + (20 \sin 60° \times 1) + (V_P \times 3) = 0$$
$$\therefore \quad -4 + 17.32 + 3V_P = 0$$
$$\therefore \quad V_P = -13.32/3 = -4.44 \text{ kN (downwards)}$$

Note that the negative sign means that this reaction is acting in the opposite direction to that assumed, i.e. it is acting downwards. In taking moments did you take correct account of the 'signs' of the moments? The load of $20 \times \sin 60°$ has a clockwise turning effect about B. The 4 kN resultant of the uniformly distributed load acts at 1 m to the left of B and has an anti-clockwise turning effect about B: hence the signs in the above equation. This can be seen by considering the figure below:

The calculation of the two vertical reactions can be checked by considering the vertical equilibrium of the right-hand part of the structure, PC. Resolving vertically:

$$\Sigma V = 0: \quad V_P + V_B - (4 \times 1) - 20 \sin 60° = -4.44 + 25.76 - 4 - 17.32$$
$$= 0 \quad \text{which is correct.}$$

Calculate the value of the horizontal reaction H_P.

44

$$\boxed{10.0 \text{ kN}}$$

You should have obtained this by considering horizontal equilibrium:

$$\Sigma H = 0: \qquad H_\text{P} - 20\cos 60° = 0 \qquad \therefore \quad H_\text{P} = 10.0 \text{ kN}$$

(i) Sketch the free-body diagram for the left-hand side of the structure, AP, and show the correct direction of the reactive forces H_P and V_P and
(ii) Calculate the reactions V_A, H_A and M_A at the built-in support.

45

Resultant of
UD load 3 x 1= 3 kN

4.44 kN

M_A

10.0 kN

H_A A   P

V_A 1.5 m

3.0 m

$$H_\text{A} = 10.0 \text{ kN} \quad V_\text{A} = -1.44 \text{ kN} \quad M_\text{A} = 8.82 \text{ kN m}$$

Check your working:

$$\Sigma H = 0: \qquad H_\text{A} - 10.0 = 0$$
$$\therefore \quad H_\text{A} = 10.0 \text{ kN}$$

$$\Sigma V = 0: \qquad V_\text{A} - (3 \times 1) + 4.44 = 0$$
$$\therefore \quad V_\text{A} = -1.44 \text{ kN (downwards)}$$

$$\Sigma M_\text{A} = 0: \qquad M_\text{A} + [(3 \times 1) \times 1.5] - (4.44 \times 3) = 0$$
$$\therefore \quad M_\text{A} + 4.5 - 13.32 = 0$$
$$\therefore \quad M_\text{A} = 8.82 \text{ kN m}$$

Note that M_A has a positive value indicating that the fixing moment at the support is in the positive clockwise direction, as originally assumed.

You should take particular note of the way that we have solved this problem by isolating the free-body diagrams of each part to either side of the pinned connection and applying the equations of equilibrium to each part in turn. We can check the accuracy of the solution by now checking the overall vertical equilibrium of the whole structure as follows:

$$\Sigma V: \quad V_A + V_B - (7 \times 1) - 20 \sin 60° = -1.44 + 25.76 - 7.00 - 17.32$$
$$= 0 \quad \text{which is correct.}$$

Now try some problems for yourself to practise the application of the three equations of equilibrium.

46

$$\boxed{\text{PROBLEMS}}$$

1. Determine the values of reactions at A and B for figures Q1, Q2 and Q3.

Ans. (Q1: $V_A = +8.33$ kN, $V_B = +5.67$ kN   Q2: $V_A = +12.0$ kN, $V_B = +29.0$ kN
Q3: $V_A = 13.25$ kN, $V_B = +8.75$ kN)

Q1   Q2   Q3

2. Determine the values of the reactions and fixing moments at A for figures Q4 and Q5.

Ans. (Q4: $V_A = +6.0$ kN, $H_A = 0$, $M_A = 9.0$ kN m anticlockwise.
Q5: $V_A = 0$, $H_A = -5.0$ kN, $M_A = 10.0$ kN m anticlockwise)

Q4   Q5

3. Determine the reactions at A and B for Figures Q6 and Q7.

Ans. (Q6: $V_A == +3.64$ kN, $V_B = +7.88$ kN, $H_A = +8.48$ kN Q7: $V_A = +5.0$ kN, $H_A = -5.0$ kN, $H_B = +5.0$ kN)

Q6

Q7

4. Determine the values of the reactions and fixing moments at A and B for figures Q8 and Q9

Ans. (Q8: $H_A = -10.61$ kN, $V_A = +17.96$ kN, $V_B = +6.65$ kN, $M_A = 44.89$ kN m anticlockwise Q9: $H_B = -10.0$ kN, $V_B + 19.16$ kN, $V_A = +10.16$ kN, $M_B = 87.96$ kN m clockwise)

Q8

Q9

47

If you had difficulty getting the correct
answers, revise the previous frames

THREE-PINNED ARCHES

A common and efficient use of a pinned connection to join two parts of a structure is in the three-pinned arch. Figure (a) shows a typical three-pinned arch, which is statically determinate.

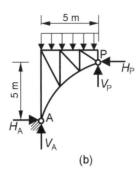

(a)

(b)

To illustrate the solution method we will analyse the arch in figure (a). Firstly all the unknown reactions should be indicated on the diagram. This has already been done and you will notice that the horizontal reactions are assumed to be in opposite directions. This will have an effect on the signs in the following equations. Look out for this.

We will use the three conditions for equilibrium applied to the whole structure and then write down a moment equation for all forces to one side of pin P.

For the whole structure:

$\Sigma V = 0 \qquad\qquad +V_A + V_B - (50 \times 12) = 0$

$\Sigma H = 0 \qquad\qquad +H_A - H_B = 0$

$\Sigma M_A = 0 \quad +(50 \times 12 \times 12/2) + H_B(7 - 5) - (V_B \times 12) = 0$

Now write down the equation for moments about P of all forces to the right of P (see figure (a)).

48

$$+(50 \times 7 \times 7/2) + (H_B \times 7) - (V_B \times 7) = 0$$

Did you remember to take the UD load acting only over a length of 7 m and not over the whole span?

Solving the four equations we obtain the values:

$$V_A = 275 \text{ kN}, \ V_B = 325 \text{ kN}, \ H_A = H_B = 150 \text{ kN}$$

These values can be checked by taking moments about pin P of all forces to the left of P (see figure (b)). Let's do this.

Sum of moments about P of forces to the left:

$$= (V_A \times 5) - (H_A \times 5) - (50 \times 5 \times 5/2)$$

and substituting our values we have:

$$(275 \times 5) - (150 \times 5) - (50 \times 5 \times 50/2) = 0: \text{ thus values check.}$$

In this example we discovered that there were horizontal reactions even though the loading was entirely vertical. We have not seen this in previous examples. Why is this example different?

49

In our previous examples only one support has been pinned, consequently the structures have been free to move horizontally at the roller support under the action of a load system. In the case of the arch analysed above, the load tends to flatten the structure and if one support was on a roller the horizontal distance between the supports would increase. The use of two pins prevents this horizontal movement, consequently horizontal reactions are developed.

50

MASS STRUCTURES

Mass structures differ from the types we have been looking at so far in that they depend upon their own mass for equilibrium. The sketch below of a typical wall for retaining earth enables us to discuss the basic principles for the equilibrium of such structures.

First let us consider the load of 15 kN acting on the face of the wall at $10°$ to the horizontal. This load represents the resultant effect of the pressure exerted on the wall by the retained earth and can be resolved into a vertical component $15 \sin 10° = 2.60$ kN acting downwards, which could be resisted by a vertical upward reaction from the ground upon the bottom of the wall, and a horizontal component $15 \cos 10° = 14.77$ kN, which would tend to push the wall to the left.

What reactive force is available to prevent such movement?

51

You will remember from an earlier example (frame 19) that the frictional force between two surfaces is limited to μN where μ is the coefficient of friction and N is the normal force between the surfaces. In this case the normal force is the weight of the wall plus the vertically downward component of the load. Hence if we let the frictional force be F, then for horizontal equilibrium:

$$\Sigma H = 0$$
or $\quad F - 14.77 = 0$
thus $\quad\quad F = 14.77$ kN

But the frictional force cannot be greater than μ times the normal force, and in this case the normal force is $31 + 2.60 = 33.60$ kN. Therefore for equilibrium to be possible $\mu \times 33.60 > 14.77$ and so μ must be greater than 0.44. If μ is not sufficiently large, some other means of resisting the horizontal load must be adopted such as sinking the wall into the ground so that a passive resistance P of the ground upon the wall can develop (see figure (b)).

What about rotational equilibrium? If we look at figure (a) and consider the effect of the 15 kN load acting alone, you can probably visualise the load tending to rotate the wall as a whole anticlockwise about the point O.

What force is available to prevent this rotation?

52

> The weight of the wall

The weight of the wall provides a restoring clockwise moment about O which counteracts the overturning anticlockwise moment of the load. If overturning is taking place, the weight and the load are the only forces acting on the wall because there is no longer any contact between the base and the ground, consequently there is no normal reaction at base level (see figure (c)).

The overturning moment about O is equal to the anticlockwise moment due to the horizontal component of the load minus the clockwise moment due to the vertical component. Hence:

Overturning moment $= (14.77 \times 1.0) - (2.60 \times 1.0) = 12.17$ kN m anticlockwise

The restoring moment due to the weight of the wall is given by:

Restoring moment $= 31 \times 0.65 = 20.15$ kN m clockwise

The moment of the force tending to overturn the wall is less than the moment which is available to resist overturning. If this were not the case then failure by overturning would occur. In this case, however, we can say that there is a *factor of safety* against overturning given by:

$$\text{Factor of safety} = \frac{\text{restoring moment}}{\text{overturning moment}} = \frac{20.15}{12.17} = \underline{1.66}$$

The provision of a factor of safety is important in design and allows for an adequate margin of safety against failure. You will appreciate that, for safe design, any factor of safety should always be greater than 1.

53

SPACE FRAMES

Before leaving this programme it will be useful briefly to consider the conditions for the equilibrium of space frames. You will remember from Frame 2 that these are three-dimensional frame structures. Consider the three-dimensional frame shown in figure (a). The arrangement is that of a scotch derrick, a particular form of crane. The jib which carries the rope and crane hook is free to rotate about the vertical mast AD, consequently loads can be lifted from anywhere within the sector indicated on the plan (figure (b)). The other views show the elevation on plane XY (figure (c)) and the elevation on plane ZY (figure (d)). For any fixed position of the jib, the structure may be analysed using the rules for static equilibrium.

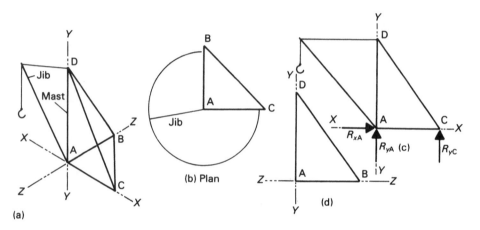

Consider the situation when the jib is aligned in the XY plane.

If we look at figure (c) we see what looks like a simple frame. If it were a simple plane frame what would be the conditions for equilibrium?

54

$$\Sigma V = 0, \ \Sigma H = 0 \text{ and } \Sigma M = 0$$

Or using the notation on the diagrams, the first two of these conditions can be written as:

$$\Sigma \text{ (forces in the } Y \text{ direction)} = 0 \quad \text{that is} \quad \Sigma Y = 0$$

and similarly

$$\Sigma X = 0$$

These two conditions ensure that the frame does not move in the Y or in the X direction, but it could still move in the Z direction at right angles to the XY plane.

What additional condition has to be specified to prevent movement in the Z direction?

$$\boxed{\Sigma Z = 0}$$

Look again at figure (c). For there to be no rotation in the XY plane, the sum of the moments about any point must equal zero. Thus the sum of the moments about A must equal zero but A is an end view of the ZZ axis. Consequently we can ensure freedom from rotation in the XY plane by specifying the condition $\Sigma M_{ZZ} = 0$. You should realise from a study of figures (a) and (c) that this is a logical conclusion. You should also realise from figure (a) that even though ΣM_{ZZ} equals zero, the frame would still be free to rotate about the XX and YY axes.

What other conditions must be specified to ensure complete equilibrium?

$$\boxed{\Sigma M_{XX} = 0 \text{ and } \Sigma M_{YY} = 0}$$

where ΣM_{XX} and ΣM_{YY} denote the sum of the moments about the X–X and the Y–Y axis with the structure viewed in the YZ and the XZ planes respectively. For complete equilibrium of a space frame, six conditions must be satisfied:

$$\Sigma X = 0, \ \Sigma Y = 0, \ \Sigma Z = 0, \ \Sigma M_{XX} = 0, \ \Sigma M_{YY} = 0 \text{ and } \Sigma M_{ZZ} = 0$$

We have seen that in the case of a plane frame, three conditions have to be satisfied for equilibrium and that a minimum of three support restraints are required. It follows that for complete equilibrium of a space frame a minimum of six support restraints are required. Possible types of support include:

Ball joints permitting rotation in all directions but not free to move linearly in any direction. These provide reactions in the X, Y and Z directions. See figure (a).

Pinned supports permitting rotation in all directions and fixed linearly in two directions but free to move linearly in one direction. These provide two reactions. See figure (b).

Simple supports permitting rotation in all directions but free to move linearly in two direction. These provide one reaction. See figure (c).

(a)　　　　(b)　　　　(c)

57

It follows that if in our example we provide a ball joint at A, a pin free to move in the X direction at C and a simple support at B, we have provided the essential six reactions for equilibrium. All six reactions are shown on the diagrams below, each shown acting in the positive direction of the relevant axis. Note that R_{XA} signifies the reaction at A in the X direction, and so on.

We will now calculate the reactions for the situation when the jib lies in a plane at 45° to the XY plane, as indicated in figure (a), and the crane is supporting a load of 150 kN. Figure (b) shows the jib and rope arrangement and we see that there is a force of 150 kN in the rope pulling horizontally at the top of the mast. This must be resolved into the X and Z directions before we can substitute in the basic equations. The necessary components are $150 \sin 45°$ and $150 \cos 45°$, that is, 106 kN in both directions. These are shown acting on the frame in figure (c).

Similarly at the bottom of the mast are two forces of 106 kN but acting in the opposite directions to those at the top. A downward load of 150 kN is also transmitted to the base of the mast.

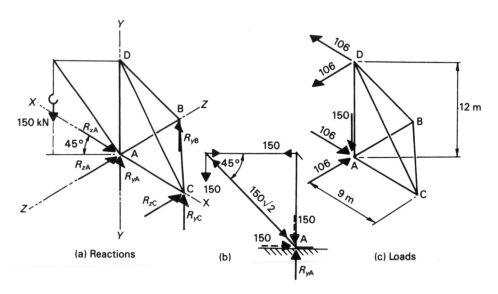

(a) Reactions   (b)   (c) Loads

We can now apply the basic equations for equilibrium:

$\Sigma M_{ZZ} = 0 \therefore -(106 \times 12) - (R_{YC} \times 9) = 0 \therefore R_{YC} = -141$ kN (141 kN downwards)

$\Sigma M_{XX} = 0 \therefore -(106 \times 12) - (R_{YB} \times 9) = 0 \therefore R_{YB} = -141$ kN (141 kN downwards)

$\Sigma M_{YY} = 0 \therefore \qquad\qquad (R_{ZC} \times 9) = 0 \therefore R_{ZC} = 0$

$\Sigma X = 0 \therefore \qquad R_{XA} + 106 - 106 = 0 \therefore R_{XA} = 0$

$\Sigma Z = 0 \therefore \qquad R_{ZA} + 106 - 106 = 0 \therefore R_{ZA} = 0$

Now calculate the value of R_{YA}.

58

$$\boxed{432 \text{ kN upwards}}$$

$$\Sigma Y = 0 \quad \therefore \ R_{YA} - 150 - 141 - 141 = 0 \quad \therefore \ R_{YA} = 432 \text{ kN}$$

59

$$\boxed{\text{TO REMEMBER}}$$

For equilibrium of plane structures $\Sigma H = 0$, $\Sigma V = 0$ and $\Sigma M = 0$.

Three support restraints are essential for a plane structure to be in equilibrium (unless loads are entirely vertical, when two will suffice).

A plane structure with more than three support restraints is externally statically indeterminate.

The addition of one internal pin provides one additional condition for equilibrium and enables one additional unknown support restraint to be determined.

A pinned support provides two components of reaction.

A roller support provides one reaction.

A fixed support provides two components of reaction and a fixing moment.

The moment of a couple is the same about any point.

A factor of safety against overturning is given by:

$$(\text{restoring moment})/(\text{overturning moment})$$

Six conditions of equilibrium must be satisfied for a space frame to be in equilibrium.

A space frame with more than six support restraints is externally statically indeterminate.

60

FURTHER PROBLEMS

1. A vertical load of 50 kN acts 0.8 m from end A of a simply supported beam AB which has a span of 3.0 m. Calculate the reactions at the ends of the beam.

Ans. ($V_A = 36.67$ kN, $V_B = 13.33$ kN)

2. A load of 50 kN acts on a beam AB which is supported by two beams CD and EF which rest on supports at C, D, E and F as in figure Q2. Calculate the loads on the four supports.

Ans. (C 22.0 kN, D 14.67 kN, E 8.0 kN, F 5.33 kN)

Q2

3. Determine the reactions at A and B for the frame in Q3.

Ans. ($R_A = 11.18$ kN ($V_A = +11.0$ kN; $H_A = +2.0$ kN) $R_B = 2.83$ kN at 45° to the horizontal)

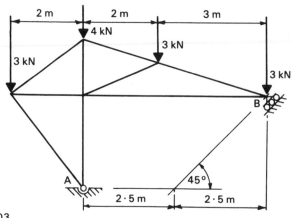

Q3

4. The beam AB in figure Q4 is pinned at A and on a roller support at B. Determine the reactions at A and B.

Ans. ($V_A = +31.22$ kN, $V_B = +36.78$ kN)

5. The beam in figure Q5 is rigidly fixed at A. Determine the reactions and fixing moment at A.

Ans. ($V_A = +10.0$ kN, $H_A = +12.0$ kN, $M_A = -42.0$ kN m)

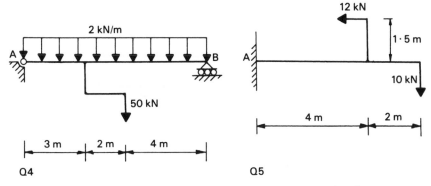

6. The beam ABC in figure Q6 has a pinned support at A and roller supports at B and C. P is an internal pin. Determine the reactions at A, B and C and at the pin P when the beam is loaded as shown.

Ans. ($V_A = +3.87$ kN, $H_A = +7.5$ kN, $V_B = +13.12$ kN, $V_C = +6.0$ kN, reaction at pin $= 4.0$ kN vertical)

7. The arch rib ABC has pinned supports at A and C. B is an internal pin. Determine the reactions at A and C when the arch is loaded as in figure Q7.

Ans. ($V_A = +19.97$ kN, $H_A = +24.86$ kN, $V_C = +7.53$ kN, $H_C = -11.87$ kN)

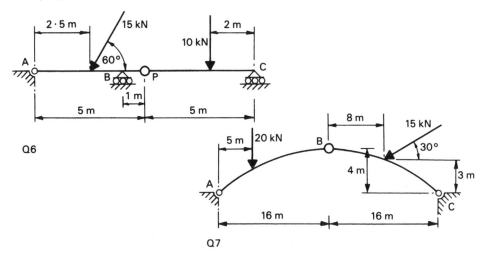

8. The mass concrete wall in figure Q8 is subjected to a force of 12 kN per metre length of wall as shown. If the density of concrete is 2400 kg/m^3 and the coefficient of friction between the base and the ground is 0.25, determine whether the wall is safe against (i) sliding and (ii) overturning. If the wall is safe, calculate the factor of safety in each case.

(*Hint:* you will need to calculate the weight per metre length.)

Ans. ((i) safe 1.49 (ii) safe 1.52)

9. The mass concrete wall in figure Q9 is to be loaded with a horizontal force W acting 1.5 m above the base as indicated. The density of the concrete is 2400 kg/m^3. If the factor of safety against overturning is to be 2.0, determine the maximum value of W.

Ans. (11.77 kN)

Q8   Q9

10. The space frame shown in figure Q10 has a ball jointed support at A. The support at B is free to move in any direction on the horizontal plane and the support at C is free to move in direction AC only. Calculate the value of the reactions at A, B and C.

(*Hint:* take A as the origin of the three axes—XX in direction AC, ZZ at right angles to XX in the horizontal plane and YY vertical.)

Ans. ($R_{XA} = +10.0$ kN, $R_{ZA} = +3.75$ kN, $R_{YA} = +10.0$ kN, $R_{ZC} = -3.75$ kN, $R_{YB} = +5.0$ kN, $R_{YC} = -5.0$ kN

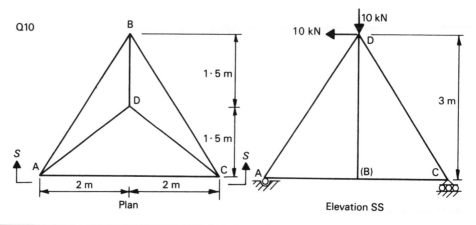

Programme 3

PIN-JOINTED
FRAME
STRUCTURES

On completion of this programme you will be able to:

- Identify the nature and form of pin-jointed frame structures
- State the conditions for internal statical determinacy in a pin-jointed plane frame
- Calculate the magnitude and nature of member forces by resolving at joints
- Identify the direction of member forces by inspection without calculation
- Identify situations where resolving in the horizontal and vertical direction may not always be most appropriate
- Apply the method of sections to calculate the magnitude and nature of member forces
- Recognise when to apply the method of sections or the method of resolution at joints
- Determine member forces by graphical methods

1

In the previous programme we studied the conditions necessary for the equilibrium of structures as whole bodies under the action of external forces. We can now begin to learn how structures react to those external forces, and how to calculate the values of the internal forces that develop within a structure as a consequence of the loads upon it. This programme will introduce you to frame structures and instruct you in methods of analysing such frameworks in order to determine the forces in the members. The term frame structure is applied to a number of different forms of construction. Thus the rigid assembly of columns and beams forming the skeleton of a large building may be described as a reinforced concrete frame. A lightweight pin-jointed roof truss may be described as a frame structure. Our studies will, however, be confined to a specific type of frame structure as specified below, and it is plane frames complying with the following definition that we shall be considering in this programme.

2

A pin-jointed frame structure may be defined as a structure built up of a number of straight members connected together at their ends by frictionless pinned joints to form a stable geometrical arrangement which is capable of carrying loads applied at some or all of the joints.

It is important to realise that since each member has a pin joint at each end and since a pin cannot transmit a moment, then no member can have a moment transmitted to it from the rest of the structure.

Pins can, however, transmit a force. Thus a member may have a force transmitted to it from the rest of the structure. Such a force could be resolved into a component (P) along the longitudinal axis of a member of a frame and a component (F) at right angles to the member as shown. Taking moments about the end A, we have:

$$\Sigma M_A = F \times L.$$

But, as A is a pin, for equilibrium the value of the total moment about A must be zero. Thus:

$$\Sigma M_A = 0$$
$$\therefore \quad F \times L = 0$$
$$\therefore \quad F = 0$$

Therefore there can be no component of force at right angles to the member and *the only possible force in a pin-jointed member is an axial one.* If this axial force is tensile, the member will increase in length under load and is called, *a tie.* If the force is compressive, the member will shorten and is called a *strut.*

3

The only forces in the members of a pin-jointed plane
frame are *axial* forces along the longitudinal axis.

When a frame is loaded, the members change in length and rotate until a state of equilibrium is reached. Look at the simple pin-jointed frame shown in figure (a).

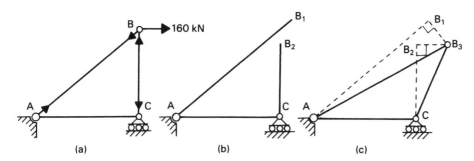

In the unloaded condition, the frame is a right-angled triangle. When the load is applied at joint B, member AB is subjected to a tensile force and consequently extends in length, while member CB is subjected to a compressive force and shortens. You will learn how to determine the magnitude and direction of these forced later in this programme. No force develops in AC, consequently that member does not change in length. Figure (b) shows what would happen to the two members AB and CB if they were not connected together at B. AB would lengthen, thus the end B would move to B_1. CB would shorten, thus the end B would move to B_2. But we know that the members are joined together at B and you can see in figure (c) that for this to be the case member AB must rotate about A and member CB must rotate about C until B_1 and B_2 come together at B_3. At this stage the movements of the two members are said to be compatible. You can also see in figure (c) the shape of the frame when deflected under the action of the external load at B. Remember that the actual values of the changes in length are very small, for example the extension of member AB might be approximately equal to its original length/1000. You should appreciate then that if figure (c) was drawn to scale it would not be possible to distinguish between B and B_3, and the shape of the frame would look the same as ABC in figure (a). The movements shown in figures (b) and (c) are grossly exaggerated. It is, however, useful to sketch such diagrams because they help us to appreciate what is actually happening in frames when subjected to external loadings.

4

When the frame in the figure shown is loaded as indicated, members AB and BC are subjected to compressive forces while member AC is subjected to a tensile force.

Sketch the shape of the deflected frame.

5

We conclude that:

 (i) For a frame to remain whole and stable under the action of loads the displacements of the members must be compatible.

(ii) Although internal movements take place, the geometrical configuration of the frame is effectively the same before and after loading. To indicate what is meant by 'effectively the same' look at the diagram above. If we were calculating the value of the reaction V_C by taking moments about A and used 4 m for the length AC' which might typically be of the order of 4.004 m, then the error involved would be insignificant.

6

INTERNALLY STATICALLY DETERMINATE FRAMES

If a frame has just sufficient members to be stable, it is said to be a *perfect frame* and the internal member forces may be determined by repeatedly using the three equations for statical equilibrium. Such a frame is referred to as being *internally statically determinate*. You will remember that in Programme 2 a frame in equilibrium under the action of external loads and support reactions and in which the values of the support reactions could be determined by using the three equations for statical equilibrium was referred to as being *externally statically determinate*. The simplest internally determinate frame consists of three members in the form of a triangle as in the figure below. You should recognise that this is a stable frame and that the removal of any one of the three members would result in the collapse of the other two. Three members are therefore essential for stability.

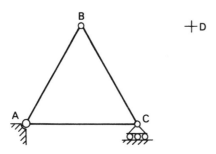

How could the additional joint D be joined into the frame in such a way that the resulting enlarged frame was stable?

7

> The additional joint (D) will require two additional members (BD and CD) to be added to the frame.

Look at the diagrams below. In figure (b) we see how the additional joint (D) can be joined into the frame using two members. In figure (c) a second joint is joined in using another two members. All the frames shown are stable structures and all are internally statically determinate.

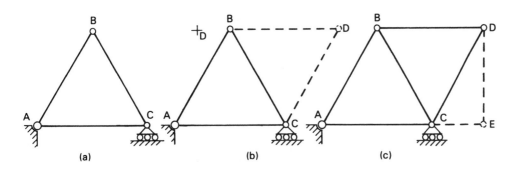

Using the above figures as a guide, can you derive an equation relating the number of members (M) and the number of joints (J) which would enable you to check whether a frame had sufficient members to be internally statically determinate?

8

$$\boxed{M = 2J - 3}$$

In the basic triangle in figure (a) of frame 7, $M = 3$ and $J = 3$. For every extra joint added to the frame, J increases by 1 and M increases by 2. Thus:

$$M - 3 = 2(J - 3)$$

and hence
$$M = 2J - 3 \qquad\qquad (3.1)$$

Equation (3.1) relates the necessary number of members to the number of joints in a perfect frame.

If a frame has too few members in relation to the number of joints, $M < (2J - 3)$, the frame is unstable and it, or part of it, will collapse.

If a frame has too many members, $M > (2J - 3)$, it is *imperfect* and cannot be completely analysed by the use of the equations for statical equilibrium. It is said to be *internally statically indeterminate*.

Use the relationship $M = 2J - 3$ to check which of the following frames are internally statically determinate.

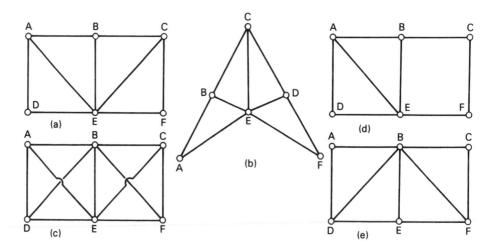

9

$$\boxed{\text{a, b and e}}$$

You should have decided that:

(i) Figures a, b and e are *perfect* frames with just sufficient members for stability.

(ii) Frame c is internally statically indeterminate. It has two more members than the minimum necessary for stability $(2J - 3 = (2 \times 6) - 3 = 9, M = 11)$. We say that it has two *redundant members*.

(iii) Frame d has insufficient members. You should be able to see that the right-hand panel will collapse as indicated in figure (f) below.

(f)

(g)

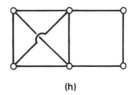
(h)

Are frames (g) and (h) statically determinate?

10

Frame (g) is determinate: frame (h) is not.

Both frames satisfy the expression $M = 2J - 3$ but you should have noticed that the right-hand panel of the frame (*h*) is similar to the right-hand panel of the previous example, figure (d), of Frame 8, and we saw that such a panel would collapse. It lacks stability. The problem with frame (h) is that, although it has the necessary number of members to satisfy equation (3.1), not all the members are in the necessary correct position. One of the diagonals in the left-hand panel should have been positioned in the right-hand panel to prevent collapse of this part of the structure.

As drawn, the left-hand panel is indeterminate since it contains a redundant member. Note that, for this panel $(2J - 3) = (2 \times 4) - 3 = 5$, but $M = 6$. In the absence of any other information, either of the diagonals in the left-hand panel could be considered as the redundant member.

You should now realise that for a frame to be statically determinate not only must $M = 2J - 3$ but *the members must be correctly positioned in the frame*. There is no simple rule to ensure that members are correctly positioned but in general if a frame is triangulated (that is, composed of triangular panels) it will be determinate.

Note that for any framework problem to be capable of complete solution using the equations of statical equilibrium, the frame must be both externally and internally determinate.

We have studied the way in which a pin-jointed frame responds to the application of a load.

Do you think that a real frame will behave in the same way as a theoretical pin-jointed one?

11

$$\boxed{\text{Not sure?}}$$

You probably felt obliged to give an indefinite answer. The question was perhaps unfair, but nevertheless a question you should consider.

In practice very few frames are likely to be constructed with perfectly pinned joints. It is in fact difficult to construct a frame with perfectly pinned joints. In many frames the members will be connected together by bolts, probably several bolts at each joint, or they may be welded together. In neither case are the joints likely to act like frictionless pins. Experience has shown that when a frame does not contain pinned joints the distribution of internal forces does not differ significantly from that of a similar frame with pinned joints. For most practical purposes, whatever the actual construction, it is normally convenient and sufficiently accurate to analyse frameworks on the assumption that the joints are pinned.

In the rest of the programme we will look at three methods for determining the forces in the members of pin-jointed frame structures.

12

METHOD OF RESOLUTION AT JOINTS

If a frame is in equilibrium, then every part of it is in equilibrium and *every joint is in equilibrium*. Consider the frame shown below. There is a pinned support at D and a roller support at F.

With one pin and one roller support the frame is externally determinate. There are 9 members and 6 joints, thus $M = 2J - 3$ and the frame is internally determinate. We may proceed.

First calculate the reactions by taking moments about D:

$$\Sigma M_\mathrm{D} = 0 \qquad \therefore \quad (20 \times 2) + (20 \times 6) + (20 \times 10) - (V_\mathrm{F} \times 8) = 0$$

$$\therefore \quad V_\mathrm{F} = +45 \text{ kN}$$

and by taking moments about F:

$$\Sigma M_F = 0 \quad \therefore \quad V_D \times 8 + (20 \times 2) - (20 \times 6) - (20 \times 2) = 0$$
$$\therefore \quad V_D = +15 \text{ kN}$$

check $\quad \Sigma V = 0 \qquad\qquad\qquad\qquad 45 + 15 - 60 = 0$

Now consider the internal axial forces in the members of the frame. Let these forces be F_{AB}, F_{BC}, F_{CF} etc. as shown. We do not know whether these forces are tensile or compressive, so for calculation purposes let's assume they are tensile. The arrow heads on figure (a) are shown accordingly. If we end up with a negative value for a force it will simply mean that the force is compressive and not tensile as initially assumed. We can see that all the joints are acted upon by a number of forces acting along the longitudinal axis of the members. Figure (b) shows one of the joints (E) together with the forces acting on it. This joint is in equilibrium and hence we can apply the rules for statical equilibrium. We can *resolve horizontally* and *vertically*:

substituting in $\Sigma H = 0$ we have $\quad F_{EF} + F_{BE} \cos 60° - F_{AE} \cos 60° - F_{DE} = 0$

and in $\qquad\qquad \Sigma V = 0 \qquad\qquad\qquad F_{BE} \sin 60° + F_{AE} \sin 60° = 0$

Can these equations be solved for the values of the forces? If not, why not?

13

NO, there are insufficient equations.

Since there are four unknown forces, we need four equations to be able to calculate the values of the forces. At any joint it is only possible to write down two unique equations; that is, $\Sigma H = 0$ and $\Sigma V = 0$. Consequently in order to obtain a solution we must look for a joint with only *two* unknown forces. The frame has been redrawn below and if you look at it you will see that there are in fact only two joints meeting that requirement (C and D). This does not look too promising but let's continue:

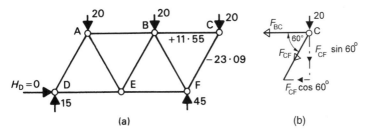

(a)   (b)

We will start at joint C, which is shown in the free-body diagram, figure (b), above, and resolve vertically:

$$\Sigma V = 0 \quad \therefore \quad -F_{CF} \sin 60° - 20 = 0 \quad \text{and} \quad F_{CF} = -23.09 \text{ kN}$$

Now resolve horizontally to determine the value of F_{BC}.

14

$$\boxed{F_{BC} = +11.55 \text{ kN}}$$

$$\Sigma H = 0 \qquad \therefore \quad -F_{BC} - F_{CF} \cos 60° = 0$$
$$\therefore \quad F_{BC} = -F_{CF} \cos 60°$$
$$= -(-23.09) \cos 60°$$
$$= +11.55 \text{ kN}$$

The force in member BC= $F_{BC} = 11.55$ kN tension (tension because the answer is positive). The force in member CF= $F_{CF} = 23.09$ kN compression (compression because the number is negative).

Having determined the value of these two forces, write them on the diagram adjacent to the members in which they act. You will see that this has already been done on figure (a) in Frame 13. You should also see that there are now other joints capable of solution.

Which two joints are now capable of solution?

15

D and F, because there are only two unknown
member forces at each of those joints

Considering joint F and resolving vertically:

$$\Sigma V = 0 \therefore \quad +45 + F_{BF} \sin 60° + F_{CF} \sin 60° = 0 :$$
$$\text{but we know that } F_{CF} = -23.09 \text{ kN}$$
$$\therefore \quad +45 + (-23.09) \sin 60° + F_{BF} \sin 60° = 0$$
$$\underline{F_{BF} = -28.87 \text{ kN}}$$

resolving horizontally:

$$\Sigma H = 0 \therefore \quad F_{CF} \cos 60° - F_{BF} \cos 60° - F_{EF} = 0$$
$$\therefore \quad (-23.09) \cos 60° - (-28.87) \cos 60° - F_{EF} = 0$$
$$\underline{F_{EF} = +2.89 \text{ kN}}$$

Which joint could be solved next?

16

B

Since the force in member BF is now known, the number of unknowns at joint B is reduced to two. Joint D, of course, is still capable of solution, but since we started at the right-hand end and are working towards the left, it is perhaps better to continue working systematically from right to left. When tackling this type of problem, it is always wise to adopt a systematic procedure.

Now resolve vertically and horizontally at joint B and determine the forces F_{AB} and F_{BE}.

17

$$F_{AB} = 5.78 \text{ kN compression}: F_{BE} = 5.78 \text{ kN tension}$$

If you did not get these values, check your equations against the following:

$\Sigma H = 0; \qquad +F_{BC} + F_{BF} \cos 60° - F_{BE} \cos 60° - F_{AB} = 0$

$\Sigma V = 0; \qquad -20 - F_{BF} \sin 60° - F_{BE} \sin 60° = 0$

Now complete the analysis by resolving the forces at joints E and A. In each case draw a free-body diagram of each joint before attempting to write down the equations at each joint.

18

The diagrams below show the forces in all the members of the frame. In figure (a), the algebraic values are given and the arrow heads left as originally inserted on the assumption that all forces were tensile. We now know that not all the forces are tensile and figure (b) shows the frame with the arrow heads correctly inserted to show the actual sense of each force. When you are starting to gain experience in the analysis of frames, it is a good idea to sketch diagrams similar to figure (b) each time you complete a problem. Such diagrams will help you to understand the part played by each member towards the overall stability of the frame, and will help you to recognise which members are struts (arrows point towards the joints) and which are ties (arrows point away from the joints). Carefully study the diagrams and relate the information given about the forces to the values derived in the previous few frames of this text.

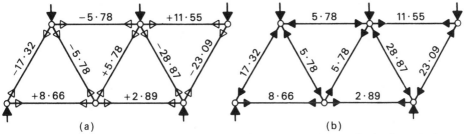

(a)   (b)

If you had to analyse a pin-jointed plane frame using the method of resolution at joints, how would you select the starting point?

19

> By locating a joint at which there are no more
> than two unknown member forces

In the example just completed, we were able to start at the loaded joint C and could have done so even if the values of the reactions were not known. When we reached joint F however, it was necessary to know the value of the reaction at that support. In some problems it is possible to complete the analysis without calculating the reactions; in other cases it may be necessary to calculate the reactions before a start can be made on the analysis. Always inspect the frame in a given problem to decide whether or not it is necessary to determine the reactions before starting to resolve the forces at the joints. You should also mark on a diagram of the frame the components of the reactions at all supports. If you do not do this, you may forget to take account of the reactions when resolving at a joint at a support.

20

PROBLEMS

Determine the forces in all the members of the following frames:

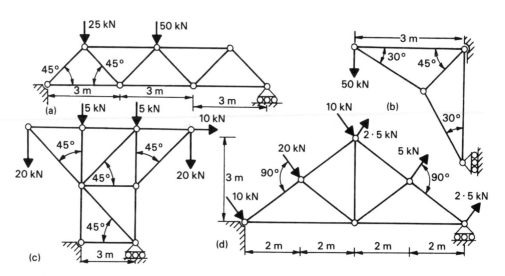

The solutions are given on the diagrams in the next frame.

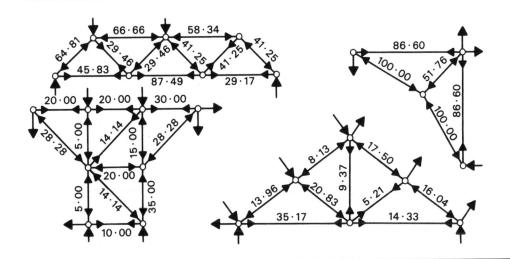

22

Now let's look at some special cases:

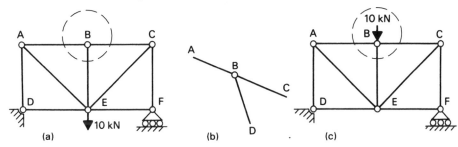

(a)   (b)   (c)

Consider joint B in figure (a). If you try to resolve vertically at this joint you discover that there is only one member (BE) which might be able to provide a vertical force. The other members are horizontal and the forces in them cannot have vertical components.

What do you think is the value of the force in BE?

23

> zero

This is because there is no other member at B capable of providing a vertical force to balance the force in BE acting on joint B. Nor is there any external load at B which can balance an internal force in BE.

It follows that, if three members, two of which are collinear, meet at a joint at which no external load acts, then the force in the third, non-collinear, member must be zero. For example in figure (b), forces in members AB and BC cannot resist any components of force at right angles to AC, consequently there can be no force in member BD.

What is the force in the member BE in figure (c)?

81

24

10 kN

You should have obtained this value by resolving vertically at B. We see that, although there are more then two unknown forces acting at B it is possible to determine the force in BE by the use of one equation (in this case by resolving vertically) or simply by inspection.

So far we have resolved horizontally and vertically at the joints. It is not essential to do this. You should remember from Programme 1 that you can resolve forces in *any* two directions at right angles to each other. It is in fact often convenient and simpler to select directions other than the vertical and the horizontal. In the figure below, for example, the calculations will be made easier if you resolve in a direction *parallel* to the rafter AD and at *right angles* to AD when determining the forces in the members at joint C.

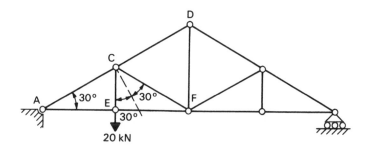

By resolving at joint E and then at joint C, determine the values of the forces in members CE and CF.

25

force in CE = 20 kN tension: force in CF = 20 kN compression

By inspection of joint E it should be obvious that the force in CE is 20 kN tension. Then by resolving at C in a direction at right angles to AD:

$$F_{CE}\cos 30° + F_{CF}\cos 30° = 0 \quad \text{(taking forces in directions away from joint C as positive)}$$

$$\therefore \quad +20\cos 30° + F_{CF}\cos 30° = 0$$
$$\therefore \quad 20 + F_{CF} = 0$$
$$\therefore \quad F_{CF} = -20 \text{ kN}$$

Now to the next frame.

26

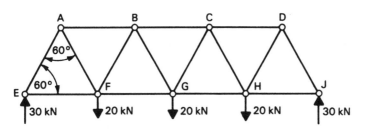

If we start at the left-hand end of the above structure and resolve vertically at E, then the force in member AE is given by:

$$F_{AE} \sin 60° + 30 = 0$$
$$\therefore \quad F_{AE} = -30/0.866$$
$$= -34.64 \text{ kN (compression)}$$

then by resolving vertically at joint A we obtain the force in AF by:

$$-F_{AF} \sin 60° - F_{AE} \sin 60° = 0$$
$$\therefore \quad -F_{AF} \sin 60° - (-34.64) \sin 60° = 0$$
$$\therefore \quad F_{AF} = +34.64 \text{ kN (tension)}$$

Now start from J and determine the value of the force in member DH.

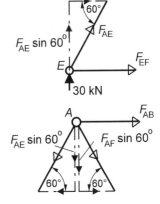

27

$$\boxed{F_{DH} = +34.64 \text{ kN}}$$

Did you see a quicker way of arriving at this result? In fact the force in DH is the same as the force in AF, and we could have anticipated this result because the frame and the loading on it are symmetrical. *In any geometrically symmetrical frame which is subjected to symmetrically placed loading, then the forces in any two members symmetrically positioned within the frame will be identical.*

The figure below is the same frame with all internal forces shown. Note how the values comply with the above statement.

Now to the next frame.

28

When faced with the task of analysing a frame, you should look to see whether any of the special cases mentioned in the previous frames are involved. You may be able to make your task easier.

It is essential for your future work that you should be able rapidly to analyse any frame. Your ability to do this quickly and accurately is increased if you can visualise the way in which the frame will deflect under load and if you can recognise by inspection which members are struts and which are ties. You should aim to achieve the ability to look at a frame and by inspection (and experience) identify the struts and ties. You may not immediately be able to do this instinctively, but to begin with let's mentally work through a logical sequence for a typical problem. Looking at the figure below you might reason as follows:

The reaction at F is pushing up, thus the force in AF must be pushing down at F. Put an arrow on AF (near to F) pointing towards F. At the other and of AF put an arrow pointing in the opposite direction, towards A. The force in FG is zero. Why? See Frame 23 if you are not sure. Move to joint A. The force in FA is pushing up on A (as shown by the arrow head) consequently the force in AG must be pulling down. Put an arrow on AG (near to A) pointing down (away from A). At the other end of AG put an arrow pointing in the opposite direction (that is, away from G). Looking back at A, the force in AG is pulling A to the right, thus the force in AB must be pushing A to the left. Put an arrow on AB. Working in this fashion through the entire frame, you will eventually have arrows on all members and can then see which are struts and which are ties.

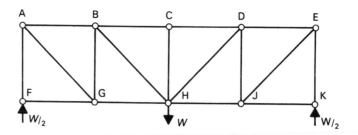

In this particular case you need only work halfway through the frame. Why?

29

Because the frame and loads are symmetrical

Complete the process of putting the arrows on the members of the frame and check against the diagram on the next page.

30

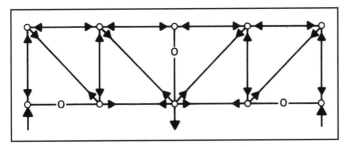

Once you have mastered this technique of identifying the direction of member forces by *inspection* of the frame the mathematics involved in the subsequent solution of the joint equilibrium equations will be much simplified. This is because an initially correct evaluation of the *direction* of member forces should result in equations of equilibrium which will then solve to give positive values of *magnitude* of force. Hence there will be no need to 'carry forward' negative values in the calculations at the remaining joints. You will discover, however, that in some cases it may not be possible to assess by *inspection* the direction of force in all the members of a frame, in which case you will have to assume the direction of force and determine the correct direction through the process of calculation.

31

PROBLEMS

Work through the following figures putting arrows on the members to indicate the actual direction of the internal forces and mark the struts with an S and the ties with a T. Check your answers against the sketches in the next frame.

32

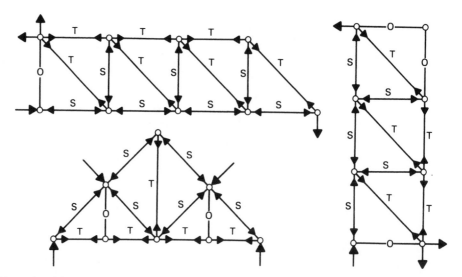

You should have noticed that in the work you have just completed a number of members had no force in them. This does not imply that they can be removed. A change in the load system or a slight movement will make their presence essential to prevent collapse of the frame. They will also be needed in practice to support the dead weight of other members, a factor which we are neglecting.

33

METHOD OF SECTIONS

Consider the frame below and assume that we wish to know the values of the forces in members BC, BG and GH. Let the forces in those members be F_{BC}, F_{BG} and F_{GH} respectively. Since we do not know the sense of these forces, let's assume they are all tensile as indicated by the arrows. Now imagine that these three members are cut and that the two parts of the frame are pulled apart as shown in figure (b), the cut members being replaced by forces F_{BC}, F_{BG} and F_{GH} equal in magnitude to, and in the same direction as, the forces in the original members before they were cut. The two parts of the frame are now effectively two independent frames each acted upon by five external forces.

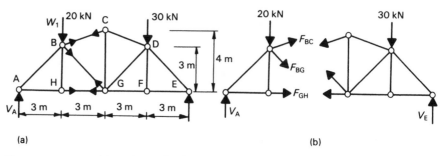

Can you say anything about the forces acting on either part of the frame?

34

> They must be in equilibrium.

If a frame is in equilibrium then the forces acting on any part of the frame must also be in equilibrium. Considering the left-hand part, the value of load W_1 is known and the reaction V_A can be calculated by considering the forces acting on the whole frame before, in our imagination, we cut the frame in two. For the part frame therefore, there are only three unknown forces and since we can write three equations for statical equilibrium the unknown forces can be determined.

The three equations of equilibrium are $\Sigma V = 0$, $\Sigma H = 0$ and $\Sigma M = 0$.

If we are going to use $\Sigma M = 0$ to determine the value of F_{BC}, can you suggest the most convenient point about which to take moments?

35

> G

By taking moments about the point of intersection of the line of action of two of the forces, those forces do not appear in the moment equation and we are left with one unknown only in one equation. G is the point of intersection of F_{BG} and F_{GH}, thus taking moments about G will enable F_{BC} to be readily calculated. The value of the reaction V_A is shown on the figure as is the dimension p_1 which can be calculated from the geometry of the frame.

$$\Sigma M_G = 0 \qquad \therefore \quad (V_A \times 6) + (F_{BC} \times p_1) - (W_1 \times 3) = 0$$
$$(22.5 \times 6) + (F_{BC} \times 3.79) - (20 \times 3) = 0 \quad \therefore \; F_{BC} = -19.79 \text{ kN}$$

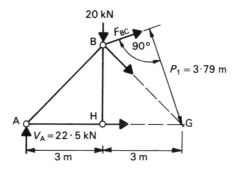

Which are the most appropriate points to take moments about in order to determine (a) F_{GH} and (b) F_{BG}?

(a) B, which is the intersection of F_{BC} and F_{BG}
(b) the intersection of F_{BC} and F_{GH}

The intersection of the line of action of F_{BC} and F_{GH} will be outside the frame and is shown as I in the sketch below.

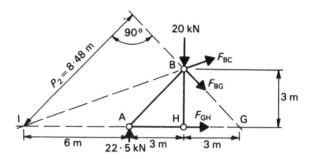

The appropriate equation to determine the values of F_{BG} will be given by:

$$\Sigma M_I = 0 \qquad \therefore \quad (F_{BG} \times p_2) + (W_1 \times 9) - (V_A \times 6) = 0$$
$$(F_{BG} \times 8.48) + (20 \times 9) - (22.5 \times 6) = 0$$
$$\therefore \quad \underline{F_{BG} = -5.31 \text{ kN}}$$

and to determine F_{GH} by taking moments about B:

$$\Sigma M_B = 0 \qquad \therefore \quad (V_A \times 3) - (F_{GH} \times 3) = 0$$
$$\therefore \quad \underline{F_{GH} = +22.5 \text{ kN}}$$

You will see that by a suitable choice of section, it should be possible to determine the value of force in any member. The only limitation is that, because only three unique equations can be written, *the section must not pass through more than three members*. The procedure for using the method of sections is often expressed as follows:

To determine the force in a member of a frame, cut the frame by a section which passes through that member and no more than two other members. Then considering one part of the frame, replace the cut members by external forces equal in magnitude to, and having the same line of action as, the forces in the cut members. The value of the required force may then be obtained by taking moments of all external forces on the part frame about the intersection of the line of action of the forces in the other two cut members.

The above explanation of the method of sections suggests, however, that the solution of the part frame is always effected by taking moments. This is, of course not so. The solution is generally conveniently obtained by taking moments, but in some cases it is more convenient to resolve forces vertically or horizontally. Indeed, in some problems it would be impossible to take moments about the intersection of the forces in two of the cut members if the members are parallel and do not intersect!

Could you have determined the values of F_{BC}, F_{BG} and F_{GH} by considering the right-hand part of the frame?

37

> yes: both parts of the frame must be in equilibrium

You have learnt how to determine the forces in members of a frame by cutting the frame by a section through the members concerned and then solving either part for the 'external' forces acting on it. This is a useful technique when a complete analysis of the frame is not required. Try this technique on the following problems.

38

PROBLEMS

1. The roof truss in figure Q1 is pinned at A and has a roller support at E. Determine the values of the forces in members BC, CG and FG. State whether the members are struts or ties. (*Hint*: having chosen the section at which you are going to cut the frame, sketch the part of the frame you are going to consider and make sure you put on the sketch *all* the forces, loads and reactions that act on that part.)

Ans. ($F_{BC} = 50.0$ kN (strut), $F_{CG} = 17.32$ kN (tie), $F_{FG} = 34.63$ kN (tie))

2. The frame in figure Q2 is pinned at A and has a roller support at B. Use the method of sections to determine the force in member BC. (*Hint*: this is a case where you need to resolve instead of taking moments.)

Ans. ($F_{BC} = 28.28$ kN (strut))

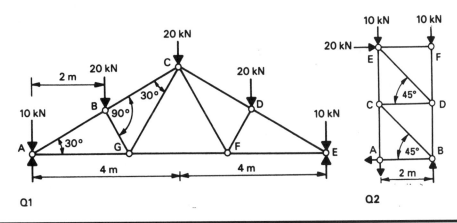

Q1

Q2

39

GRAPHICAL METHOD: FORCE DIAGRAMS

The graphical method which will be described in this and the following frames is now largely superseded by other methods. You may, however, find it of use if you have to analyse a frame which is geometrically complicated and thus tedious to solve mathematically. Practise of the method is also useful in helping students to understand the inter-relation of the forces in the members of a frame.

When using a graphical method of analysing a frame, it is convenient to use *Bow's notation* for referring to forces. This involves lettering the spaces between the members of a frame, as shown below, instead of lettering the joints. The spaces between the external forces are also lettered. In the diagram below the joints and members are also numbered. This is purely for reference purposes in this text and would not be necessary in practice.

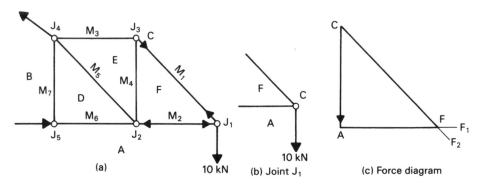

(a) (b) Joint J_1 (c) Force diagram

Now consider joint J_1 which is shown in figure (b). The forces acting on the joint are denoted by the letters of the spaces either side of them. Thus the load is force CA, the force in the bottom frame member is AF and the force in the diagonal is FC. Note that we have progressed in a clockwise direction round the joint when using this convention to denote the forces. Since these forces are concurrent and in equilibrium, a triangle of forces can be drawn from which the magnitude and direction of the unknown forces (AF and FC) can be determined. You should remember from Programme 1 how to do this but let's remind ourselves. First draw vector *CA* parallel to the load and to some scale representing 10 kN (see figure (c)). Then from A draw a line AF_1 parallel to member M_2 and from C draw a line CF_2 parallel to member M_1. The intersection of these lines locates F on the force diagram. CAF is the force diagram for joint J_1 and the lengths of the sides of the triangle represent the magnitude and direction of the member forces.

We know that force CA acts downwards. You will also remember that when vectors represent forces in equilibrium the direction of those vectors is such that they follow one another round the force diagram. By looking at the force diagram (c), determine the direction of forces AF and FC at joint J_1.

40

> Force AF acts to the right.
> Force FC acts up and to the left.

Put arrows on the ends nearest joint J_1 of members M_2 and M_1 in the free-body diagram to indicate these directions. Put arrows in the opposite direction at the other ends of members M_2 and M_1. (You can see that this has already been done on the figure (a) on the previous page.)

You now see that at joint J_3 in the free-body diagram there are only two unknown forces. Thus we can now draw a force diagram for joint J_3 and determine the magnitude and direction of the forces in members M_4 and M_3.

Draw the force diagram for joint J_3 and determine the direction of the forces in members M_4 and M_3.

41

> The force in member M_4 is upwards at joint J_3.
> The force in member M_3 is to the left at joint J_3.

Look at the diagrams below to see how we arrive at this conclusion. Figure (a) shows the forces acting at joint J_3 and figure (b) is the force diagram.

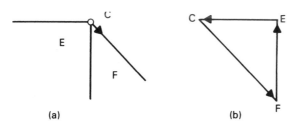

(a)   (b)

When attempting the question in frame 40 you may have experienced some difficulty and eventually realised that the procedure for 'naming' the forces is important. When we were considering joint J_1 we 'named the load CA: we could have called it AC! We had in fact chosen to work clockwise round the joint when naming the forces. Having thus chosen we must be consistent and work clockwise round *all* joints. Thus at joint J_3 the forces are CF, FE and EC. If you are not consistent in this respect you will not be able to read the true direction of action of the forces from the completed force diagrams.

Put arrows on members M_4 and M_3 in the free-body diagram to show the direction of the member forces. Then sketch the force diagram for joint J_2.

42

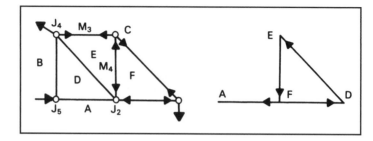

Having now drawn the force diagrams for joints J_1, J_3 and J_2, let's consider joints J_4 and J_5. The force diagrams for these joints are given below.

Force diagram for joint J_5

Force diagram for joint J_4

From these two diagrams we can obtain not only the forces in the remaining members of the frame but also the values of the two external reactions.

If we put arrows on the members in the free-body diagram every time we complete the force diagram for a joint, then by the time we have drawn all the force diagrams all members in the free-body diagram will be marked with arrows and we will immediately be able to distinguish between struts and ties.

What we have done so far is acceptable but is in fact wasteful of time and effort.

Look at all the five force diagrams that we have sketched. Do you notice any common features?

43

Each line appears in at least two diagrams.

For example, the line CF appears in the force diagrams for joint J_1 (as vector *FC*) and joint J_3 (vector *CF*). It follows that the diagram for joint J_3 could be superimposed on to that for joint J_1 if the line CF is taken as a line common to both diagrams. In fact it should be possible to combine all the diagrams into one force diagram for the entire frame.

Sketch the composite force diagram incorporating the individual diagrams drawn for the five joints.

44

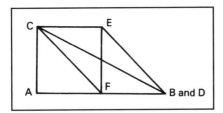

In practice we would not draw the individual diagrams for each joint, but would construct the composite diagram straight away, building it up step by step as we worked progressively through the frame from joint to joint.

Warning: do not put arrows on the composite force diagram. Why not? Because the vectors act in different directions depending upon which joint is being considered. You would thus end up with arrows in different directions on the same vector and confusion would result. In order to interpret the force diagram to give direction of force, proceed as follows:

Suppose we wish to determine the sense of the force in member M_5. Is that member a strut or a tie? The force acts in different directions at each end, so let's choose the end at joint J_2. Reading the letters round joint J_2 in a *clockwise* sense the force we want is DE. On the force diagram vector *DE* implies a direction up and to the left (to move from D to E you must move up and to the left). Thus the force in member M_5 at the end nearest joint J_2 acts in a direction away from the joint. Similarly the force in member M_5 at the end nearest joint J_4 is seen to act away from that joint.

Bearing in mind what we have just decided, is member M_5 a strut or a tie?

45

A tie: because the internal member force is tensile.

In many problems it is necessary to determine the reactions before a force diagram can be commenced. There are graphical methods for obtaining the reactions and you may find such methods described elsewhere. In practice, however, it is probably better to calculate the reactions mathematically before proceeding to draw the force diagram as outlined above.

We have used a simple frame to learn how to construct and interpret a force diagram. The real advantage of the method is only appreciated however when a geometrically complicated framework has to be analysed. You may practise this method by doing problems Q7 and Q8 at the end of this programme. Complete solutions are provided.

46

SPACE FRAMES

A space frame can be analysed in the same way as a plane frame using the method of resolution at joints, the only difference being that more equations are involved and that the resolution of the forces is more complicated. Manual solution methods are not so convenient for the analysis of space frames which can be more readily analysed using computer methods. Software packages are available for this purpose.

47

TO REMEMBER

For a plane frame to have a sufficient number of members to be statically determinate:

$$M = 2J - 3$$

Members must be correctly positioned in the frame.

To solve a frame by resolution at joints, start at any joint at which there are no more than two unknowns. It is not always necessary to determine the reactions before starting to resolve at the joints.

To determine the value of a force in a particular member, use the method of sections. Cut the frame by a section through the member and no more than two others. Solve either part as a frame in equilibrium.

A graphical method of solution is useful to solve frames which have a complicated geometry.

48

PROBLEMS

1. Use the method of resolution at joints to determine the forces in all members of the frames in figures Q1, Q2 and Q3.

Q1

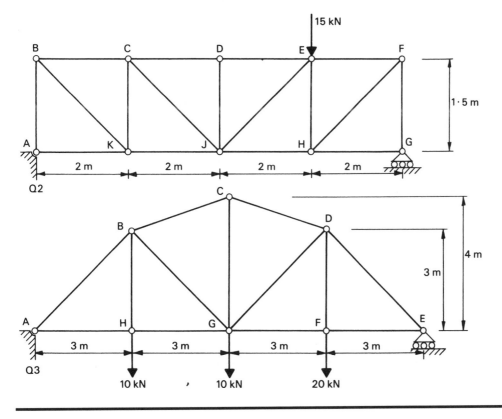

Q2

Q3

10 kN   10 kN   20 kN

Ans.

Member	Force	Member	Force	Member	Force
Q1 BC	−10.00	CD	−10.00	AG	+7.50
GF	+12.50	FE	+7.50	AB	−16.77
BG	+5.59	GC	−5.59	CF	−5.59
FD	+5.59	DE	−16.77		
Q2 BC	−5.00	CD	−10.00	DE	−10.00
EF	−15.00	AK	0	KJ	+5.00
JH	+15.00	HG	0	AB	−3.75
KC	−3.75	JD	0	HE	−11.25
GF	−11.25	BK	+6.25	CJ	+6.25
EJ	−6.25	FH	+18.75		
Q3 AB	−24.75	BC	−19.76	CD	−19.76
DE	−31.82	AH	+17.50	HG	+17.50
GF	+22.50	FE	+22.50	BH	+10.00
CG	+12.50	DF	+20.00	BG	+1.77
DG	−5.30				

2. Use the method of sections to determine the forces in members marked X, Y and Z in the frames in figures Q4, Q5 and Q6. (*Hint:* you may need to consider more that one section.)

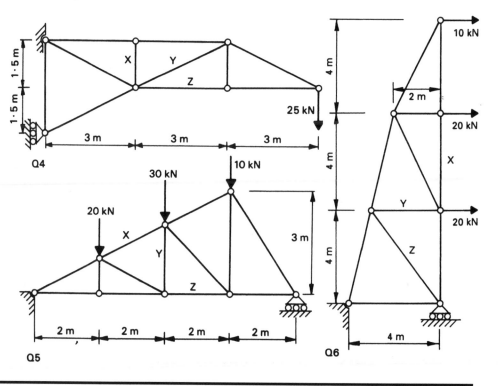

Ans.

Frame	X	Y	Z
Q4	0.00	−55.90	−50.00
Q5	−50.31	+10.00	+45.00
Q6	−20.00	+36.67	−45.83

3. Construct the force diagrams for the frames in figures Q7 and Q8, and determine the magnitude and the sense of the forces in all members. Note that in these problems the spaces have been lettered. Bow's notation may be used to denote both a member and the force in that member. The force diagrams are given on the next page.

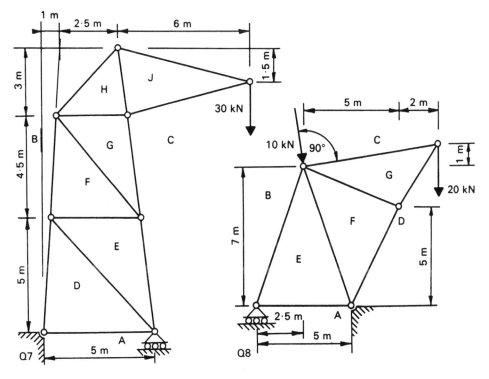

Ans.

Member	Force	Member	Force	Member	Force
Q7					
AD	−2.2	BD	27.1	BF	38.4
BH	73.9	BJ	59.9	CE	−68.7
CG	−85.2	CJ	−60.2	DE	15.3
EF	−11.3	FG	21.0	GH	−60.0
HJ	−69.6				
Q8					
AE	−5.4	BE	16.0	CG	14.9
DF	−26.1	DG	−26.6	EF	−22.9
FG	−3.3				

Force diagrams for problems Q7 and Q8

Q7

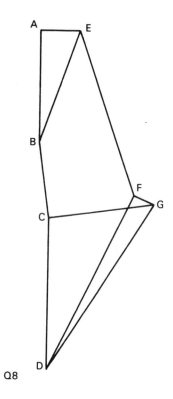

Q8

Programme 4

SHEARING FORCES
AND
BENDING MOMENTS

On completion of this programme you will be able to:

- Define the term 'bending moment'
- Define the term 'shearing force'
- Draw deflected shape diagrams for simple beams
- Calculate the value and sense of shearing force at different sections along a beam
- Calculate the value and sense of bending moment at different sections along a beam
- State the formula for maximum bending moment in a simply supported beam carrying a central point load
- State the formula for maximum bending moment in a simply supported beam supporting a uniformly distributed load (UDL) over the whole span
- Draw shearing force diagrams for simply supported beams, with or without cantilevers, subject to combinations of point loads and UDLs
- Draw bending moment diagrams for simply supported beams, with or without cantilevers, subject to combinations of point loads and UDLs
- Identify and calculate the position of points of contraflexure
- Draw bending moment diagrams for simple portal frame structures

1

In the previous programme we studied the effect of external loads upon the members of pin jointed frames. We saw that, as a result of the loading, such members were subjected to direct forces acting along their longitudinal axes and increased or decreased in length according to the sense of those forces. In practice however, the majority of structures are not pin jointed and often loads are applied at points other than the joints between members. You will appreciate, for example, that the self-weight of members will act along their entire length. In this programme we will investigate the way in which structural members respond to such loads. The reasoning we will adopt could be applied to any member of a structure regardless of whether it be a horizontal beam, a vertical column or a member inclined at an angle other than 90° to the horizontal. For convenience, however, we will initially confine ourselves to a consideration of horizontal beams.

In Programme 3 we studied the behaviour of pin-jointed frames. Why are the members of such frames subjected only to longitudinal axial forces?

2

> (1) Each member has a pinned joint at each end.
> (2) Loads are only applied at the joints of the frames

BENDING MOMENTS

The effect of the pins at the ends of the members combined with joint loading is to permit axial deformations and axial forces only within the members. The members we are now going to consider may be loaded anywhere along their lengths and consequently other forms of deformation will take place resulting in internal forces other than simply axial forces.

We start by considering a simply supported beam AB as shown. The beam is loaded only by a single concentrated load W acting at the mid span point C. We will for the moment ignore the self-weight of the beam. You should be able to visualise that the beam is likely to respond to the load by bending as in figure (b). The underside would be stretched and would be in tension.

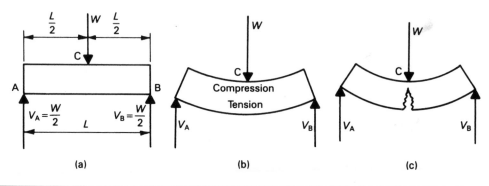

(a)   (b)   (c)

If the load was large enough the beam *could* fail in bending owing to the development of excessive tension resulting in tearing at C as indicated in figure (c). The bending is of course initiated by the load but all the forces, including the reactions, should be taken into consideration when analysing the response of the beam to loading. Thus if you look at the part of the beam to the right of C you could visualise that, relative to C, the reaction is apparently forcing the beam to bend concave upwards.

You will realise that in order to carry out a numerical analysis it is necessary to be able to quantify the bending effect at any point in a beam. Let's see how we do this. Look again at the part of the beam to the right of C. To quantify the bending effect at the point C we can say that the *bending moment* at C is the moment about C of the reaction V_B. The value of this *bending moment* is $W/2 \times L/2 = WL/4$ anticlockwise.

Looking to the left of C what is the bending moment at C due to the reaction V_A?

3

$$W/2 \times L/2 = WL/4 \ clockwise$$

You will notice that this is the same numerically as the bending moment about C due to the reaction V_B but the sense is reversed. Hence we can quantify the magnitude of the bending effect at C by calculating the moments of all the forces to the left or right of C, but we cannot base a sign convention upon the sense of the moments because the sense is different depending on which side we consider. What we can do is to say that bending moments that cause *sagging* of a beam (see figure (a) below) are positive, while bending moments causing *hogging* of a beam (figure (b)) are negative. Note that when the beam is subjected to positive bending the bending moment to the right of the point under consideration is anticlockwise, the bending moment to the left of the point under consideration is clockwise. The reverse is true for a beam which is hogging.

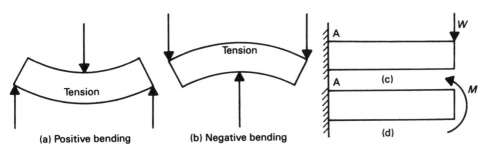

(a) Positive bending (b) Negative bending (c) (d)

End A of each of the two beams shown in figures (c) and (d) is rigidly fixed to a wall. Beam (c) is loaded by a single concentrated load and beam (d) is subjected to an external couple of moment M. By visualising the way in which the beams will bend, determine whether the bending is positive or negative.

4

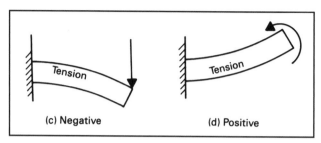

(c) Negative   (d) Positive

In case (d) the effect of the couple will be to stretch the bottom of the beam and compress the top. The only way such deformations can take place is if the beam bends upwards as indicated.

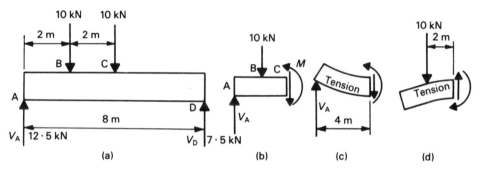

(a)   (b)   (c)   (d)

Now look at the beam AD with a concentrated load of 10 kN at the mid-span point C and another concentrated load of 10 kN at B. In this example the values of the two reactions have already been calculated to be as indicated. In future problems you may have to start by determining the values of the reactions as explained in Programme 2.

In order to calculate the bending moment at C try to visualise the left-hand half of the beam as in figure (b). Figure (b) also shows the force and moment which must be developed within the beam at C to maintain the free body of this part of the beam in equilibrium. This moment is the bending moment at C. To the left of C and viewed relative to C the reaction V_A, if considered to be acting alone, would apparently tend to bend the beam upwards as in figure (c) and cause tension on the bottom face. This shape of bending has been defined as positive, thus V_A produces a positive bending moment about point C, the numerical value being $V_A \times 4$ kN m. In a similar way, the load at B, if considered to be acting alone, produces a negative bending moment about C of numerical value 10×2 kN m and causes tension on the top face. The total net bending moment at C due to the forces to the left of C is given by the algebraic sum of the separate bending moments. Thus:

$$\text{Bending moment at C} = +(V_A \times 4) - (10 \times 2)$$
$$= +(12.5 \times 4) - (10 \times 2)$$
$$= +30 \text{ kN m}$$

In the calculation just completed, we considered the left-hand part of the beam. Now calculate the bending moment at C by considering the part of the beam to the right of C.

5

$$+30 \text{ kN m}$$

V_D is the only force to the right of C and relative to C this reaction will apparently tend to bend the beam in a similar way to V_A (see figure (c) of Frame 4). This is positive (sagging) bending and thus V_D produces a bending moment at C of value $+(V_D \times 4) = 7.5 \times 4 = +30$ kN m. You should have anticipated this result from the work done in Frame 3. Summarising what we have learnt so far about the bending moment in a beam, we can state:

The bending moment at any section in a beam is the algebraic sum of the moments about that section of all the forces to the right, or to the left, of that section.

It has been suggested in the previous few frames that you try to visualise the way in which a beam bends under the action of loads. You will find later on in your studies that the ability to do this, and to be able to visualise the way in which more complicated structures deflect, leads to the early identification of those zones within a structure where tension is likely to develop and is thus a great help in structural analysis and design. You should aim to develop this ability. Make a start by tackling the following problems. You may find it helpful to use a flexible plastic rule which you can hold and bend to simulate the effect of loading and support restraint.

6

PROBLEMS

The following sketches show a series of beams subjected to different loading conditions. In each case visualise the way in which the beam will bend and sketch the shape of the deflected beam. Show the loads and the reactions on your sketches and mark the lengths of the beam (top or bottom) where tension develops owing to bending.

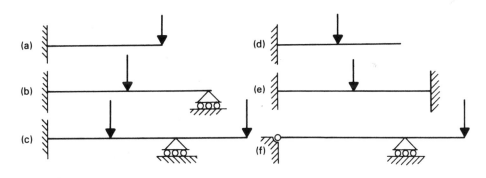

7

Check your solutions against the following diagrams.

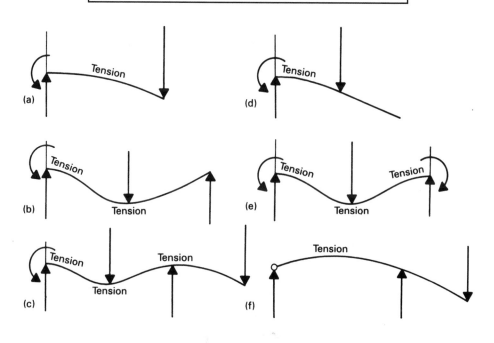

(a)   (d)

(b)   (e)

(c)   (f)

8

SHEARING FORCES

Now, let's consider another way in which the beam of Frame 2 could fail. For convenience the beam is shown below.

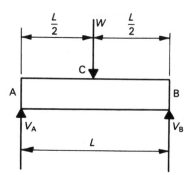

Can you suggest another possible mode of failure?

9

You might have suggested that the beam could fail as indicated in the sketch above. Such a failure is as if the beam had been cut in two by a guillotine. The failure is described as a *shear failure* and the forces producing such failures are described as *shearing forces*. In the case of horizontal beams carrying vertical loads, the shearing forces will be vertical.

Look at figure (a) below and try to visualise the shearing effect at C of all the forces to the left of C.

(a) (b) Positive shear (c) Negative shear

The reaction V_A, if acting alone, will tend to push the left-hand part of the beam upwards as in figure (b). Relatively, the right-hand part of the beam will tend to be pushed downwards. Shearing forces that cause upward shear deformation to the left (or downwards to the right) of the section under consideration are defined as positive. Hence a positive shearing force (S_1) is developed at C owing to the reaction V_A as shown in figure (b). For vertical equilibrium of the length AC, this shearing force must equal the reaction V_A. That is, $S_1 = V_A = +12.5$ kN.

In the same way, the load at B, if considered to be acting alone, will tend to push the left-hand part of the beam downwards. Thus the load at B produces a negative shearing force (S_2) at C of value -10.0 kN. The total net shearing force at C is given by the algebraic sum of the separate shearing forces. Thus: shearing force at $C = +12.5 - 10.0 = +2.5$ kN.

What value of shearing force do you get if you look at the forces to the right of C?

10

$$-7.5 \text{ kN}$$

The negative sign is as a result of our sign convention given in Frame 9. When you did the same sort of calculation for the bending moment you got the same answer working from the left as from the right. Why is it not the case for the calculation of shearing force?

11

In order to understand the apparent anomaly in the previous frame you must remember that although we talk about concentrated loads acting at a point, they do in fact act over a definite area or finite length of beam. In the figure below it is assumed that the 10 kN load at C is applied uniformly over a length of 100 mm.

What is the value of the shearing force at section Y_1–Y_1?

12

$$+2.5 \text{ kN}$$

The shearing force at the section Y_1–Y_1 is the algebraic sum of all the forces to the left of Y_1–Y_1: that is, $+12.5 - 10.0 = 2.5$ kN. If we now move 10 mm to the right, to section Y_2–Y_2, you will see that the forces to the left of section Y_2–Y_2 have been changed by the addition of part of the 10 kN load. This is tending to push the left-hand part of the beam downwards and is thus providing an additional negative shearing force at section Y_2–Y_2. If the 10 kN load is uniformly applied over a length of 100 mm, then in a length of 10 mm a load if 1kN will be applied and the total net shearing force at section Y_2–Y_2 will be $+2.5 - 1.0 = +1.5$ kN.

What will be the shearing force at the section Y_3–Y_3?

13

$$+2.5 - 10.0 = -7.5 \text{ kN}$$

This now agrees with the value in Frame 10. You will have realised that the shearing force in a beam changes abruptly at the point of application of a concentrated load. It is necessary to calculate the shearing force just to the left and just to the right of such a load. Thus:

The shearing force just to the left of C = the algebraic sum of all
forces to the left of C
$= +12.5 - 10.0 = \underline{+2.5 \text{ kN}}$

The shearing force just to the right of C = the algebraic sum of all
forces to the left of C
plus the load force at C
$= +12.5 - 10.0 - 10.0 = \underline{-7.5 \text{ kN}}$

Summarising what we have learnt so far about the shearing forces in a beam, we can state:

The shearing force at any section in a horizontal beam is the algebraic sum of all the vertical forces to the left, or to the right, of that section.

Now to the next frame.

14

BENDING MOMENT AND SHEARING FORCE DIAGRAMS

You should now understand the meaning of *bending moment* and *shearing force* and realise that the values of these vary along the length of a beam. It is useful to show this variation by plotting graphs in which the length of the beam represents the horizontal axis and the values of shearing force or bending moment are plotted vertically from this axis. Let's plot such graphs for the beam shown below.

What are the values of the reactions V_A and V_B?

15

> Since the load W is acting at mid span,
> the reactions are equal and of value $W/2$.

We will start with the *shearing force diagram* which is the graph of shearing force variation. Since we are going to plot on a horizontal axis representing the length of the beam, it is convenient to draw the shearing force diagram directly below and as a projection from the diagram of the loaded beam (see the figure below).

Starting at the left-hand end we see at A a concentrated load in the form of the reaction V_A. If we then consider a section just to the right of A and look at forces to the left of this section, then reaction V_A provides a positive shearing force of value $+W/2$. Plot this as line ad on the diagram. If you look along the beam to the right of A you will realise that the value of the shearing force cannot change until point C. Why not? Because no forces act on the beam between A and C and hence the shearing force is constant over this length. At point C, remembering the argument in Frame 13, we reason as follows:

Just to the left of C the shearing force $= +V_A = +W/2$ kN

Plot this as line fe

Just to the right of C the shearing force $= +V_A - W = -W/2$ kN

Plot this as line fg

You will appreciate that the line efg should, strictly speaking, not be vertical. The distance over which the load is applied is however very small compared with the length of the beam, consequently it is sufficiently accurate to draw efg vertical. If you cannot understand this, refer back to Frames 11 and 12. Points d and e are joined by a straight line because the shearing force is constant over the length AC.

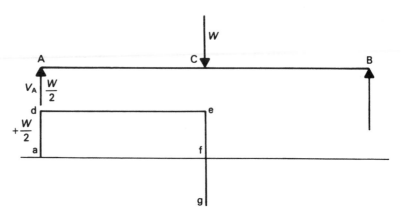

Complete the construction of the shearing force diagram.

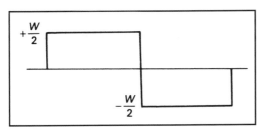

Having completed the shearing force diagram, we will now plot the bending moment diagram for the same beam. Look at the diagram below and consider a point X on the beam distance x from the end A. If you look to the left of X you will see one force only, the reaction V_A, which produces a bending moment at X of value $M_x = +(W/2) \times x$. The moment has a positive sign as it is a sagging moment (see Frame 3 if in doubt).

This is the equation for a straight line and is valid for the length of beam between A and C, for all values of x between zero and $L/2$. If, however, we wish to calculate the value of bending moment at a point such as X_1 which is to the right of C, the equation must be modified by the addition of a term related to the load W. Thus the bending moment at X_1:

$$= +(W/2) \times x_1 - W \times (x_1 - L/2)$$

This also is the equation for a straight line which is in this case valid for the length CB of the beam. We now know that the bending moment diagram will consist of straight lines and by substituting values of x and x_1 in the appropriate equations we can derive values of bending moment to plot the required diagram. Thus:

At A: $x = 0$ and from the first equation $M_x = 0$.

Plot this at point a

At C: $x = L/2$ ∴ from the first equation $M_x = +(W/2) \times (L/2) = +WL/4$

Plot this at point b

Note that if we substitute $x_1 = L/2$ in the second equation:

$$M_x = +(W/2) \times (L/2) - W \times (L/2 - L/2) = +WL/4$$

The value of bending moment at C must be the same whichever equation we use, thus the calculation we have just done provides a useful check on our work.

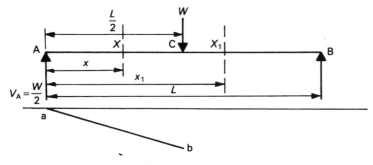

Complete the bending moment diagram.

17

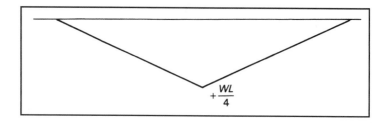

Note that we have plotted positive bending moments on the underside of the base line. If you imagine the base line to represent the beam, the positive values are plotted *on the side of the beam which is in tension* as the beam bends. You will discover later on in your studies that this convention for plotting is particularly useful when designing reinforced concrete members. When considering more complex situations, sketching the shape of the deflected beam, as in Frame 6, to locate those zones in which tension will develop will give you guidance as to which side of the base line to draw the bending moment diagram. *Always remember to draw the bending moment diagram on the side of the beam which is in tension.*

In the diagram below, the shearing force (SF) and the bending moment (BM) diagrams have been drawn below and in projection with the free body diagram of the beam. Plotting in this way enables the values of shearing force and bending moment at any point in the beam to be rapidly determined by vertical projection. Thus at Q distance $L/4$ from B, you can see that the shearing force is $-W/2$ and the bending moment is $+WL/8$.

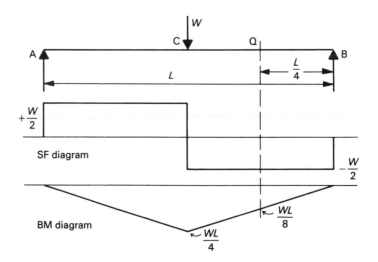

What is the value of the shearing force at the point in the beam where the bending moment is a maximum?

18

ZERO at C

If you didn't give the answer of zero, look back at frames 11 and 12. When plotting bending moment diagrams it is a useful fact that at beam sections where the shearing force is zero, the bending moment will always be either a maximum or a minimum. We will now show that the relationship is always true.

Figure (a) below is the free body diagram for a beam simply supported at A and B and carrying a non-uniform distributed load indicated by the irregular figure drawn on top of the beam. You have seen similar figures in Programme 2. Let the intensity of loading be w kN/m (note that this is not necessarily constant).

Figure (b) shows an enlarged view of a small segment abcd of the beam at a section X–X, the length cd being δx. To the right of the face bc there is a downward-acting load, the value of which is given by the area of the load distribution diagram.

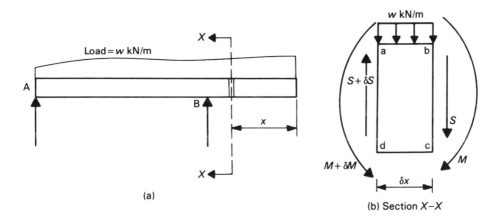

(a)

(b) Section X–X

The load to the right of section X–X produces a positive shearing force S at the face bc and to resist this there must be an upward shearing force $S + \delta S$ on the face ad. This shearing force on face ad is bigger than the shearing force on face bc because there is a load $w\delta x$ on the length of beam between those two faces.

Now consider the bending moments. The load to the right of section X–X will produce a bending moment at the face bc. Let's call this moment M. To resist this moment there will be a bending moment $M + \delta M$, at the face ad in the opposite direction.

Look at figure (b) above and by taking moments about the left face (ad) write down an equation relating the moments and shearing forces acting on the portion of beam abcd.

19

$$M + \delta M = M + S\delta x + (w\delta x)\delta x/2$$

From the above relationship and by ignoring second-order terms (δx^2) we obtain:

$$M + \delta M = M + S\delta x$$

thus: $\qquad\qquad\qquad\qquad \delta M/\delta x = S$

and as $\delta x \to 0$ then: $\qquad\quad \mathrm{d}M/\mathrm{d}x = S$

The bending moment M is a maximum or minimum if $\mathrm{d}M/\mathrm{d}x$ equals zero. It hence follows that as $\mathrm{d}M/\mathrm{d}x = S$, then the bending moment is a maximum or a minimum if $S = 0$.

Can you suggest why a knowledge of this relationship will be of assistance to you when plotting bending moment diagrams?

20

Determining the point of zero shearing force will locate the point of maximum bending moment: a critical point in the bending moment diagram.

SHEARING FORCE AND BENDING MOMENT DIAGRAMS FOR UNIFORMLY DISTRIBUTED LOADS

The figure at the top of the next page shows a beam AB simply supported at A and B and carrying a uniformly distributed load of w kN/m. This is a very common loading situation. For example, the self-weight of a beam usually constitutes such a load although the cross-section of a beam may not always be constant throughout its length. If this is the case, the value of w will vary along the length of the beam. To draw the SF diagram, consider a section X–X at a distance x from A:

The total load on the beam $\qquad\qquad\qquad\qquad\qquad W = wL$

and by symmetry the rections are equal and $\qquad\qquad = wL/2$

The shearing force at X–X $\qquad = +wL/2$ due to the reaction V_A

$\qquad\qquad\qquad\qquad\qquad\qquad\quad = -wx$ due to the load to the left of X–X

Thus the net shearing force: $\quad \mathrm{SF} = +wL/2 - wx$

This is the equation of a straight line with a slope of $-w$ and is applicable to all points between A and B. To plot the SF diagram only two points are required:

when $x = 0$ $\qquad\qquad\qquad \mathrm{SF} = +wL/2$

when $x = L$ $\qquad\qquad\qquad \mathrm{SF} = +wL/2 - wL = -wL/2$

These values are plotted to give the shearing force diagram as shown.

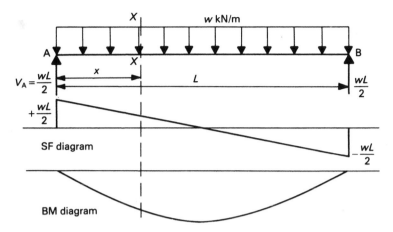

Now let's consider bending moments. The bending moment at a section X–X distrance x from A is obtained by taking moments of forces to the left of X–X:

$$= +V_A \times x \text{ due to the reaction } V_A$$

and

$$= -(wx) \times x/2 \text{ due to the load to the left of } X\text{–}X$$

Thus the net BM at X–X $\quad = +(wL/2)x - wx^2/2$

This is the equation of a parabola and will apply for all values of x between zero and L. Why? Because there is no change in the load system along the length of the beam. In order to plot the bending moment diagram it would be useful to locate the point of maximum bending moment. You can probably guess, correctly, that this will be at the mid-span position. If you did not guess this, look at the shearing force diagram and identify the point at which the shearing force is zero.

To plot the bending moment diagram we need the value of the maximum bending moment. What is the value of the maximum bending moment?

21

$$\boxed{wL^2/8}$$

Substituting $x = L/2$ into the equation for bending moment derived in Frame 20, the bending moment at mid span is given by:

$$M = +(wL/2)L/2 - w(L/2)^2/2$$
$$= +wL^2/4 - wL^2/8$$
$$= +wL^2/8$$

Thus the maximum bending moment in a beam simply supported at each end and carrying a uniformly distributed load is $wL^2/8$ (or $WL/8$, where $W = w \times L$ equals the *total* load on the beam).

Do you remember the expression for the maximum bending moment in a similar simply supported beam with a single central concentrated load?

22

$$\boxed{WL/4}$$

The expressions $WL/4$ and $WL/8$ (or $wL^2/8$) for the maximum bending moment in a simply supported beam subject to either a single central concentrated load or to a uniformly distributed load respectively are in common use and should be remembered.

Now to the next frame.

23

In the next few frames we shall construct shearing force and bending moment diagrams for beams with multiple loads. We shall develop a standard procedure which can be used to solve any problem.

The beam shown below is pinned at A and has a roller support at D. It carries concentrated loads at B, C and E.

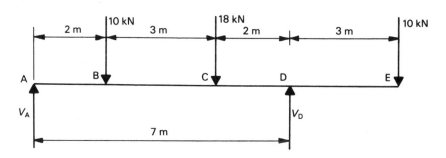

The first step is to determine the values of the reactions.

Calculate the values of V_A and V_D.

24

$$\boxed{\begin{aligned} V_A &= 8 \text{ kN} \\ V_D &= 30 \text{ kN} \end{aligned}}$$

If you did not get these values, revise the methods used in Programme 2 for calculating beam reactions.

Now let's derive the information to plot the shearing force diagram. We will work from left to right. To follow the calculations you will need to refer to the diagram below. In this and subsequent problems you may find it useful to cover the free-body diagram with a piece of paper and move it slowly from left to right progressively to uncover those forces causing shear at the section being considered.

$$
\begin{aligned}
\text{SF just to the right of A} &= +V_A & &= +8 \text{ kN} \\
\text{left of B} &= +V_A & &= +8 \text{ kN} \\
\text{right of B} &= +8 - 10 & &= -2 \text{ kN} \\
\text{left of C} &= +8 - 10 & &= -2 \text{ kN} \\
\text{right of C} &= +8 - 10 - 18 & &= -20 \text{ kN} \\
\text{left of D} &= +8 - 10 - 18 & &= -20 \text{ kN} \\
\text{right of D} &= +8 - 10 - 18 + 30 & &= +10 \text{ kN} \\
\text{left of E} &= +8 - 10 - 18 + 30 & &= +10 \text{ kN}
\end{aligned}
$$

These values are plotted to give the shearing force diagram. Note that the plotted points are joined by horizontal straight lines between the load positions with 'steps' in the diagram at the point of application of each concentrated load. This is always the case when only concentrated loads act on a beam. When distributed loads are acting, however, the lines of the shearing force diagram will be sloping, as for the example given in Frame 20.

Now for the bending moment diagram. We will take moments of all forces to the left of the point considered and calculate the values of bending moment at the load points, since we know from previous work that when a beam is subjected to concentrated loads only the bending moment diagram consists of straight lines with critical points (changes of direction) occurring at load positions.

$$
\begin{aligned}
\text{BM at A} & & &= 0 \\
\text{at B} &= +(8 \times 2) & &= +16 \text{ kN m} \\
\text{at C} &= +(8 \times 5) - (10 \times 3) & &= +10 \text{ kN m} \\
\text{at D} &= +(8 \times 7) - (10 \times 5) - (18 \times 2) & &= -30 \text{ kN m} \\
\text{at E} &= +(8 \times 10) - (10 \times 8) - (18 \times 5) + (30 \times 3) & &= 0
\end{aligned}
$$

These values are plotted to give the BM diagram.

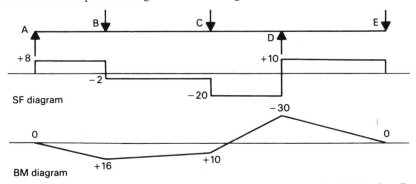

Note that there is a maximum sagging bending moment under the load at B and a maximum hogging bending moment at the reaction point at D.

Repeat the above calculations but take moments of all forces to the right of each point.

25

You should have obtained the same values.

You should now realise that, when calculating bending moments, you can take moments of all forces to the left or right of the point considered. You will also appreciate that in the previous frame if we had taken moments to the left for part ABC of the beam and to the right for part ED we would have simplified the calculations. However, when calculating shearing forces it is better to work systematically from left to right. Now for another example.

Beam ABCD, on the next page, is pinned at A and on rollers at C. It carries concentrated loads at B and D, a uniformly distributed load of 2 kN/m between A and C, and a uniformly distributed load of 1 kN/m between C and D.

By taking moments about A, the reactions are determined as:

$$V_A = 13.2 \text{ kN} \quad \text{and} \quad V_C = 25.8 \text{ kN}$$

Then for the SF diagram working from left to right:

$$
\begin{aligned}
\text{SF just to the right of A} &= V_A & &= +13.2 \text{ kN} \\
\text{to the left of B} &= +13.2 - (2 \times 4) & &= +5.2 \text{ kN} \\
\text{to the right of B} &= +13.2 - (2 \times 4) - 10 & &= -4.8 \text{ kN} \\
\text{to the left of C} &= +13.2 - (2 \times 10) - 10 & &= -16.8 \text{ kN} \\
\text{to the right of C} &= +13.2 - (2 \times 10) - 10 + 25.8 & &= +9.0 \text{ kN} \\
\text{to the left of D} &= +13.2 - (2 \times 10) - 10 + 25.8 - (1 \times 4) & &= +5.0 \text{ kN}
\end{aligned}
$$

These values are plotted to give the shearing force diagram. Note that the lines joining the plotted points are sloping because the beam is, in this case, carrying distributed loads over its whole length.

Now for the bending moment diagram: BM at A = 0.
What is the value of the bending moment at B?

26

+36.8 kN m

Taking moments of all forces to the left of B:

The bending moment at $B = +(13.2 \times 4) - (2 \times 4) \times 4/2 = +36.8$ kN m

Did you remember to take into account only that part of the UD load to the left of B?

Continuing: the BM at $C = +(13.2 \times 10) - (2 \times 10) \times 10/2 - (10 \times 6)$

$$= -28.0 \text{ kN m}$$

or alternatively taking moments to the right:

BM at $C = -(5 \times 4) - (1 \times 4) \times 4/2 = -28.0$ kN m

The shearing force is zero at B and C, thus the values of BM calculated at those points are maximum values.

The calculated values of bending moment are plotted and joined by parabolic curves. The lines are curved because distributed loads are acting on the beam (see Frame 20). You may find it necessary to calculate intermediate values in order to plot the curves more precisely. For example, you could calculate the value of bending moment at a point half way between A and B in order to determine the precise shape of the curve between A and B.

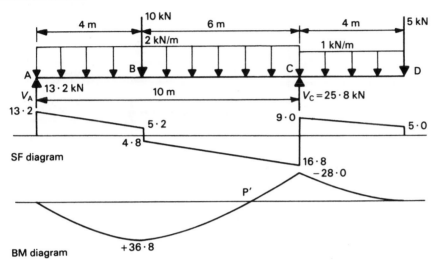

Now sketch the shape of the deflected beam and mark the lengths of the beam (top or bottom) where tension develops.

27

You should have sketched the beam as shown above being curved downwards at the left end and curved upwards at the right-hand end. The direction of curvature changes at the point P to the left of the support C. Point P is called a **point of contraflexure** and coincides with point P' on the BM diagram where the bending moment changes sign from positive bending (sagging) to negative bending (hogging). Note that the deflected shape of the beam approximates to the shape of the BM diagram. You should always sketch the deflected shape *before* doing any calculations and use this as a guide to the correct shape of the BM diagram. The BM diagram should always be drawn on the side that is in tension at all points along the beam span.
Now calculate the distance from A to the point of contraflexure.

28

$$\boxed{8.12 \text{ m}}$$

The bending moment at a point of contraflexure is zero. Thus you should have tackled this question by writing the equation for bending moment at a point between B and C and distance x from A, then equating to zero and solving for x. See below.

Bending moment at P is given by:

$$M_P = +(13.2 \times x) - 10 \times (x - 4) - (2 \times x) \times x/2$$
$$= +13.2x - 10x + 40 - 2x^2/2$$
$$= +3.2x + 40 - x^2$$

then if
$$M_P = 0$$
$$x^2 - 3.2x - 40 = 0$$

from which
$$x = 8.12 \text{ m}$$

Note that to perform this calculation it was necessary to determine, by inspection, the approximate region of the beam where the point of contraflexure occurs so that the correct bending moment equation could be written down and equated to zero.
Now calculate the position of the point of contraflexure in the beam of Frame 23.

29

$$\boxed{5.5 \text{ m from A}}$$

You should have derived the equation:

$$8x - 10 \times (x - 2) - 18 \times (x - 5) = 0$$
$$8x - 10x + 20 - 18x + 90 = 0$$
$$\therefore \quad -20x + 110 = 0$$

and
$$x = 5.5 \text{ m from A}$$

30

PROBLEMS

In the following problems you are required to sketch the shearing force and bending moment diagrams, marking the values of all important points. First sketch the shape of the deflected beams and hence sketch the shape of the bending moment diagrams before attempting any calculations. For question Q5, work from right to left.

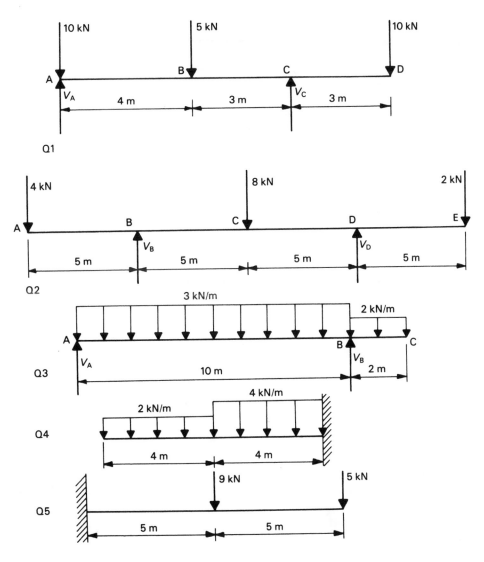

Check your solution against the sketches in the following frame.

31

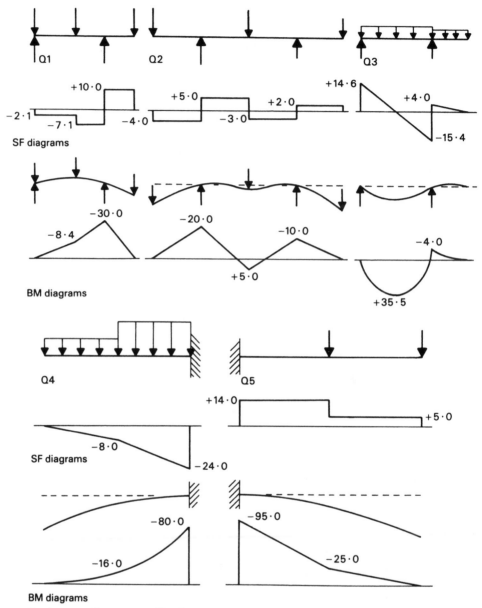

SF diagrams

BM diagrams

Q4   Q5

SF diagrams

BM diagrams

In the next frame we will solve a more complex example.

32

The beam shown at the top of the next page has a pinned support at A, roller supports at C and F, and a pin joint (hinge) at D.

First we determine the values of the reactions at A, C and F, using the technique that we learnt in Programme 2 for analysing beams containing internal pins.

Left-hand part of beam

Considering the part DF and taking moments about the pin D of all forces to the right of D:

$\Sigma M_D = 0$: $\qquad (10 \times 2) + (2 \times 5) \times 5/2 - V_F \times 5 = 0$

$\qquad \qquad \qquad \qquad \qquad \qquad \therefore\ V_F = (20 + 25)/5 = 9.0\ \text{kN}$

Resolving vertically for the forces acting on the length of beam DF

$\Sigma V = 0$: $\qquad \qquad V_D + V_F - (\text{total load on DF}) = 0$

$\qquad \qquad \qquad V_D + 9 - (10 + 2 \times 5) = 0 \quad \therefore\ V_D = 11.0\ \text{kN}$

An equal and opposite reaction will be exerted at D on the left-hand part AD.
Considering the left-hand part AD and taking moments about A:

$\Sigma M_A = 0$: $\quad (10 \times 3) + (2 \times 6) \times 6/2 + (11 \times 6) - (V_C \times 4) = 0$

$\qquad \qquad \qquad \qquad \qquad \qquad \therefore\ V_C = (30 + 36 + 66)/4$

$\qquad \qquad \qquad \qquad \qquad \qquad \qquad = 33.0\ \text{kN}$

$\Sigma V = 0$: $\qquad \qquad V_A + V_C - (\text{total load on AD}) = 0$

$\qquad \qquad \qquad V_A + 33 - (10 + 12 + 11) = 0 \quad \therefore\ V_A = 0\ \text{kN}$

We can now calculate the values of shearing force at the critical points:

$$\begin{array}{llll}
\text{Just to the right of A} & \text{SF} = V_A & = & 0\ \text{kN} \\
\text{to the left of B} & \text{SF} = 0 - (2 \times 3) & = & -6.0\ \text{kN} \\
\text{to the right of B} & \text{SF} = 0 - (2 \times 3) - 10 & = & -16.0\ \text{kN} \\
\text{to the left of C} & \text{SF} = 0 - (2 \times 4) - 10 & = & -18.0\ \text{kN} \\
\text{to the right of C} & \text{SF} = 0 - (2 \times 4) - 10 + 33 & = & +15.0\ \text{kN} \\
\text{to the left of D} & \text{SF} = 0 - (2 \times 6) - 10 + 33 & = & +11.0\ \text{kN}
\end{array}$$

Calculate the shearing forces at D, E and F by considering the free-body diagram of that part of the beam to the right of the internal pin.

33

Right of D +11.0 kN	Left of E +7.0 kN
Right of E −3.0 kN	Left of F −9.0 kN

Did you remember that the pin at D transmits a vertical upward force of 11.0 kN to the part of the beam to the right of D, and hence has the same effect as if a simple support was provided at D?

34

The shearing force diagram is plotted below, in projection with the free-body diagram which, for convenience, has been redrawn. Note that although the part AD transmits a reaction of 11.0 kN through the pin to part DF, this does not appear as a jump in the SF diagram since it is cancelled out by an equal but opposite reaction of 11.0 kN transmitted from the part DF to the part AD.

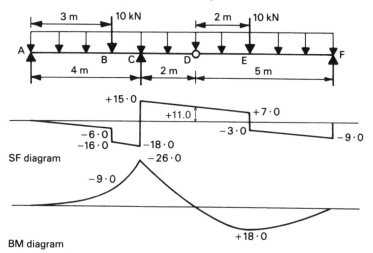

SF diagram

BM diagram

Now let's calculate the bending moments.
First for part AD, taking moments of forces to the left:

$$\text{BM at A} = 0$$
$$\text{at B} = +(0 \times 3) - (2 \times 3) \times 3/2 = -9.0 \text{ kN m}$$
$$\text{at C} = +(0 \times 4) - (2 \times 4) \times 4/2 - 10 \times 1 = -26.0 \text{ kN m}$$
$$\text{at D} = 0 \text{ (because the moment at a pin must be zero)}$$

Then for part DF, taking moments of forces to the right:

$$\text{BM at F} = 0$$
$$\text{at E} = +(9 \times 3) - (2 \times 3) \times 3/2 = +18.0 \text{ kN m}$$
$$\text{at D} = 0$$

You will remember that when loads are uniformly distributed the bending moment diagram is parabolic. You might, however, be unsure whether the curve is convex upwards or downwards. If you are not sure of the shape of any part of the curve, check by calculating intermediate values. Thus to check the shape of the curve between E and F, calculate the value of the BM half way between E and F. That is:

$$\text{BM} = +(9 \times 1.5) - (2 \times 1.5) \times 1.5/2 = +11.3 \text{ kN m}$$

If you plot these values, you will find that the curve between E and F is convex downward as shown. With practice, you will eventually realise that for vertically downward UD loads acting on a horizontal beam, the shape of the BM curve is always convex downwards.

Sketch the deflected shape of the beam.

Note that again it would have been advantageous to have sketched the shape of the deflected beam before carrying out the calculations. This will identify which parts of the beam are in tension and, as the BM diagram is drawn on the side of tension, this will indicate the correct shape of the diagram. Note also that the curvature (bending) of the beam changes direction at the pin D.

In the next frame we will see how to analyse a beam loaded by a couple.

36

The beam shown has a pin support at A, and is on a roller support at E. Concentrated loads act at B and D and an external couple applies a clockwise moment of 10.0 kN m to the beam at the mid point C. If you are not sure how couples act, refer back to Programme 2.

First we determine the reactions.

By moments about A: $\Sigma M_A = 0$: $-(V_E \times 6) + (5 \times 4) + (5 \times 2) + 10 = 0$

$$\therefore\ V_E = 40/6$$

$$= 6.67 \text{ kN}$$

By moments about E: $\Sigma M_E = 0$: $+(V_A \times 6) - (5 \times 4) - (5 \times 2) + 10 = 0$

$$\therefore\ V_A = 20/6$$

$$= 3.33 \text{ kN}$$

Calculate the shearing forces at A, B, D and E.

37

Right of A	+3.33 kN	Left of B	+3.33 kN
Right of B	−1.67 kN	Left of D	−1.67 kN
Right of D	−6.67 kN	Left of E	−6.67 kN

Check your calculations:

$$\text{Just to the right of A} \quad SF = +V_A \qquad\qquad = +3.33 \text{ kN}$$

$$\text{to the left of B} \quad SF = +V_A \qquad\qquad = +3.33 \text{ kN}$$

$$\text{to the right of B} \quad SF = +3.33 - 5.0 \qquad = -1.67 \text{ kN}$$

$$\text{to the left of D} \quad SF = +3.33 - 5.0 \qquad = -1.67 \text{ kN}$$

$$\text{to the right of D} \quad SF = +3.33 - 5.0 - 5.0 = -6.67 \text{ kN}$$

$$\text{to the left of E} \quad SF = +3.33 - 5.0 - 5.0 = -6.67 \text{ kN}$$

The values are plotted on the next page. Note that the shearing force does not change at the point C. The couple applies a moment but no vertical force.

Now for the bending moment diagram. Taking moments of forces to the left:

$$\text{BM at A} \qquad\qquad\qquad\qquad = \quad 0$$

$$\text{BM at B} = +(3.33 \times 2) \qquad\qquad = \quad +6.66 \text{ kN m}$$

$$\text{BM just to the left of C} = +(3.33 \times 3) - (5 \times 1) \qquad = \quad +5.0 \text{ kN m}$$

$$\text{BM just to the right of C} = +(3.33 \times 3) - (5 \times 1) + 10.0 = +15.0 \text{ kN m}$$

$$\text{BM at D} = (6.67 \times 2) \qquad\qquad = +13.33 \text{ kN m}$$

If you are still having trouble with the signs, look at the figures below. When calculating the bending moment at C you can see in figure (a) that, relative to C, reaction V_A (3.33 kN) tends to bend the beam upwards (positive bending) and in figure (b) you can see that the couple is also tending to bend the beam upwards (positive bending). The bending moment values are also plotted on the next page. Note how, at the point C, the value of the bending moment suddenly changes; the external couple is inducing a moment of 10.0 kN m in the beam at that point.

38

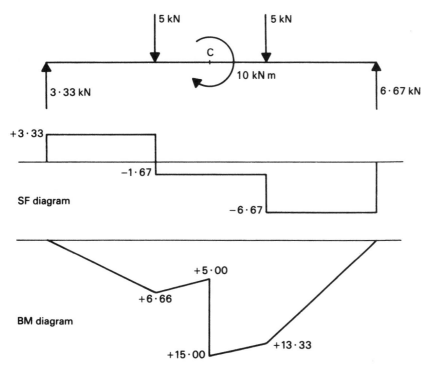

Now try the following problems.

39

PROBLEMS

Construct the shearing force and bending moment diagrams for the two beams shown below. The solutions are given in the next frame.

40

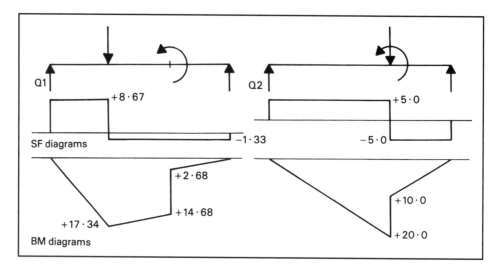

When tackling the second question, did you remember that the loading could be considered to be replaced by a concentrated load of 10 kN acting vertically downwards at B together with an anticlockwise moment of 10×1 kN m applied to the beam at B. If you have difficulty visualising this, refer back to Programme 2.

41

THE PRINCIPLE OF SUPERPOSITION

You should by now be capable of constructing shearing force and bending moment diagrams for beams subject to any form of loading. It is appropriate at this stage to introduce a concept, known at the *Principle of Superposition* which can help in the more rapid construction of such diagrams when a beam is subjected to complex loading. Phrasing the Principle of Superposition to suit the present context we may say:

If a structure is made of linear elastic materials and is loaded by a combination of loads which do not strain the structure beyond the linear elastic range, then the resulting shearing forces and bending moments are equal to the algebraic sum of the shearing forces and bending moments which would have been produced by each of the loads acting separately.

Let's see how we might use the Principle of Superposition. Suppose we have a simply supported beam AB having a single concentrated load W at mid span. If the self-weight of the beam is w kN/m, then the total load system is as shown in figure (a) on the next page. To apply the Principle of Superposition we consider the loads to be applied separately to the beam as in figures (b). Note how the third sketch in figure (b) is a combination of the first two sketches.

In figure (c) we have plotted the shearing force diagrams for the separate loadings. The third sketch, which is the shearing force diagram for the total load system is produced by adding together the ordinates of the first two sketches. Thus: $y_3 = y_1 + y_2$.

Similarly in figure (d) we have produced the bending moment diagram for the total load system by adding together the ordinates of the bending moment diagrams for each load treated separately. Thus: $z_3 = z_1 + z_2$.

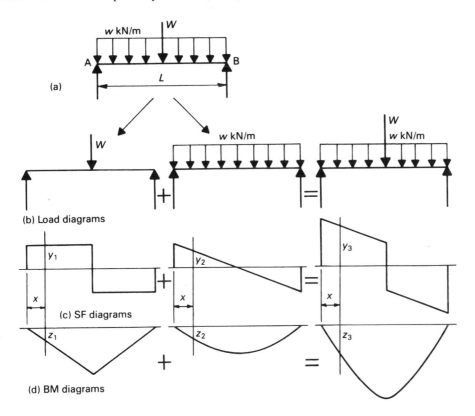

(a)

(b) Load diagrams

(c) SF diagrams

(d) BM diagrams

The Principle of Superposition can be applied to structural behaviour properties other than shearing forces and bending moments. For example, you will realise in later programmes that the principle can be used to help determine the deflection of structures under complicated load systems.

In the definition of the principle which was given on the previous page we used the term 'linear elastic'. Do not worry if you do not see the full implication of this term, as it will be explained in the next programme. For the present you can assume that you can use the Principle of Superposition provided the total load on a structure is significantly less than the critical value at which the structure would fail.

What is the value of the maximum bending moment in the beam of the above example under the action of the total load system?

42

$$WL/4 + wL^2/8$$

The maximum bending moment under the action of the total load system is, by inspection, at the mid-span position and equals the maximum bending moment ($WL/4$) under the action of the concentrated load plus the maximum bending moment ($WL^2/8$) under the action of the uniformly distributed self-weight.

Let's do another example. The beam ABCD has a pin support at A, and is on a roller support at C. Length AC weighs 3 kN/m and length CD 1 kN/m. Concentrated loads of 10 kN and 2 kN act at B and D respectively. The diagrams below show how shearing force diagrams for each of the four loads acting separately may be combined to give the shearing force diagram for the total load system.

Now use the Principle of Superposition to derive the bending moment diagram for the total load system.

43

Check your answer against the figure below.

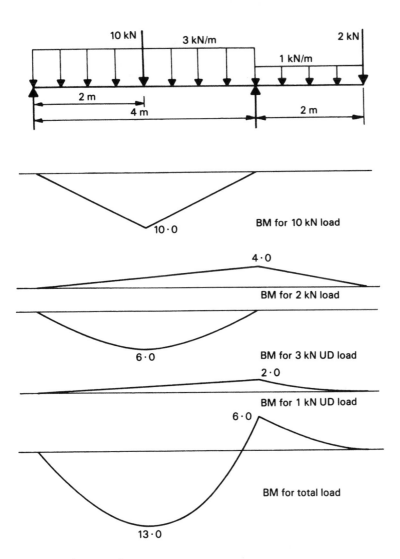

BM for 10 kN load

BM for 2 kN load

BM for 3 kN UD load

BM for 1 kN UD load

BM for total load

Now move on to the next frame.

44

In this programme, we have so far confined our studies to horizontal beams. The concepts of shearing forces and bending moments may, however, be applied equally well to other structures and structural members. As an example, we will construct the bending moment diagram for the three-pinned portal frame shown below. A and F are pinned supports, D is a pin (hinge), and B and E are rigid joints.

Calculate the reactions at A and F.

45

$$V_A = 8.75 \text{ kN}; \ V_F = 1.25 \text{ kN}; \ H_A = H_F = 1.25 \text{ kN}$$

You studied the method for calculating the support reactions for three-pinned structures in Programme 2. Refer back to Programme 2 if in doubt about the calculations below.

Considering the right-hand part only (DF) and taking moments about D:

$\Sigma M_D = 0$: $\qquad\qquad\qquad H_F \times 3 - V_F \times 3 = 0$

$$\therefore \ H_F = V_F$$

Considering the whole frame and taking moments about A:

$\Sigma M_A = 0$: $\qquad\qquad (10 \times 1) - H_F \times (5 - 3) - V_F \times 6 = 0$

$$\therefore \ 10 - 2H_F - 6V_F = 0$$

But $H_F = V_F$ $\qquad\qquad\qquad\qquad\qquad\qquad\qquad \therefore \ 8V_F = 10$

$$\therefore \ V_F = 1.25 \ \text{kN}$$

$$\therefore \ H_F = 1.25 \ \text{kN}$$

and by resolving horizontally, $\qquad\qquad\qquad\qquad\qquad H_A = H_F = 1.25 \ \text{kN}$

Considering the whole frame and taking moments about F:

$\Sigma M_F = 0$: $\qquad V_A \times 6 - H_A \times (5 - 3) - (10 - 5) = 0$

$$\therefore \ 6V_A - 1.25 \times 2 - 50 = 0$$

$$V_A = 52.5/6 = 8.75 \ \text{kN}$$

Then working clockwise round the frame starting at A:

BM at A $\quad\quad\quad\quad\quad\quad\quad = 0$ (A is a pin)

The bending moment at B can be determined by considering column AB and taking moments of all forces *below* B. Hence:

BM at B $\quad\quad\quad\quad\quad\quad = H_A \times 5 = 1.25 \times 5 = 6.25$ kN m

By inspection, H_A will tend to bend AB so that it is in tension on the left-hand side. Consequently if we wish to plot on the tensile side of the members, we will plot the value 6.25 kN m on the left-hand side of AB (see the figure below).

The bending moment at C is obtained by taking moments to the *left* of C:

BM at C $= V_A \times 1 - H_A \times 5 = (8.75 \times 1) - (1.25 \times 5) = 2.5$ kN m

BM at D $\quad\quad\quad\quad\quad\quad = 0$ (D is a pin)

Then working from the other end:

The bending moment at E can be determined by considering column EF and taking moments of all forces *below* E. Hence:

BM at E $\quad\quad\quad\quad\quad\quad = H_F \times 3 = 1.25 \times 3 = 3.75$ kN m

By inspection, H_F will tend to bend EF so that it is in tension on the right-hand side. Consequently if we wish to plot on the tensile side of the members, we will plot the value 3.75 kN m on the right-hand side of EF (see the figure below). The complete bending moment diagram is shown below and consists of straight lines, as there are only point loads acting on the structure.

Note that since joint B is rigid and since the material on the outer face of column AB at end B is in tension, then the material on the outer (top) face of beam BD at end B will also be in tension. You can visualise the layers (or fibres) on the outer face of the portal frame as being continuous round the corner at B, thus the tension continues round the corner. Consequently when we are drawing the bending moment diagram for the horizontal member BDE, we plot the value of bending moment at B (6.25 kN m) on the top face.

Similarly at joint E the value of bending moment (3.75 kN m) is plotted on the outer face of the column FE and on the upper face of the beam BDE.

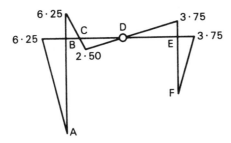

In the next frame we summarise what we have learnt in this programme.

46

The bending moment at a section in a beam is the algebraic sum of the moments about that section of all the forces to the right, *or to the left*, of that section.

Bending moments which cause a beam to sag are taken as positive.
Bending moments which cause a beam to hog are taken as negative.

Bending moment diagrams showing the variation of bending moment along a beam are plotted on the tensile side of the beam.

Bending moment diagrams resulting from the action of concentrated point loads consist of straight lines with changes of direction at the point of application of the loads.

Bending moment diagrams resulting from the action of uniformly distributed loads consist of parabolic curves.

The shearing force at a point in a beam is the algebraic sum of all the vertical forces to the right, *or to the left*, of that point.

Shearing force diagrams resulting from the action of concentrated point loads on horizontal beams consist of horizontal straight lines with vertical jumps at load positions.

Shearing force diagrams resulting from the action of uniformly distributed loads on horizontal beams consist of sloping straight lines.

The maximum bending moment occurs at a point where the shearing force is zero.

The maximum bending moment in a simply supported beam of span L carrying a single concentrated load W at mid span is $WL/4$.

The maximum bending moment in a simply supported beam of span L carrying a uniformly distributed load of w per metre length is $wL^2/8$ (or $WL/8$).

The bending moment at a pin is zero.

PROBLEMS

In the following problems, try to sketch the deflected shapes and hence the shape of the BM diagram before attempting any calculations of bending moment values. The solutions to all problems are given at the end of the programme.

1. In figure Q1 the beam ABCD has a pinned support at B and is on a roller support at D. Draw the shearing force and bending moment diagrams, marking all important values.

Q1

2. In figure Q2 beam ABCDEF has a pinned support at C and is on a roller support at D. Draw the shearing force and bending moment diagrams. Determine the position of the maximum positive bending moment and determine its magnitude.

Q2

3. The beam ABCDE in figure Q3 has a pinned support at A, is on roller supports at D and B, and has a pin (hinge) at C. Draw the shearing force and bending moment diagrams. Locate the points of contraflexure by determining their distance from D.

4. The beam ABCDE in figure Q4 has a pinned support at E, is on roller supports at A and D, and has a pin at C. Draw the shearing force and bending moment diagrams and locate the point of contraflexure.

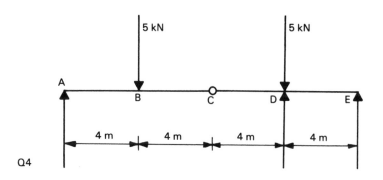

Q4

5. The beam ABC in figure Q5 has a pinned support at A and is on a roller support at B. It has a self-weight of 2 kN/m and a couple applies a clockwise moment of 4 kN m at C. Draw the shearing force and bending moment diagrams.

Q5

6. The beam ABC in figure Q6 has a pinned support at A and is on a roller support at C. The bracket at B carries a rail for an overhead crane, the load of 6 kN being offset by 0.5 m as shown. Draw the shearing force and bending moment diagrams.

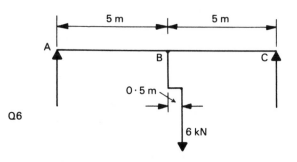

Q6

7. The frame ABCDEFG has pinned supports at A and G, a pin at D and has rigid joints at B and F. Draw the bending moment diagram, marking the important values.

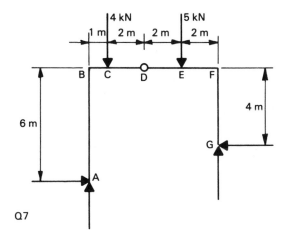

Q7

8. The frame ABCDE in figure Q8 has pinned supports at A and E, a pin at B and a rigid joint at D. Draw the bending moment diagram, marking all important values. (*Hint*: column AB has a pin at either end. What does this tell you about the bending moments in this member of the frame?)

Q8

SOLUTIONS TO PROBLEMS

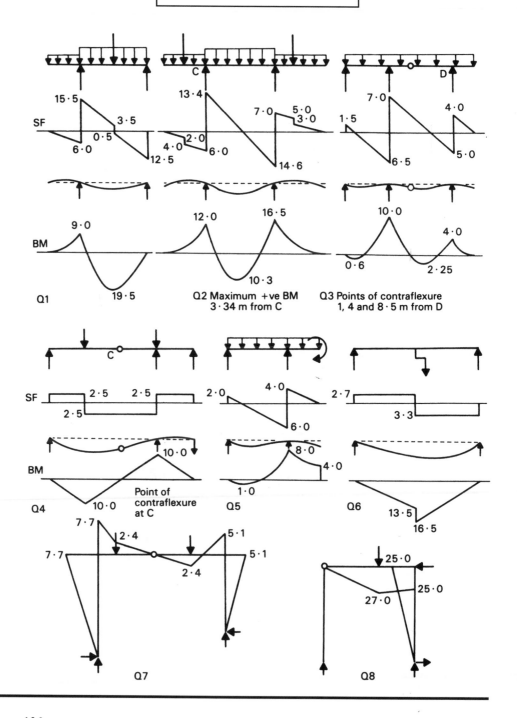

SF

15·5
3·5
0·5
6·0
12·5

13·4
7·0
5·0
3·0
4·0
2·0
6·0
14·6

7·0
1·5
4·0
6·5
5·0

BM

9·0
19·5

12·0
16·5
10·3

10·0
4·0
0·6
2·25

Q1

Q2 Maximum +ve BM
3·34 m from C

Q3 Points of contraflexure
1, 4 and 8·5 m from D

SF

2·5   2·5
2·5

2·0   4·0
6·0

2·7
3·3

BM

10·0
10·0

8·0
4·0
1·0

13·5
16·5

Q4   Point of
contraflexure
at C

Q5

Q6

7·7
2·4
5·1
7·7
2·4
5·1

25·0
25·0
27·0

Q7

Q8

Programme 5

STRESS ANALYSIS
(DIRECT STRESS)

On completion of this programme you will be able to:

- Define 'stress'
- Calculate tensile stress values for a range of simple axial stress problems
- Calculate compressive stress values for a range of simple axial stress problems
- Define 'strain'
- Calculate strain values for a range of simple axial stress problems
- Describe the concept of linear elastic behaviour
- State the relationship between stress and strain
- Define and calculate lateral strain
- Explain the terms 'Young's Modulus', 'Poisson's ratio' and 'Bulk Modulus'
- Calculate volumetric strain
- Extend the principles of one-dimensional stress systems to simple two- and three-dimensional stress systems
- Calculate stress in thin-walled cylinders

1

In Programme 3 we saw how external loads on a pin-jointed frame structure induced axial forces in the individual members of the frame. In Programme 4 we saw how external loads induced shearing forces and bending moments in beams. We now begin to look more closely at the way in which the material of which the structure is made reacts to the effects of loading. We start in this programme by considering in detail the behaviour of structural members subjected only to axial tensile or compressive forces. The effect of shearing forces and bending moments will be considered in subsequent programmes.

We have already noted in previous programmes that members subjected to tensile forces increase in length and members subjected to compressive forces decrease in length. It is true to say that whenever a force is applied to a piece of material, that material will change shape. The change may only be microscopic but in all cases a loading is accompanied by deformation. Thus all structures when loaded will deform, the amount of deformation depending on factors such as the magnitude of the load and the type of material used in the structure. In this programme we will learn how to calculate the effect of axial loading on some simple structures and structural elements.

2

DIRECT TENSILE STRESS

Consider a structural member having a cross-sectional area A and which is in equilibrium under the action of external forces P which are applied at the ends of the member in a direction parallel to the longitudinal axis of the member and thus at right-angles to the cross-section. The internal force developed in the member is also P as shown on the free-body diagram (figure (b)) which is obtained by taking any section through the member and considering the equilibrium of the part of the member to the left of that section.

The force P is transmitted along the length of the member through the material fibres of which the member is made. The material is resisting the tendency to pull apart and is said to be in a *state of stress*. The small arrows shown on the section Q–Q indicate those internal stresses which develop over the entire area of a cross-section and which act in a direction at right angles to the plane of the cross-section. In order to quantify the state of stress, normally denoted by σ (sigma), we calculate the force acting on each unit of area of the cross-section and say that the average *magnitude of stress* is given by: $\sigma = P/A$. The term *magnitude of stress* is abbreviated to *stress*.

Thus for any section Q–Q, we can say that there is a *direct (tensile) stress* acting on the section given by:

$$\text{average stress } \sigma = P/A$$

The term *direct* implies that the member is subject to an axial force only and that the stress is normal (at right angles) to the plane on which it acts. In this programme we will denote tensile stresses as positive.

We have derived the value of *average* stress by dividing the total force by the total area. Does this mean that the stress is necessarily uniform over the entire area? It does not. We have no justification to make such an assumption. A uniform distribution of stress would however be desirable since all fibres would then be evenly stressed to resist the external force and the material would be utilised in the most efficient manner.

Can you suggest how the external force should be applied to the member to ensure that the stress in the member is uniformly distributed across a cross-section?

3

> The force should be applied so that it acts along
> an axis through the centroid of the cross-section.

You may have intuitively arrived at this conclusion. If not, look at the figure below. The stress on section QQ is shown as being uniform across the section and of value σ. The resultant total internal tensile force will thus be given by $P = \sigma \times A$

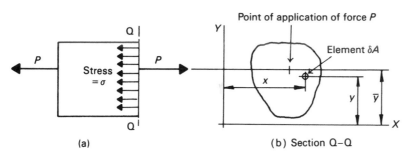

(a) (b) Section Q–Q

To determine the line of action of the resultant force P, we calculate the moment of P about the X-axis and equate to the sum of the moments about the X-axis of the forces ($\sigma \times \delta A$) on each elemental area (δA) of the cross-section. Thus, referring to figure (b):

$$P \times \bar{y} = \Sigma(\sigma \times \delta A) \times y$$

but $P = \sigma \times A$

$$\therefore \quad \sigma \times A \times \bar{y} = \sigma \Sigma y \times \delta A$$

$$\therefore \quad A\bar{y} = \Sigma y \delta A$$

Similarly it can be shown that: $\quad \bar{x} \quad = \Sigma x \delta A$

You should remember these relationships from Programme 1. What do they suggest to you?

4

> You should recognise these relationships from Programme 1
> as defining the position of the centroid of an area.

It follows that, if the direct stress in a member is uniformly distributed across any section, then the resultant internal tensile force acts at the centroid of the cross-section. For equilibrium, the external force P must be collinear with the internal force. Thus, if the stress is to be uniformly distributed, the external force must be applied at the centroid of the cross-section.

What would happen if the external force was not applied at the centroid of the cross-section?

5

> The stress distribution across the section would not be
> uniform. The member would be subject to bending in
> addition to direct loading.

The first answer follows from the reasoning of the previous frames, but you may not have made the second observation. Figure (a) below shows a tie on which the loading is eccentric (that is, the load is not collinear with an axis through the centroid). The member bends as shown by the broken lines and the stresses on a cross-section are a combination of the results of direct tension and bending. We will study the effects of combined tension and bending in Programme 7. For the time being, however, we will confine our attention to members subject to direct loading only and in which the stress is uniformly distributed across the cross-section. We now know that this means we are considering situations where loads are applied at the centroids of cross-sections as shown in figures (b), (c) and (d).

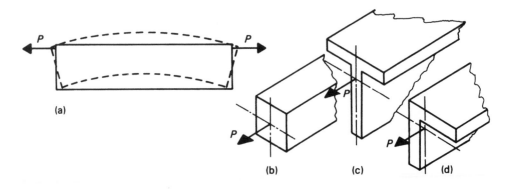

(a)

(b)   (c)   (d)

Note that although in figures (b) and (c) the line of action of the force P passes through the member, in figure (d) the line of action of force P lies outside the member. In practice it is not always easy to provide end connections to ensure that external forces (loads) are transmitted to members at the centroids of their cross-sections. This is particularly so in the case of frameworks where members are often fabricated from angles with cross-sections as in figure (d). It is nevertheless normally sufficiently accurate to assume that the members of frameworks are subject only to direct stress (no bending). Now let's look at some examples of members subject to direct tensile stress:

(e)

(f)

In figure (e) a vertical bar of circular cross-section 25 mm diameter has its central length reduced to 12 mm diameter. If a load of 18 kN is attached to the lower end, determine the tensile stresses in the bar.

6

> Stress in the central length $= 159.16$ N/mm^2
> Stress in the end lengths $= 36.67$ N/mm^2
> Note the units of stress!

The cross-sectional area in the central length $= \pi 12^2/4 = 113.09$ mm^2
and at the ends $= \pi 25^2/4 = 490.87$ mm^2
The tensile stress $\sigma =$ force/area
$\therefore$ the stress in the central length $= 18\,000/113.09 = 159.16$ N/mm^2
and in the end lengths $= 18\,000/490.87 = 36.67$ N/mm^2

The central length is the most heavily stressed part of the bar and if the load is increased sufficiently, fracture will take place somewhere along that central length. The strength of the bar is determined by the strength of the central length since that length is the weakest part of the bar.

Now try the following problem.

The suspension link, figure (f), is fabricated from steel plate 8 mm thick. If the tensile stress is not to exceed 165 N/mm^2, determine the maximum allowable value of load W. The load is transmitted to the link by steel pins passing through the 12 mm diameter holes.

7

$$\boxed{17.16 \text{ kN}}$$

The cross-sectional area of the link at its narrowest part

$$= 15 \times 8$$
$$= 120 \text{ mm}^2$$

Now stress σ = load/area $\therefore$ load (W) = stress × area

Thus if the stress in the link at its narrowest part is not to exceed 165 N/mm^2, then the maximum load on the link is given by:

$$W = 165 \times 120/10^3 = 19.80 \text{ kN}$$

But, in this example, the narrowest part is not the weakest part of the link. The weakest part is at the ends where the drilling of the holes has considerably reduced the area of material under stress. At a section through the centre line of the hole, the effective width of the link is reduced to:

$$25 - 12 = 13 \text{ mm}$$

Thus the effective area of cross-section = $13 \times 8 = 104$ mm^2 and if the stress at this section is not to exceed 165 N/mm^2, then the maximum load on the link is given by:

$$W = 165 \times 104/10^3 = \underline{17.16 \text{ kN}}$$

8

DIRECT COMPRESSIVE STRESS

Now look at the figure below showing a column 300 mm square resting upon a base 500 mm square. A load of 1500 kN acts vertically on the centre of a steel plate 150 mm square set symmetrically on the top of the column.

   Since the load is axial along an axis of symmetry through the centroid of cross-section, then the stress will be uniformly distributed across any section. Thus, in the column, the direct compressive stress is given by:

$$\sigma = \text{load/area} = -1500 \times 10^3/(300 \times 300)$$
$$= \underline{-16.67 \text{ N/mm}^2}$$

Similarly, the stress in the base is given by:

$$\sigma = \text{load/area} = -1500 \times 10^3/(500 \times 500)$$
$$= \underline{-6.0 \text{ N/mm}^2}$$

What is the direct compressive stress

(i) in the steel plate? and

(ii) in the column immediately below the steel plate?

9

$$\boxed{\begin{array}{l}\text{(i)} \ -66.67 \ \text{N/mm}^2 \\ \text{(ii)} \ -66.67 \ \text{N/mm}^2\end{array}}$$

$$\sigma = \text{load/area} = -1500 \times 10^3/(150 \times 150) = -66.67 \ \text{N/mm}^2$$

The stress in the concrete immediately below the steel plate will be the same as in the steel plate because the load is applied to the concrete over an area of $150 \times 150 \ \text{mm}^2$ only. The load, however, rapidly spreads over the entire area of the column. In design, it is often assumed that the broken lines, drawn at an angle of $45°$, indicate the way in which the load is distributed into the column. Thus at a distance of 75 mm below the top of the column, the entire cross-sectional area is under stress and the direct compressive stress in the concrete at that level will be:

$$\sigma = \text{force/area} = -1500 \times 10^3/(300 \times 300) = -16.67 \ \text{N/mm}^2$$

The figure below shows a strut subjected to a compressive force P applied through steel pins which pass through 12 mm diameter holes drilled in the strut.

If the maximum permissible compressive stress in the strut is 150 N/mm², calculate the maximum value of P.

10

$$\boxed{90 \ \text{kN}}$$

Force $(P) = \text{stress} \times \text{area} = 150 \times (50 \times 12) = 90\,000 \ \text{N} = 90 \ \text{kN}.$

Note that, unlike in the case of the tie, the holes drilled in the strut do not reduce its load-carrying capacity. This is because the contact surface between pin and strut, through which the load is transmitted, is on the side of the pin towards the centre of the strut. If you take a section through the strut which also cuts the pin, you will see that this reduced area of section is not stressed and therefore does not reduce the strength of the strut.

11

PROBLEMS

1. A bar with a circular cross-section of diameter 12 mm is subjected to a longitudinal tensile force of 10 kN. Determine the value of the direct tensile stress in the bar.

Ans. (88.42 N/mm²)

2. A concrete cube 150 mm × 150 mm × 150 mm is loaded in a compression testing machine. If the compressive force acting normal to one face of the cube is 250 kN, calculate the compressive stress in the concrete.

Ans. (−11.11 N/mm²)

3. A length of steel having a rectangular cross-section 50 mm × 12 mm is used as a tie in a plane frame. If the tensile force in the member is 60 kN, calculate the tensile stress in the steel: (a) if the ends of the member are welded to the rest of the frame, and (b) if the member is bolted to the rest of the frame by one bolt through a 10 mm diameter hole at each end. The hole is drilled through the 50 mm width of the section.

Ans. ((a) 100 N/mm² (b) 125 N/mm²)

4. A tie member of a plane frame is made from a length of L shaped steel angle section having a cross-sectional area of 1200 mm². If the permissible tensile stress in the steel is not to exceed 165 N/mm², calculate the maximum permissible force in the member. The connection at each end of the tie consist of one bolt through a 13 mm diameter hole drilled in one of the two 6 mm thick legs of the angle.

Ans. (185.13 kN)

12

STRAIN

Now we will consider the deformation of materials under stress. You already appreciate that materials in tension increase in length in the direction of the applied force and that materials in compression decrease in length. For example, figure (a) on the next page shows a bar of length L_a which extends a distance δL_a due to the application of a tensile axial load. Figure (b) shows a bar of original length L_b which shortens by a distance δL_b due to the application of a compressive axial load. Let's think about the relationship between the change in length and the original length of bars which are stressed in this way.

(a)   (b)

Figure (c) shows two bars made from the same material, having the same cross-sectional area, and acted upon by forces of equal magnitude. One bar is however twice the length of the other.

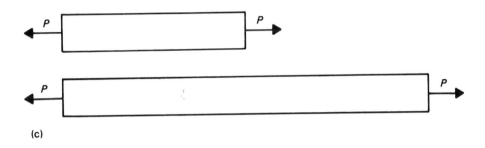

(c)

Can you guess what is the relationship between the magnitude of the extension of the longer bar and the extension of the other bar?

13

> The longer bar would extend
> twice as much as the shorter.

Intuitively you should realise that the longer bar will extend further than the shorter. The extension of a given bar under a given load will depend upon the original length and is in fact directly proportional to the original length. If a bar of original length L changes in length by an amount δL, then the fractional change in length is $\delta L/L$.

This dimensionless parameter is used to define the strain in the material. That is:

$$\text{Strain} = \frac{\text{change in length}}{\text{original length}}$$

Strain is normally denoted by the symbol ε (epsilon).

You should now appreciate that the effect of a load acting on a structural member is to cause both stress and strain in that member. In the next frame we will study the relationship between stress and strain.

14

THE STRESS–STRAIN RELATIONSHIP

If specimens of material are tested in a laboratory such that corresponding values of strain and stress can be recorded, then graphs may be plotted to show the relationship between strain and stress. The following figures show graphs typical of those which result from tensile tests on commonly used metals.

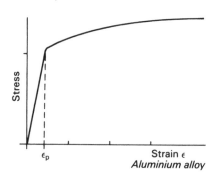

The important feature of each graph is the straight line with which each starts. Stress is seen to be directly proportional to strain provided that the strain does not exceed some value ε_p, the limit of proportionality. You can see the limit of proportionality marked on the graphs shown above. If the materials had been tested in compression instead of tension, the first part of the resulting graphs would similarly have been straight lines although the later part of each graph would differ from that of the tensile test. We see then that, provided the limit of proportionality is not exceeded, *stress is proportional to strain*. This relationship is known as Hooke's law and is written as:

$$\sigma = E\varepsilon \text{ (or stress } \sigma/\text{strain } \varepsilon = E)$$

where E is known as *Young's Modulus of Elasticity* and is usually quoted, for a given material, in units of kN/mm^2.

ELASTICITY

Another common feature of the behaviour of metals, and some other materials, under test is that when the stress is reduced the strain also reduces. This phenomenon of recoverable strain is known as *elasticity*. You will be familiar with elastic bands and appreciate their behaviour when stretched and then released.

It can be shown experimentally that provided an elastic material is not strained beyond the limit of proportionality, then when the load is reduced to zero the strain likewise reduces to zero and the material reverts to its original length. If a material is loaded such that the strain is in excess of the limit of proportionality, then the material suffers a permanent deformation.

LINEAR ELASTICITY

When a metal is loaded such that the strain does not exceed the limit of proportionality it behaves in a *linear elastic* fashion; strain is completely recovered on unloading, and stress is directly proportional to strain. Such behaviour is termed *linear elasticity*. In addition to metals, other common structural materials such as timber, masonry and concrete under some conditions may be considered to act in a *linear elastic* fashion.

Our future studies in this book will be based on *linear elasticity*.

15

CALCULATION OF LINEAR DEFORMATION

Figure (a) below shows a steel tie subjected to a tensile force of 100 kN. Let's determine the increase in length of the tie under load.

(a)

(b)

If the original, unloaded, length of the tie is 3 m, the cross-sectional area 645 mm^2 and if E for steel is 200 kN/mm^2 then:

The stress in the tie $\sigma = \text{force}/\text{area} = 100 \times 10^3/645 = 155$ N/mm^2

$\quad$ Stress/strain $= E$ $\quad$ that is $\quad \sigma/\varepsilon = E$

$\qquad\qquad\qquad$ or $\quad \varepsilon = \sigma/E$

$\therefore \quad$ in this example the strain $\varepsilon = 155/(200 \times 10^3) = 0.775 \times 10^{-3}$

$\qquad\qquad$ But, the strain $\varepsilon = \text{change in length}/\text{original length}$

$\therefore \quad$ the change in length (extension) $= \varepsilon \times \text{original length}$

$$= (0.775 \times 10^{-3}) \times (3 \times 1000)$$

$$= \underline{+2.33 \text{ mm}}$$

The positive sign indicates an extension in length corresponding to a (positive) tensile stress. Throughout these calculations did you note the factors used to convert all units to either Newtons or mm? Correct conversion to consistent units is important!

Figure (b) above shows an aluminium strut 1.2 m long which is part of a lightweight roof truss. When the roof truss is loaded the strut is subjected to a compressive force of 25 kN. Calculate the strain in the strut and determine the change in length.

(Area of cross-section = 220 mm^2, E for aluminium = 70 kN/mm^2)

16

$$\varepsilon = -1.6 \times 10^{-3}$$
$$\text{decrease in length} = 1.94 \text{ mm}$$

Stress $\sigma = \text{force/area} = -25 \times 10^3/220 \quad = -113.64 \text{ N/mm}^2$
Strain $\varepsilon = \text{stress}/E \quad = -113.64/(70 \times 10^3) = -1.62 \times 10^{-3}$
Change in length $= \varepsilon \times \text{original length} \quad = -(1.62 \times 10^{-3}) \times (1.2 \times 10^3)$
$= \underline{-1.94 \text{ mm}}$

LATERAL STRAIN

Now look at the following figures showing (a) a bar of circular cross-section and (b) a member having a rectangular cross-section, both being subject to direct tensile forces. We know that the bars will extend in length under the action of the external forces P.

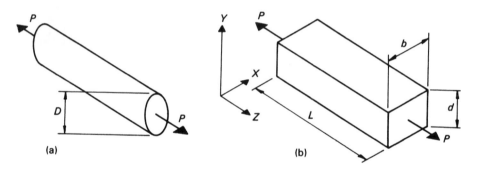

(a)   (b)

Can you visualise the effect of the forces P on the lateral dimensions b, d and D?

17

The lateral dimensions will decrease.

Common experience, perhaps the stretching of a piece of elastic with a rectangular cross-section, should have led you to make the above general observation. It is indeed true that all lateral dimensions decrease if any member is stretched longitudinally. Similarly it is true that all lateral dimensions increase if a member is compressed longitudinally. Thus longitudinal strain is always accompanied by lateral strain. Experiments show us that lateral strain (ε_L) is directly proportional to the longitudinal stress (σ) in the same way that longitudinal strain (ε) is directly proportional to (σ).

The lateral strain is thus proportional to the longitudinal strain and we say that:

$$\text{Lateral strain} = -v \times \text{longitudinal strain}$$

where v is a constant for a particular material and is known as *Poisson's Ratio*.

Experimental evidence also shows that the lateral strain is the same in all directions for materials that are homogeneous and isotropic (having the same elastic properties in all directions). Many structural materials are isotropic or can be assumed to be so.

Thus in figure (b) of Frame 16, if the breadth b of the member decreases by δb, the strain in the X direction $= -\delta b/b$ and this will be numerically the same as $-\delta d/d$, the lateral strain in the Y direction. The longitudinal strain, in the Z direction, is $+\delta L/L$ thus:

$$\delta b/b = \delta d/d = -v(\delta L/L)$$

The negative sign implies that if the longitudinal strain is an extension, then the lateral strains are contractions and vice versa.

18

Now look at the following figure which shows a structural member of rectangular cross-section subjected to a tensile force of 78 kN. Assuming that Young's Modulus of Elasticity (E) = 200 kN/mm^2 and that Poisson's ratio $v = 0.33$, we will calculate the changes in the linear dimensions of the member.

The cross-sectional area $= 50 \times 10 = 500$ mm^2

∴ the direct tensile stress, $\sigma = 78 \times 10^3/500 = 156$ N/mm^2

The longitudinal strain is given by $\varepsilon = \sigma/E$

$$= 156/(200 \times 10^3)$$
$$= 0.78 \times 10^{-3}$$

and the lateral strain $= -v \times \varepsilon = -0.33 \times 0.78 \times 10^{-3} = -0.257 \times 10^{-3}$

∴ the change in length $= \varepsilon \times$ original length $= +0.78 \times 10^{-3} \times 1.6 \times 10^3 = +1.25$ mm
the change in width $= \varepsilon_L \times$ original width $= -0.257 \times 10^{-3} \times 50 \quad = -12.85 \times 10^{-3}$ mm
the change in depth $= \varepsilon_L \times$ original depth $= -0.257 \times 10^{-3} \times 10 \quad = \underline{-2.57 \times 10^{-3}}$ mm

19

$$\boxed{\text{PROBLEMS}}$$

1. A steel tie is 1.4 m long, has a cross-sectional area of 110 mm^2 and carries a tensile load of 10.5 kN. If the value of Young's Modulus of Elasticity (E) is 200 kN/mm^2 and Poisson's ratio $v = 0.3$, calculate: (i) the direct tensile stress, (ii) the longitudinal strain, (iii) the lateral strain and (iv) the change in length.

Ans. ((i) 95.45 N/mm^2, (ii) 0.48 × 10^{-3}, (iii) −0.14 × 10^{-3}, (iv) 0.67 mm)

2. A hollow cylindrical steel tube with an outer diameter of 300 mm is to be used as a column to carry a vertical load of 2000 kN. If the direct stress in the steel is not to exceed 120 N/mm^2, calculate: (i) the thickness of metal required in the wall of the tube (*Hint*: calculate the internal diameter) and (ii) the change in external diameter under load.

Assume that $E = 200$ kN/mm^2 and $v = 0.3$.

Ans. ((i) 18.87 mm, (ii) +0.054 mm)

3. A steel bar of 25 mm diameter has its central length reduced in diameter to 12 mm. If $E = 200$ kN/mm^2 and $v = 0.33$ determine the reduction in diameter of (i) the ends and (ii) the central length of the bar under the action of a tensile force of 25 kN.

Ans. ((i) 2.10 × 10^{-3} mm, (ii) 4.38 × 10^{-3} mm)

4. A tie bar is 1.0 m long, has a circular cross-section of 25 mm diameter and is subjected to a force of 125 kN. If $E = 200$ kN/mm^2 and $v = 0.3$, determine the change in volume of the bar when loaded. (Hint: calculate the original volume, the changes in dimensions and hence the volume after loading.)

Ans. (249 mm^3)

20

VOLUMETRIC STRAIN

You may have been surprised by the answer to the fourth problem in Frame 19. The volume of the bar changed under load. We will now make a more detailed study of this effect. Consider the element of material shown in the figure on the previous page.

The longitudinal stress $\sigma = P/bd$

The longitudinal strain $\varepsilon = +\delta L/L$ where δL is the increase in length

The lateral strain $\varepsilon_L \quad = -\delta b/b$ and $= -\delta d/d$ where δb and δd are the decreases in b and d respectively

The original volume $\quad = L \times b \times d$

The volume after loading and the subsequent deformation
$$= (L + \delta L)(b - \delta b)(d - \delta d)$$
$$= L(1 + \varepsilon)b(1 + \varepsilon_L)d(1 + \varepsilon_L)$$
$$= Lbd(1 + \varepsilon)(1 + 2\varepsilon_L + \varepsilon_L^2)$$

If we expand the brackets and neglect second-order terms in ε and ε_L because they are small in magnitude then:

the volume after loading $= Lbd(1 + \varepsilon + 2\varepsilon_L)$
the change in volume is thus $=$ final volume $-$ original volume
$$= Lbd(1 + \varepsilon + 2\varepsilon_L) - Lbd = Lbd(\varepsilon + 2\varepsilon_L)$$
$$= Lbd\varepsilon(1 - 2v) \text{ since } \varepsilon_L = -v\varepsilon$$

For the majority of structural materials v has values between 0.35 and 0.25. Thus in the case of most structural applications there will be a change in volume as a result of loading. The ratio (change in volume/original volume) is termed the *volumetric strain*.

$$\therefore \quad \text{Volumetric strain } = \delta V/V = Lbd\varepsilon(1 - 2v)/Lbd = \varepsilon(1 - 2v)$$

The expression $\delta V/V = \varepsilon(1 - 2v)$ may be used to determine the volumetric strain for any body for any shape of cross-section. Thus for problem 4 of Frame 19

the longitudinal stress $\sigma = $ load/area $= (125 \times 10^3)/(\pi 25^2/4)$
$$= 254.65 \text{ N/mm}^2$$
the longitudinal strain $\varepsilon = \sigma/E = 254.65/200 \times 10^3 = 0.001\,27$
the volumetric strain $\quad = \varepsilon(1 - 2v) = 0.001\,27(1 - 2 \times 0.3) = 0.000\,508$
The original volume $\quad = (1 \times 10^3) \times \pi 25^2/4 = 490.87 \times 10^3 \text{ mm}^3$
$\therefore \quad$ the change in volume $\quad =$ volumetric strain $\times$ original volume
$$= 0.000\,508 \times 490.87 \times 10^3 = 249 \text{ mm}^3$$

This agrees with your previous answer to this problem.

21

BULK MODULUS

In the last few frames we have considered problems involving stress applied in one direction only. Now imagine a body completely immersed in a liquid. The effect of the liquid is to exert a uniform pressure, or stress, on the entire surface of the body and the body is compressed. The ratio of stress to volumetric strain is found to be constant and is known as the *Bulk Modulus of Elasticity (K)*. This is analogous to Young's Modulus of Elasticity (E). In the next frame we will solve a problem involving the use of the Bulk Modulus.

22

A steel plate 300 mm × 300 mm × 100 mm is completely immersed in water and is subjected to a uniform hydrostatic pressure of 3 N/mm² in all directions. The compressive stress on all faces of the plate is thus 3 N/mm². Determine the change in volume if the Bulk Modulus K is 167×10^3 N/mm².

$$\text{Stress/Volumetric strain} = K$$
$$\therefore \quad \text{Volumetric strain} = \text{Stress}/K = 3/(167 \times 10^3) = 18.0 \times 10^{-6}$$
$$\therefore \quad \delta V/V = 18.0 \times 10^{-6}$$
$$\text{But original volume } V = 300 \times 300 \times 100 \qquad = 9.0 \times 10^6 \text{ mm}^3$$
$$\therefore \quad \delta V = 18.0 \times 10^{-6} \times 9.0 \times 10^6 \quad = \underline{162 \text{ mm}^3}$$

The situation in which a member is subjected to a uniform stress in all directions is not commonly encountered in practice. A situation in which there are two or three stresses of different value acting in two or three mutually perpendicular directions is however quite likely to be encountered. Consider the figure below.

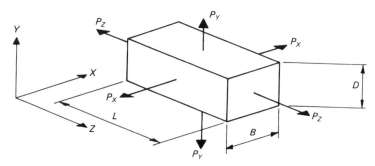

The member is subjected to a longitudinal force P_Z in the Z direction and to lateral forces P_X and P_Y in the X and Y directions respectively. The stresses in the three directions are given by:

In the X direction $\qquad \sigma_X = P_X/(L \times D)$

in the Y direction $\qquad \sigma_Y = P_Y/(L \times B)$

and in the Z direction $\quad \sigma_Z = P_Z/(B \times D)$

The strain in any direction will consist of three components. Thus the strain in the Z direction will be

$$\varepsilon_Z = +\sigma_Z/E \text{ due to the force } P_Z$$

together with $\qquad -v\varepsilon_X = -v\sigma_X/E \text{ due to the force } P_X$

and $\qquad -v\varepsilon_Y = -v\sigma_Y/E \text{ due to the force } P_Y$

The total strain in the Z direction is thus given by:

$$\varepsilon_Z = +\sigma_Z/E - v\sigma_X/E - v\sigma_Y/E$$

Similarly the strain in the X direction is given by:

$$\varepsilon_X = +\sigma_X/E - v\sigma_Y/E - v\sigma_Z/E$$

and the strain in the Y direction is given by:

$$\varepsilon_Y = +\sigma_Y/E - v\sigma_Z/E - v\sigma_X/E$$

These relationships can conveniently be displayed in tabular form as shown on the next page.

23

Strain	due to		
	σ_X	σ_Y	σ_Z
ε_X	$+\sigma_X/E$	$-v\sigma_Y/E$	$-v\sigma_Z/E$
ε_Y	$-v\sigma_X/E$	$+\sigma_Y/E$	$-v\sigma_Z/E$
ε_Z	$-v\sigma_X/E$	$-v\sigma_Y/E$	$+\sigma_Z/E$

Now let's solve a problem using these relationships. The figure shows a member of rectangular cross-section which slots into a rectangular hole in a slab. The dimensions of the hole are slightly larger than the dimensions of the member, thus allowing a tolerance, or clearance, of 0.01 mm in one direction and 0.02 mm in the other. As a compressive force (P_Z) is applied to the member, the lateral dimensions increase until the member is a tight fit in the hole. The problem is to determine the longitudinal stress necessary to expand the member sufficiently in the lateral directions so as completely to fill the hole.

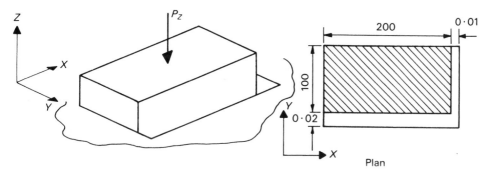

In which direction will the clearance be reduced to zero first?

24

In order that the member just fits tightly in the X direction, the strain in that direction must be 0.01/200. This is a lesser strain than 0.02/100 which is the strain necessary to eliminate the clearance in the Y direction. The strain in the X direction is given by:

$$\varepsilon_X = \sigma_X/E - v\sigma_Y/E - v\sigma_Z/E$$

and when the clearance in the X direction is just reduced to zero, $\varepsilon_X = 0.01/200$. At this stage there are no forces in the X or Y directions, thus $\sigma_X = 0$ and $\sigma_Y = 0$. Substituting in the above equation gives $0.01/200 = 0 - 0 - v\sigma_Z/E$.

If $E = 200 \text{ kN/mm}^2$ and $v = 0.3$, calculate the value of σ_Z.

25

$$\boxed{-33.33 \text{ N/mm}^2}$$

$$0.01/200 = -0.3\sigma_Z/(200 \times 10^3)$$

$$\therefore \quad \sigma_Z = -33.33 \text{ N/mm}^2$$

Thus a longitudinal stress of -33.33 N/mm^2 is needed in order to eliminate the clearance in the X direction. The negative sign implies that the longitudinal strain is in the opposite sense (compression) to the change in length (extension) in the X direction.

If the longitudinal force is increased further until the clearance in the Y direction is just reduced to zero, then $\varepsilon_Y = 0.02/100$. The figure below shows the situation when this is achieved.

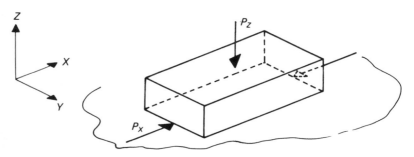

ε_X is still 0.01/200 since the external restraint along the short side of the member restricts further lateral expansion in the X direction but because of this restraint, a lateral compressive force P_X will have developed at the contact face. There will not however be a force at the contact face in the Y direction and hence $\sigma_Y = 0$. We can then write down the equation:

$$\varepsilon_X = \sigma_X/E - v\sigma_Y/E - v\sigma_Z/E$$

and substituting $\quad 0.01/200 = \sigma_X/(200 \times 10^3) - 0 - 0.3\sigma_Z/(200 \times 10^3)$

$$\therefore \quad 10 = \sigma_X - 0.3\sigma_Z \tag{5.1}$$

Also: $\qquad\qquad\qquad \varepsilon_Y = \sigma_Y/E - v\sigma_Z/E - v\sigma_X/E$

and substituting $\quad 0.02/100 = 0 - 0.3\sigma_Z/(200 \times 10^3) - 0.3\sigma_X/(200 \times 10^3)$

$$40 = -0.3\sigma_Z - 0.3\sigma_X \tag{5.2}$$

Subtracting (5.1) from (5.2) gives $30 = -1.3\sigma_X$

$$\therefore \quad \underline{\sigma_X = -23.08 \text{ N/mm}^2 \text{ (compression)}}$$

Then substituting in equation (5.1):

$$10 = -23.08 - 0.3\sigma_Z$$

$$\therefore \quad \underline{\sigma_Z = -110.27 \text{ N/mm}^2}$$

Thus a longitudinal compressive stress of 110.27 N/mm^2 would be needed to eliminate all clearance in the X and Y directions. Now to the next frame.

STRESSES IN THIN-WALLED CYLINDERS

We will conclude this programme by looking at the stresses that arise in the material of a thin-walled closed cylinder subject to internal pressure. By 'thin walled' we imply that the material of which the cylinder is made is thin relative to the diameter of the cylinder. The figure below shows a cylinder with internal diameter D of length L containing liquid or gas at a pressure p N/mm^2.

The pressure p will be uniform over the entire internal surface and will act normal, that is, at right angles, to the surface. The total longitudinal force (P) acting on one end will thus be:

$$P = \text{pressure} \times \text{area} = p \times \pi D^2 / 4$$

A force of similar magnitude acts on the other end and you can see in the free-body diagram figure (c) that the material in the wall of the cylinder is in tension and consequently resists the longitudinal tensile force P. Since the thickness (t) of the material is small compared with the diameter (D), then at any cross-section XX the area of material in tension is given by:

$$\text{area} = \pi D \times t$$

Can you write down an expression for the value of the longitudinal stress (σ_L)?

27

$$\sigma_L = \frac{\text{force}}{\text{area}} = \frac{(p \times \pi D^2 / 4)}{\pi D t} = \frac{pD}{4t}$$

If the longitudinal tensile stress was too great, the cylinder might fail by tearing into two parts as in figure (d).
Try to visualise and sketch another possible form of failure.

28

HOOP STRESS

The figure shows how the cylinder might fail by splitting along two failure lines which are diametrically opposite each other. Consider the vertical plane YY in the following figure (a). The uniform internal pressure p acting normally to the curved surface to the right of section YY effectively provides a total resultant force P. There will be an equal but opposite force P on the inner surface of the cylinder to the left of section YY and the two forces will together tend to tear the wall of the cylinder along its length at A and B. *The material in the wall of the cylinder at A and B is thus in tension* and a direct tensile stress σ_H exists (see the free-body diagram, figure (c). The direction of this stress is tangential to the cylinder. You should realise that similar tensile stresses of the same value σ_H will exist in the material at all points around the circumference. The stress situation is similar to that in a steel hoop put round a wooden barrel to hold the staves together. Hence the use of the term 'hoop stress' to describe this type of stress.

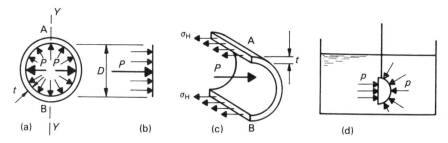

| (a) | (b) | (c) | (d) |

If the cylinder is of length (L) and the wall has a thickness (t), write down the relationship between P and σ_H. (Hint: look at the figure (c).)

29

$$P = 2\sigma_H L t$$

Check your solution on the next page.

The sectional area of the wall at both A and B $= L \times t$

Therefore the *total* tensile force in the wall at A and B $= 2(\sigma_H \times L \times t)$

And for equilibrium this must equal the force P:

$$\therefore \quad P = 2\sigma_H L t$$

To express σ_H in terms of the internal pressure we must relate the force P to the pressure p. We can do this if we remember the basic principle that the force in any direction due to a uniform pressure acting on a curved surface equals the pressure multiplied by the area of the projection of that surface on to a plane at right angles to the direction of the force.

This is an important principle and may be proved analytically. You may have been introduced to this concept in your studies of Hydrostatics. If not, and you are not convinced of the validity of the statement, look at figure (d) in Frame 28. One half of a solid sphere (for example, one half of a ball) is suspended in a container of liquid. The liquid exerts a uniform pressure p on all parts of the surface. Would you expect the body to move to the left or right? You should expect neither – the body would hang motionless. The resultant force on the hemispherical surface of the body must thus be equal to the resultant force acting on the plane surface on the left of the body. The resultant force on the plane surface is the pressure p multiplied by the area of that plane surface, and that plane surface is, you will realise, the projection of the hemispherical surface.

Looking back at our original cylinder, figure (a), you should now appreciate that the force P is equal to the pressure p multiplied by the vertical plane area indicated in figure (b).

Now develop an expression relating the hoop stress σ_H to the internal pressure p.

30

$$\boxed{\sigma_H = \frac{pD}{2t}}$$

$$P = p \times \ \text{projected area}$$
$$= p \times DL$$

But we have already seen that $\qquad P = 2\sigma_H L t$

$$2\sigma_H L t = pDL$$

$$\therefore \quad \sigma_H = \frac{pD}{2t}$$

This equation for σ_H and the equation for σ_L developed in Frame 27 may both be used to analyse problems of cylindrical structures subjected to internal or external pressure.

In the next frame we will use the two equations to solve a typical problem.

31

A cylindrical tank to be used with an air compressor is 2 m in length and 0.6 m in diameter. If the wall of the tank is 3 mm thick and if the direct stress in any direction in the wall material is not to exceed 100 N/mm^2, determine the maximum permissible pressure of the air in the tank.

Considering the longitudinal direct stress:

$$\sigma_L = \frac{pD}{4t}$$

or, rearranging

$$p = \frac{\sigma_L 4t}{D}$$

$$= \frac{100 \times 4 \times 3}{600}$$

$$= 2.0 \text{ N/mm}^2$$

Considering hoop stress

$$\sigma_H = \frac{pD}{2t}$$

or, rearranging

$$p = \frac{\sigma_H 2t}{D}$$

$$= \frac{100 \times 2 \times 3}{600}$$

$$= 1.0 \text{ N/mm}^2$$

The hoop stress is thus more critical than the longitudinal stress, and the safe maximum pressure is 1.0 N/mm^2.

Note that the material in the wall of the cylinder is in fact subjected to two direct stresses at the same time: a longitudinal stress and a hoop stress. The effect of the two stresses acting simultaneously is to produce a combined stress situation which is more complex than indicated by the simple approach adopted in this programme. In Programme 10 you will learn how to analyse material subject to combined stresses.

In the last few frames we have considered tanks subjected to internal pressure. Can you think of any constructional situation in which the material of a structure will be subjected to a hoop stress as a result of external pressure?

32

If an underground tunnel is constructed and lined with steel or concrete, the soil will exert a pressure tending to crush the tunnel. The steel or concrete will be subject to a compressive hoop stress. Buried pipelines will likewise be subject to compressive hoop stress.

33

TO REMEMBER

Direct stress is the normal force acting on a surface divided by the area of that surface ($\sigma = $ force/area) (units $= $ N/mm^2).

A direct stress is uniform over a surface (or section) only if the normal force passes through the centroid of the surface (or section).

A direct stress acts at right angles to the surface under consideration.

Tensile stresses are produced by tensile forces which tend to lengthen members.

Compressive stresses are produced by compressive forces which tend to shorten members.

Strain is the ratio of change to original length ($\varepsilon = \delta L/L$) and is a small dimensionless measurement.

Longitudinal strain is the strain in the direction of the force acting on a member.

Provided that an elastic material is not strained beyond the limit of proportionality, stress is directly proportional to strain. The strain is completely recovered when the load is removed.

Young's Modulus of Elasticity (E) = stress/strain = σ/ε (units = kN/mm^2).

Lateral strain is the strain in a direction at right angles to the direction of the force acting on a member.

Poisson's ratio (v) = $-$ (lateral strain)/(longitudinal strain) and is dimensionless.

Bulk Modulus (K) = pressure/(volumetric strain) (units = kN/mm^2).

34

1. A tie bar is 2 m in length, has a circular cross-section of 19 mm diameter and carries a longitudinal load of 35 kN. Calculate the stress in the bar and the change in length ($E = 200$ kN/mm^2).

Ans. (123.44 N/mm^2, 1.23 mm).

2. A draw bar between a tractor and a trailer is made from a length of steel with a rectangular cross-section 100 mm by 12 mm. The load is transmitted to the bar via a pin through a 25 mm diameter hole drilled through the bar at each end. If the maximum permissible stress in the steel is 150 N/mm^2, determine the maximum load that can be taken by the bar.

Ans. (135 kN).

3. Figure Q3 shows one end of a beam resting upon a bearing pad made of a rubber compound and which is 250 mm square in plan area. If the vertical reaction at the end of the beam is 3000 kN, calculate the compressive stress in the bearing pad.

Ans. (-48.0 N/mm^2)

Q3

4. A 10 m high concrete column in a large building is 300 mm square at the top and tapers uniformly to 225 mm square at the bottom. The vertical load on the top of the column is 2500 kN. Calculate the compressive stress in the concrete:

(i) at the top and
(ii) at the bottom of the column.

The weight of concrete is 24 kN/m^3
  (*Hint*: you will need to calculate the weight of the column.)
Ans. (i) -27.78 N/mm^2, (ii) -49.71 N/mm^2)

Q4

5. Figure Q5 shows a 'strong back' device for use in lifting long heavy castings. The loading is such that the load is equally shared by the lower links. The twin links connecting the device to the crane hook are each made from 75 mm by 12 mm steel strip and have 20 mm diameter holes to take the connecting pins. The lower links are each made from 50 mm by 12 mm steel and have 18 mm diameter holes. Considering only the tensile strength of the links, determine the maximum load that can be safely lifted. Assume that the maximum permissible tensile stress is 150 N/mm².

Ans. (198.0 kN)

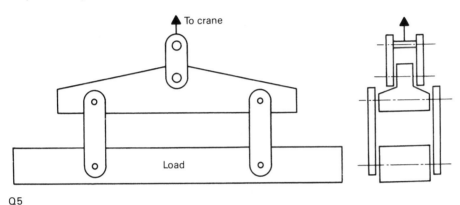

Q5

6. A compression member 0.3 m long has a rectangular cross-section 150 mm by 100 mm. It passes through a slot in a rigid block as shown in figure Q6 such that there is complete restraint in the X direction. Therefore no change of dimension can take place in the X direction. There is however no restriction of movement in the Y direction.

If a load of 3800 kN is applied to the member as shown, calculate:
(i) the stress in the X direction
(ii) the strain in the Z direction and
(iii) the strain in the Y direction.
Assume that $E = 200$ kN/mm² and Poisson's ratio $v = 0.3$.

Ans. (i) -76.0 N/mm², (ii) -1.153×10^{-3}, (iii) 0.494×10^{-3})

Q6

7. A steel tie bar 1.1 m long and 50 mm diameter is subjected to a tensile stress of 120 N/mm^2. Determine:
 (i) the extension
(ii) the change in lateral dimension and
(iii) the change in volume.
Assume that $E = 200$ kN/mm^2 and Poisson's ratio $v = 0.3$.
Ans. (i) 0.66 mm, (ii) −0.009 mm, (iii) 518 mm^3)

8. A cylindrical reinforced concrete water tank sits on solid ground and is filled with water to a height of 6 metres. The internal diameter of the tank is 10 metres. If the circumferential hoop stress is limited to 1.5 N/mm^2 determine the required thickness of the walls of the tank. (*Hint*: the pressure in the tank at depth h is ρgh where $\rho = 1000$ kg/m^3.)
Ans. (196 mm)

9. A long thin cylindrical pressure vessel consists of two halves bolted together using 20 mm diameter bolts, as shown in figure Q9. The internal diameter of the vessel is 500 mm and the wall thickness is 5 mm. If the allowable hoop stress in the walls is limited to 120 N/mm^2 calculate:
 (i) the maximum allowable internal pressure
(ii) the maximum spacing of the bolts along the length of the vessel if the tensile stress in each bolt is limited to 160 N/mm^2 when the vessel is fully pressurised.
   Ans. (2.40 N/mm^2, 83.78 mm)

Q9

20 mm bolts

500

5 mm

Programme 6

BENDING STRESS

On completion of this programme you will be able to:

- Explain the terms 'neutral axis'
- Calculate the 'second moment of area' of regular and irregular shapes
- State and apply the parallel axis theorem
- Calculate the elastic section modulus of regular and irregular shapes
- State the formulae for the second moment of area and elastic section modulus of a solid rectangular section
- Recognise the state of bending stress and strain within a beam's cross-section
- Calculate stress values at different levels in a beam's cross-section
- State the standard equation of bending
- Recognise and describe the terms in the equation of bending
- Apply the equation of bending to the solution of a range of simple beam problems
- Carry out simple design procedures for beams in bending
- Define and identify the 'principal axis' of a beam's cross-section

1

In Programme 4 you learnt how to determine the shear force and bending moment at any section of a loaded beam. In future work in structural design you will learn how to design beams capable of withstanding the effects of shear force and bending moment.

As an introduction to beam design we will, in this programme, develop and study the equations that can be used to analyse the bending stresses that result from the application of bending moment. Although the theory that we will develop can be applied to other types of structural elements, we will restrict our study to a consideration of beams only.

2

When a beam is loaded, strains and stresses are introduced internally at every cross-section and we must be able to determine their magnitude and distribution. Are the strains and stresses uniform or do they vary along the length of a beam and throughout the depth of a cross-section? Experience based on our previous work should tell you that a typical simply supported beam deflects under load as shown below.

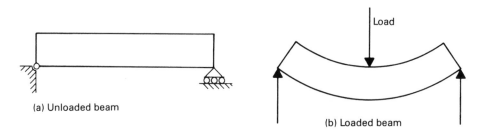

(a) Unloaded beam

(b) Loaded beam

Now look at figure (b). Which face (top or bottom) of the loaded beam is in tension and which is in compression? (Hint: remember that material which increases in length is in tension, material which decreases in length is in compression.)

3

> The top is in compression.
> The bottom is in tension.

You should have decided that the top of the beam is in compression and the bottom in tension. This implies that the strain and stress at any section in a beam will vary from compressive to tensile across the depth of the beam. To help us understand what happens in the material of a beam, it is useful to imagine the beam as made up of several layers of fibres as shown below in figures (a) and (b). When bending under load takes place, such a beam could deflect in one of two ways, either as in figure (c) or as in figure (d).

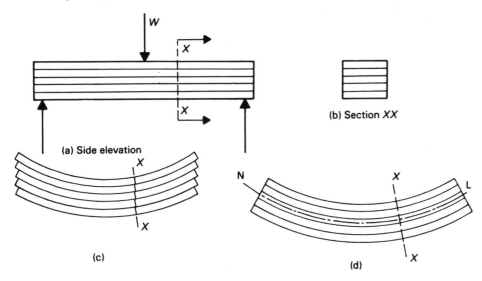

(a) Side elevation

(b) Section *XX*

(c)

(d)

In case (c) each layer deforms independently and slipping takes place along the interface between adjacent layers. Because each layer is deforming independently, compressive stresses will occur at the top surface and tensile stresses will occur at the bottom surface of each layer. This would happen if our beam was indeed made up of many separate layers with no adhesion or bond between them. If our beam was built up using separate layers and we wished to prevent it bending and slipping along the interfaces as in (c), the layers would have to be securely bonded together by glue, bolts or similar means before the load was applied.

A real beam, which is solid and assumed to be homogeneous (having the same composition throughout) would deflect as in figure (d) without any relative slipping between layers. A section *X–X* in the elevation in figure (a) is in the vertical plane at right angles to the longitudinal axis of the beam and is a plane (flat) surface. You can see in figure (d) that, after bending, section *X–X* remains plane and at right angles to the longitudinal axis although no longer vertical. It is therefore assumed that all plane sections normal to the longitudinal axis of the beam remain plane after bending.

For bending to take place as in figure (d), fibres towards the top of the beam shorten and are in compression, while those fibres towards the bottom of the beam lengthen and are in tension. Somewhere between the top and bottom is a layer which is neither compressed nor stretched, and hence the fibres along this layer do not change in length. This layer is called the Neutral layer and is shown by the line N–L in figure (d).

4

Now look at the beams sketched below:

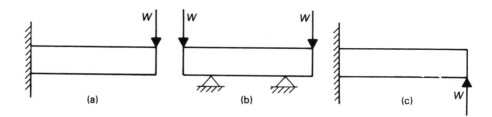

(a)   (b)   (c)

Can you say:
(i) For each beam: is the top or the bottom face in tension?
(ii) For all beams: at what level are the fibres of the beam subjected to the greatest extension or the greatest shortening?

5

> (i) (a) top (b) top (c) bottom
> (ii) at the top or bottom face of the beam

In all cases the layer of fibres at the top or bottom face of a beam will be subject to the greatest extension if that face is in tension. If the top or bottom face is in compression then the fibres at that level will be subjected to the greatest shortening. You will find it helps to see this if you visualise the way in which the beam deflects. The following figures show the deflected shape of the beams considered in Frame 4.

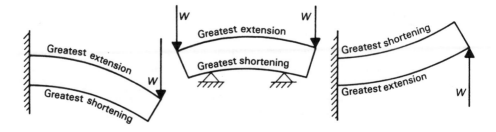

If a beam under load deflects in the shape of a circular arc, we can identify a centre of curvature O and a radius of curvature R as shown in the next figure (a). We will define R as the radius from the centre of curvature to the neutral layer.

In practice the radius of curvature of a deflected beam will not necessarily be the same throughout the length of the beam. We will discover why later in this programme. We can, however, consider a beam as comprising several elements of such a short length that the radius of curvature can be considered to be constant over an element. Such an element is shown in figures (b) and (c).

(a) (b) Undeformed element (c) Deformed element

In figure (b), PQ is the neutral layer and ST is a typical layer at a distance y from the neutral layer and on the tension side of the beam. In figure (c) these layers are shown as P'Q'and S'T'respectively.

Can you state the relationship between the length of PQ and P'Q'?

6

$$\text{length PQ} = \text{length } P'Q'$$

In Frame 3 we defined the neutral layer as the layer where the fibres do not change in length when the beam deflects. Hence as PQ is the neutral layer, its length when the beam is unloaded must be the same as when it is loaded.

Let the length of $PQ = s$

Then from figure (c) PQ = P'Q' $= s = R \times \theta$

where θ is the angle subtended at O as shown.

All layers of fibres in the element from the top face ab to the bottom face cd are of the same length ($s = R \times \theta$) when the beam is unloaded. If we now consider the layer ST we can see that, whereas the length of this is s in the unloaded beam, its length is greater in the loaded beam and equal to the length of the arc S'T'in figure (c).

From figure (c), the length of layer S'T' $= (R+y)\theta$

The original length of ST $= s = R\theta$

Hence the increase in length of layer ST $= (R+y)\theta - R\theta = y\theta$ (6.1)

From equation (6.1) the increase in the length of layer ST is seen to be directly proportional to the distance (y) of layer ST from the neutral layer.

Can you sketch a diagram to show how the change in length of a layer of fibres varies across the section of the element from the top face ab to the bottom face cd?

7

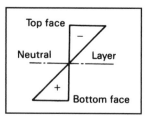

You should have realized from equation (6.1) that the change in length of any layer throughout the depth of the element is directly proportional to its distance (y) from the neutral layer. For any value of y on the tension side of the element, the fibres will extend (shown as positive on the diagram above). Likewise for values of y on the compressive side of the element, the fibres will shorten (shown as negative on the diagram above).

(a) Elevation of element   (b) Plot of change in fibre length against distance from neutral layer   (c) Strain distribution diagram

Rather than plot a diagram showing the change in length of the fibres throughout the depth of the element, it is more useful to plot a diagram showing the variation of strain throughout the depth of the element. The strain in any layer of fibres is the change in length of the layer (δs) divided by the original length of that layer. All layers of fibres in the undeformed element are, however, of the same original length (s). Hence the strain distribution diagram (figure (c)) can be obtained from figure (b) by dividing each ordinate of figure (b) by the original length (s), thus giving a strain diagram of the same shape. The maximum tensile and compressive strains are seen in figure (c) to occur in the outer fibres at the bottom and top faces of the element. As the small element that we have been considering is representative of any point along a beam, we can draw similar strain distribution diagrams to show the strain condition across the depth of any beam at any cross-section along the length of that beam.

Now sketch the strain distribution diagrams at section X–X for each of the beams shown below, and mark the zones of tensile strain and the zones of compressive strain. Indicate the position of the neutral layer in each case.

(d)   (e)   (f)

(d) + indicates tensile strain

(e) (f) − indicates compressive strain

We have already seen that for a *linearly elastic* material, stress is directly proportional to strain and:

$$\frac{\text{stress}}{\text{strain}} = \frac{\sigma}{\varepsilon} = E$$

that is:

$$\sigma = E\varepsilon$$

You will consequently realise that if we multiply the ordinates of a strain diagram by the *elastic modulus E* for the material of which the beam is made, we will obtain a *stress distribution diagram* as shown in figure (c) below. Note that, when deriving figure (c) we have assumed that the value of *E* is the same in the tension zone as in the compression zone. Also, since we have previously defined tensile strains as positive we have now, to be consistent, shown tensile stresses as positive.

(a) Elevation of element   (b) Strain distribution diagram   (c) Stress distribution diagram

For the small element that we have been considering, stresses above the neutral layer are compressive and stresses below the neutral layer are tensile. The maximum compressive and maximum tensile stress can be seen to occur in the outer fibres at the top and bottom faces of the element.

The stress diagram that we have just developed is triangular in shape. This is not always the case and later in your studies you may encounter stress diagrams of different shape. All except the final programmes in this book will however refer to triangular stress distribution. In other words our analysis is based on the assumption that materials behave in a truly linear elastic manner.

9

In this frame we will look at the stress distribution diagram in greater detail so that we can further develop the theory of bending in beams.

For simplicity we will initially look at beams which have a rectangular cross-section. Not all beams, however, will have a rectangular section and we will discover in subsequent frames how to analyse such beams.

(a) Part elevation of beam   (b) Section *XX*   (c) Stress distribution diagram

Look at the cross-section (b) above which is subjected to the stresses shown in the stress diagram in figure (c). We see that the cross-hatched area is under compressive stress. All fibres above the neutral layer are subject to compressive stress ranging from zero at the neutral layer to a maximum at the top of the beam. The fibres thus react to external loading by developing reactive forces acting in a direction parallel to the longitudinal axis of the beam. In figure (b) all fibres within the compressive zone are exerting reactive forces in the same direction. These forces can be combined into a single resultant as indicated by the arrow C in figure (c). *C* represents the resultant of all compressive forces in the beam above the neutral layer. Similarly *T* represents the resultant of all tensile forces in the beam below the neutral layer.

Now and with reference to figure (c) above, consider the following question: does your present understanding enable you to predict the position of the line of action of the resultant force C and the line of action of the resultant force T?

10

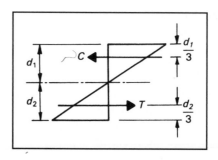

You may have predicted that C and T act through the centroid of their respective *stress blocks* as shown. This is a reasonable prediction based on work done in previous programmes and is true in this case, but would not be true if the beam was not rectangular in cross-section. Why not? Because the line of action of a resultant *force* passes through the centroid of a *force* distribution diagram whereas we have been considering *stress* distribution diagrams. The two diagrams are only the same shape if the beam is of uniform width.

Now look at figure (a) below. The line X–X, which is the line of intersection between the neutral layer and the plane of the cross-section, is known as the *neutral axis* (NA) of the cross-section. We will prove later that the neutral axis always passes through the geometric centre of the cross-section. Consequently, in the case of a rectangular section the neutral axis is at mid depth. The neutral axis is however not necessarily at mid depth in the case of more complicated shapes of cross-section.

Figure (b) below shows the stress distribution diagram for the rectangular cross-section (a), and we see that the tension and compression stress blocks are of identical shape. Now consider the part of the beam above the neutral axis. The stress varies from zero to maximum value σ_{max} so that the average stress is $\sigma_{max}/2$. This average stress acts over the hatched area $b \times d/2$ so that the resultant force C is given by:

$$C = \text{average stress} \times \text{area} = \frac{\sigma_{max}}{2} \times \frac{b \times d}{2}$$

$$= \frac{\sigma_{max}bd}{4}$$

$$\text{Similarly } T = \frac{\sigma_{max}bd}{4} \qquad \text{Hence } \underline{T = C}$$

(a) Cross-section of beam   **(b) Stress distribution diagram**

Notwithstanding the calculations that we have just done, you should realise that the tensile force (T) has to equal the compressive force (C). Why should T = C?

11

> Because for equilibrium of the section the sum of the horizontal forces acting on it must be zero, and as there is no external force acting on the section then T must equal C.

Having decided that $T = C$, we see that now we have two equal forces acting on the cross-section which form a couple. This couple exerts a moment of C (or T since they are equal) times the distance between their lines of action ($2d/3$).

This *internal* moment provides the beam's resistance to the *external* bending moment resulting from the external loads and is consequently called the *moment of resistance* (M). For equilibrium of the section, the internal moment of resistance must be equal in magnitude and opposite in sense to the external moment acting on the section.

From the stress diagram in Frame 10 it can been seen that:

$$\text{Moment of resistance} = M = C \times \frac{2d}{3}$$

$$= \frac{\sigma_{max} \times bd}{4} \times \frac{2d}{3}$$

$$= \frac{\sigma_{max}bd^2}{6} \tag{6.2}$$

Given a permissible value for the maximum stress σ_{max}, you should now be able to determine the size of the beam required to resist a given bending moment. But remember that the above derivation applies only to a beam of rectangular cross-section.

A timber beam of solid rectangular cross-section is required to withstand an external bending moment of 11.25 kN m. If the maximum permissible bending stress is 5.0 N/mm² and the beam has a breadth b = 150 mm, calculate the minimum required depth (d) of the beam.

12

> 300 mm

To resist a bending moment of 11.25 kN m the beam must be able to develop a moment of resistance of 11.25 kN m. Then using equation (6.2):

$$M = \frac{\sigma_{max}bd^2}{6} \qquad \therefore \quad 11.25 \times 10^6 = 5.0 \times \frac{150d^2}{6}$$

$$\therefore \quad d^2 = \frac{11.25 \times 10^6 \times 6}{5.0 \times 150}$$

$$\therefore \quad \underline{d = 300 \text{ mm}}$$

Now answer the following questions:

(i) Bearing in mind the shape of the stress distribution diagram, do you consider that for a beam of fixed length and for a given weight of material, a rectangular cross-section provides the greatest possible moment of resistance?

(ii) For the same length and weight of material, could a beam of different shape cross-section be stronger?

13

> (i) no   (ii) yes

In fact a rectangular section does not utilise the material of which a beam is made to the best advantage. A rectangular section may be the easiest shape to make but for a given weight of material does not give the greatest possible moment of resistance. Nor would such a shape provide the most economic beam to withstand a given loading.

In structural design we normally require to keep the weight of a structure as low as possible since this will reduce the load on the foundations and lead to the lowest cost. This will be achieved if, where feasible, we use beams in which as much of the beam section as possible is stressed as highly as possible, hence ensuring an efficient use of the available material. Therefore as much of the material as possible should be located near the top and bottom faces of the beam where the maximum compressive and tensile stresses occur.

Now assume that we have a given quantity of material to make a beam of a certain fixed length and that we have a choice of three cross-sections as shown below. (Note that the areas of all three sections are identical.)

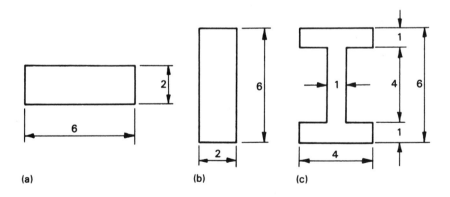

(a)   (b)   (c)

Which of the sections would give the greatest moment of resistance? (Assume that the maximum permissible stress (σ_{max}) is the same in each case.)

14

The best section is (c).

You should have reasoned as follows:

From equation (6.2) the moment of resistance $M = \dfrac{\sigma_{max} bd^2}{6}$

$\therefore$ for section (a) $\quad M = \sigma_{max} \times 6 \times 2^2/6 = 4\sigma_{max}$

for section (b) $\quad M = \sigma_{max} \times 2 \times 6^2/6 = 12\sigma_{max}$

Hence section (b) is stronger than section (a).

Although you cannot as yet calculate the moment of resistance of section (c), you can see that more material is positioned near the outer fibres and is concentrated in the regions of greatest stress. Thus section (c) is stronger that section (b).

We can now learn how to calculate the moment of resistance for a beam with a cross-section other than rectangular. To do this we consider the general case of a beam with irregular cross-section as shown below:

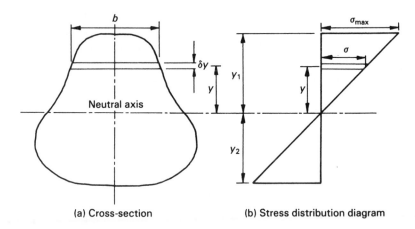

(a) Cross-section   (b) Stress distribution diagram

Consider a thin strip of the cross-section of width b, thickness δy and at distance y from the neutral axis. Let the stress in this strip $= \sigma$.

Then the longitudinal force acting on the strip $=$ stress $\times$ area

$$= \sigma \times (b\delta y)$$

And hence the longitudinal force acting on the cross-section $= \displaystyle\int_{-y_2}^{y_1} \sigma b \, dy$

But as there can be no resultant longitudinal force acting on the section, this force must be zero if static equilibrium is to be assured.

thus $\quad \displaystyle\int_{-y_2}^{y_1} \sigma b \, \mathrm{d}y = 0$ $\qquad\qquad$ (6.3)

By similar triangles in figure (b) you will see that:

$$\frac{\sigma}{y} = \frac{\sigma_{max}}{y_1}$$

thus $\quad \sigma = \dfrac{\sigma_{max}}{y_1} \times y$

and equation (6.3) becomes:

$$\frac{\sigma_{max}}{y_1} \int_{-y_2}^{y_1} by \, \mathrm{d}y = 0$$

But σ_{max} and y_1 cannot be zero, thus $\int_{-y_2}^{y_1} by \, \mathrm{d}y$ must be zero.

This integral expression is the first moment area of the cross-section about the neutral axis, and we know from Programme 1 that if the first moment of area about an axis is zero then that axis must pass through the centroid of that area.

What does this tell you about the position of the neutral axis?

15

> *The neutral axis passes through*
> *the centroid of the cross-section.*

This is a very important conclusion and enables us to identify where the neutral axis of the cross-section of a beam is located.

Now consider four horizontal beams each carrying vertical loads and having cross-sections as shown below:

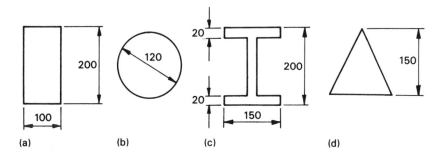

(a) $\qquad$ (b) $\qquad$ (c) $\qquad$ (d)

Identify the position of the neutral axis in each case. (Note that since the loads are vertical, the beams will bend about a horizontal axis. You are consequently looking for a neutral axis which is horizontal in each case.)

16

(a)   (b)   (c)   (d)

You should have found this straightforward since you can identify the centroids of (a), (b) and (c) by symmetry and you know that the centroid of a triangle is $\frac{1}{3}$rd of the height above the base. For some other sections, however, you may need to use the method used in Programme 1 to locate the centroid and hence the neutral axis.

SECOND MOMENT OF AREA

Now refer to the diagram in Frame 14.

The longitudinal force in the thin strip $\quad\quad\quad = \sigma b \delta y$

The moment of this force about the neutral axis $\quad = \text{force} \times \text{distance} \; (y)$

$$= \sigma b \delta y \times y$$

The total moment of the resultant force about the neutral axis $= \displaystyle\int_{-y_2}^{y_1} \sigma b y \, \mathrm{d}y$

But as before (Frame 14) $\quad\quad\quad\quad\quad\quad \sigma = \dfrac{\sigma_{max} \times y}{y_1}$

Thus the total moment about the neutral axis $\quad = \text{the Moment of Resistance}$

$$= M$$

$$= \frac{\sigma_{max}}{y_1} \int_{-y_2}^{y_1} b y^2 \, \mathrm{d}y \quad\quad (6.4)$$

The integral expression in equation (6.4) is termed the *second moment of area* of the cross-section about the neutral axis, and for convenience is given the symbol I. I has units of $(\text{length})^4$ and typically would be calculated in mm^4. Equation (6.4) can now be written more simply as

$$M = \frac{\sigma_{max} I}{y_1} \quad\quad \text{where } I = \int_{-y_2}^{y_1} b y^2 \, \mathrm{d}y$$

but $\dfrac{\sigma_{\max}}{y_1} = \dfrac{\sigma}{y}$ (see Frame 14). Thus $M = \dfrac{\sigma I}{y}$

or

$$\frac{\sigma}{y} = \frac{M}{I} \qquad\qquad (6.5)$$

Where: M is the moment of resistance

I is the second moment of area about the neutral axis

σ is the bending stress in any fibre at a distance y from the neutral axis.

The second moment of area of a beam section is 10×10^6 mm^4. If the section is subjected to a bending moment of 2 kN m, calculate the bending stress at a distance of 50 mm from the neutral axis.

17

$$\boxed{10 \text{ N/mm}^2}$$

$$\sigma = \frac{M \times y}{I} = \frac{2 \times 10^6 \times 50}{10 \times 10^6} = 10 \text{ N/mm}^2$$

Equation (6.5) can be used to calculate the stress at any level in the cross-section of a beam. The stress will be tensile or compressive depending upon which side of the neutral axis is being considered and whether the beam is being subjected to a hogging or sagging bending moment. This can always be determined by inspection. You should note that in other studies you may encounter the use of symbol I to represent moment of inertia, whereas here we are using it to represent the second moment of area. The two uses are not identical: moment of inertia has units of mass $\times$ length^2 whereas the second moment of area is a geometrical property of an area and has units of length^4.

Let's now calculate the second moment of area for a rectangular beam section. We have now shown (as previously assumed) that the neutral axis passes through the centroid of the section which is at mid depth.

For a beam b wide and d deep we can consider an elemental strip as shown in the figure below.

$$I = \int_{-y_2}^{y_1} by^2 \, \mathrm{d}y = \int_{-d/2}^{d/2} by^2 \, \mathrm{d}y = b\left[\frac{y^3}{3}\right]_{-d/2}^{d/2}$$

$$= b\left[\frac{d^3}{24} + \frac{d^3}{24}\right] = \frac{bd^3}{12} \qquad\qquad (6.6)$$

You should remember this formula, as it is widely used in structural analysis and design.

Calculate the I value for a rectangular section 50 mm wide and 100 mm deep.

18

$$\boxed{4.17 \times 10^6 \text{ mm}^4}$$

$$I = bd^3/12 = \left(50 \times 100^3\right)/12 = 4.17 \times 10^6 \text{ mm}^4$$

ELASTIC SECTION MODULUS

Frame 16 introduced the concept of second moment of area. We now consider an equally useful parameter in structural design the *elastic section modulus* (Z). This is defined as the second moment of area of a section about the neutral axis divided by the distance (y_{max}) from that axis to the furthermost fibre.

$$\text{Thus} \quad Z = \frac{I}{y_{max}}$$

For example, in the case of a *solid rectangular section b* wide and *d* deep:

$$I = bd^3/12 \text{ and } y_{max} = d/2 \qquad \therefore \quad Z = \frac{I}{y_{max}} = \frac{bd^3/12}{d/2} = \frac{bd^2}{6}$$

You may recognise this from equation (6.2) in Frame 11. You should also remember this equation as it is widely used in structural design. Note that if the neutral axis is not at mid height of the section (for example, in the case of a T section) then the section will have two Z values: one related to the top face of the section and the other to the bottom face.

Now, from the equation developed in Frame 16: $\quad \dfrac{\sigma}{y} = \dfrac{M}{I}$

and if σ_{max} is the stress in the furthermost fibre then

$$\frac{\sigma_{max}}{y_{max}} = \frac{M}{I}$$

$$\text{or} \quad \sigma_{max} = \frac{M}{I/y_{max}}$$

$$= \frac{M}{Z}$$

Which rearranges to give: $\qquad\qquad M = \sigma_{max}Z \qquad\qquad (6.7)$

Where   M   = the moment of resistance of the section
          σ_{max} = the maximum bending stress in the outer face of the section
          Z   = the elastic section modulus appropriate to the face being considered.

We can use this equation:

(i) to determine the moment of resistance of a beam since:

the moment of resistance of a beam = maximum permissible bending stress $\times Z$

or (ii) to determine the maximum bending stress induced in a beam by an externally applied bending moment since;

$$\text{maximum bending stress} = \text{bending moment}/Z$$

'Universal' steel beams of I shaped cross-section are widely used in building construction. They are usually specified by quoting their depth, breadth and mass per metre length. Their geometric properties are usually obtained from standard steel tables.

Given two universal beams (A and B) with properties as follows, which would resist the greatest bending moment? (Note: the first two numbers are overall depth and breadth in mm; the third number is the mass per metre length in kg.)

A $178 \times 102 \times 21.5$ $Z = 170.9 \ cm^3$: B $127 \times 76 \times 13.4$ $Z = 74.9 \ cm^3$

19

$\boxed{\text{beam A}}$

You should have chosen A because its Z value is the greater. This is correct provided that the maximum value of bending stress is the criterion for design. Later in your studies you will discover, however, that other criteria such as deflection or shearing stress may dictate the maximum safe load that a beam can carry.

When analysing a beam, it is important to distinguish between the terms *plane of bending* and *axis of bending*. Consider the beam shown below:

(a) Side elevation (plane *YY*)

(b) Section

The beam is bending in the Y–Y plane as evidenced by the fact that we see the shape of the deflected beam in the side elevation which is in the Y–Y plane. The beam is bending about the X–X axis passing through the centroid of the section and I_{XX} would be needed for calculations of strength and, as you will see in a later programme, deflections. In this case we are using the symbol I_{XX} to denote the second moment of area. The subscript X–X indicates that bending is taking place about the X–X axis. Similarly, if bending was taking place about the Y–Y axis, we would use I_{YY}.

So far we have considered horizontal beams subject to vertical loading. Many structural members other than horizontal beams may however be subject to bending moments. Beams may likewise be subject to loads in a direction other than vertical.

If the beam above was not loaded as shown, but was subject instead to a horizontal wind load (acting on the side face of the beam at right angles to plane Y–Y): (i) which is the plane of bending? (ii) which is the axis of bending?

20

(i) bending is in the horizontal X–X plane
(ii) bending is about the Y–Y axis

We see this by looking at the beam in plan:

You will remember from Programme 4 that the value of bending moment will usually vary along the length of the beam. Consequently since $M = \sigma \times Z$ (or $\sigma = M/Z$) the value of stress (and strain) in the fibres at any level will also usually vary with changing moment along the length of the beam. You will discover later in this programme that the radius of curvature of the deflected beam also usually varies along the length of the beam.

Now try some problems.

21

PROBLEMS

1. A timber beam has a solid rectangular cross-section 50 mm wide and 150 mm deep. Calculate I_{XX} and Z_{XX}.

Ans. (14.06 × 10⁶ mm⁴, 187.50 × 10³ mm³)

2. If the maximum permissible bending stress in the timber is 10.5 N/mm², calculate the moment of resistance of the beam in question 1.

Ans. (1.97 kN m)

3. If a timber beam 50 mm wide and 150 mm deep is used as a cantilever 0.75 m long, and if the maximum permissible bending stress is 12 N/mm², calculate the maximum single concentrated vertical load which can be supported at the end of the cantilever. (Hint: where does the maximum bending moment in a cantilevered beam occur?)

Ans. (3.0 kN)

4. A timber fence post of square cross-section 150 mm × 150 mm is securely concreted into the ground. As a result of wind loading at right angles to the fence, the post is subjected to a bending moment of 4.0 kN m. Calculate the maximum bending stress.

Ans. (7.11 N/mm²)

22

PARALLEL AXIS THEOREM

In the previous frame you have carried out calculations for beams of solid rectangular section. Many beams however are not rectangular in section. We may obtain values of I and Z for such beams from standard tables used in design offices, or we may calculate them from first principles.

For sections consisting of rectangular components, we can often speed the process of calculation by dividing an area into individual rectangular parts (for which we know $I = bd^3/12$) and then translating the value of I for each part to an I value about the axis of bending of the whole section using the theorem of Parallel Axes. You may have developed the theorem of Parallel Axes in your study of mathematics but, using the case of a rectangular section as an example, we include it here for completeness.

Consider the rectangular area shown. Let I_{CC} be the second moment of area of the section about an axis through its centroid and let I_{XX} be the second moment of area about a parallel axis at a distance h from the centroidal axis. In subsequent frames this axis $(X–X)$ will be the neutral axis of the composite section of which this rectangle is part.

The second moment of area of the elemental strip about the $X–X$ axis $= b\delta y y_1^2$

Hence the second moment of area of the whole section about the $X–X$ axis

$$I_{XX} = \int by_1^2\, dy$$

But: $y_1 = y + h$ ∴ $I_{XX} = \int b(y + h)^2\, dy = \int b(y^2 + 2yh + h^2)\, dy$

$$= \int by^2\, dy + \int b2yh\, dy + \int bh^2\, dy$$

$$= \int by^2\, dy + 2h\int by\, dy + h^2\int b\, dy$$

But: $\int by^2\, dy =$ the second moment of area (I_{CC}) about the centroidal axis

$\int by\, dy =$ the first moment of area about an axis through the centroid

$= 0$ (by definition of centroid)

and $\int b\, dy =$ the area of the rectangle $(A = b \times d)$

Thus: generally $I_{XX} = I_{CC} + Ah^2$, or for a solid rectangle: $I_{XX} = bd^3/12 + bdh^2$.
On to the next frame.

23

Now we will use the theorem of parallel axes to help us calculate the second moment of area of a beam with an I shaped cross-section. The second moment of area is calculated about the neutral axis which, as we have shown, passes through the centroid of the section. Many I beams are symmetrical. In this example however we have selected an asymmetrical beam to emphasise that the first step is to determine the position of the centroid. In a symmetrical section, of course, the centroid is located by inspection.

The section shown above is treated as if made of four rectangular parts A, B, C, D. First we determine the position of the centroid by taking the sum of the first moments of area of each part about a convenient axis (we will take the bottom face of the beam) and equating to the total area times $\bar{y}$ where $\bar{y}$ is the height of the centroid above the bottom face.

Thus
$$A \times \bar{y} = \Sigma \left(A_{\text{part}} \times y_{\text{part}} \right)$$

$$\{(200 \times 25) + (150 \times 20) + (20 \times 200) + (150 \times 20)\}\bar{y}$$
$$= (200 \times 25 \times 252.5) + (150 \times 20 \times 230) + (20 \times 200 \times 120) + (150 \times 20 \times 10)$$

Thus
$$\bar{y} = 164.2 \text{ mm}$$

Now
$$I_{XX} = I_{\text{CC}} + Ah^2$$
$$= bd^3/12 + bdh^2$$

Thus for part A:
$$I_{XX} = (200 \times 25^3)/12 + (200 \times 25)(252.5 - 164.2)^2$$
$$= 39.24 \times 10^6 \text{ mm}^4$$

and for part B:
$$I_{XX} = (150 \times 20^2)/12 + (150 \times 20)(230.0 - 164.2)^2$$
$$= 13.09 \times 10^6 \text{ mm}^4$$

and for part C:
$$I_{XX} = (20 \times 200^3)/12 + (20 \times 200)(120.0 - 164.2)^2$$
$$= 21.15 \times 10^6 \text{ mm}^4$$

and part D:
$$I_{XX} = (150 \times 20^3)/12 + (150 \times 20)(10 - 164.2)^2$$
$$= 71.43 \times 10^6 \text{ mm}^4$$

For the complete section
$$I_{XX} = (39.24 + 13.09 + 21.15 + 71.43) \times 10^6$$
$$= 144.91 \times 10^6 \text{ mm}^4$$

This is an unwieldy calculation and is best done in tabular form. We will do this in the next frame.

Now, if the beam above is to bend about the Y–Y axis (at right angles to X–X), calculate the value of I_{YY}.

24

$$\boxed{28.05 \times 10^6 \text{ mm}^4}$$

In the figure in Frame 23, axis Y–Y passes through the centroids of each part (A, B, C and D) thus the theorem of parallel axes is not needed in this instance, ($h = 0$ for each part).

$$\overset{\text{A}}{} \qquad \overset{\text{B}+\text{D}}{} \qquad \overset{\text{C}}{}$$
Thus $I_{YY} = (25 \times 200^3)/12 + 2(20 \times 150^3)/12 + (200 \times 20^3)/12 = 28.05 \times 10^6 \text{ mm}^4$

We will now set out the previous calculation for I_{XX} in tabular form. This shows a systematic and better way of presenting this type of calculation. Refer back to Frame 23 and note how the various stages of the calculation in that frame also appear in the following table.

Part	Area (A) (mm^2)	y (mm)	Ay	$I_{CC}(bd^3/12)$	$h = (y - \bar{y})$	Ah^2
A	200×25 $= 5000$	252.5	$(\times 10^3)$ 1262.5	$(\times 10^6)$ 0.26	$252.5 - 164.2$ $= 88.3$	$(\times 10^6)$ 38.98
B	3000	230.0	690.0	0.10	65.8	12.99
C	4000	120.0	480.0	13.33	44.2	7.81
D	3000	10.0	30.0	0.10	154.2	71.33
Total	15000		2462.5	13.79		131.11

Thus $\bar{y} = \dfrac{2462.5 \times 10^3}{15000}$
$= 164.17 \text{ mm}$

and $I_{XX} = \dfrac{\begin{array}{r} 13.79 \times 10^6 \\ +131.11 \times 10^6 \end{array}}{144.90 \times 10^6 \text{ mm}^4}$

Now to the next frame.

25

COMPLETION OF THE BASIC THEORY OF BENDING

We have already seen in Frame 6 that a short layer (ST) of fibres at a distance y from the neutral layer and having a length $(R+y)\theta$ in the loaded beam has a length $R\theta$ in the unloaded beam, the change in length due to bending being:

$$(R+y)\theta - R\theta = y\theta$$

If you have difficulty following this statement, refer back to Frames 5 and 6. For clarity figure (c) of Frame 5, which shows the deformed element, is reproduced below:

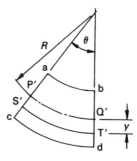

Since the change in length of $\mathrm{ST} = y \times \theta$

then the strain in the fibre $= \dfrac{\text{change in length}}{\text{original length}}$

$$= y\theta/R\theta$$

that is:

$$\varepsilon = y/R$$

Now the Modulus of Elasticity $E = \dfrac{\text{stress}}{\text{strain}} = \dfrac{\sigma}{\varepsilon} = \dfrac{\sigma}{y/R}$

Hence: $\dfrac{E}{R} = \dfrac{\sigma}{y}$

We can combine this with the relationship of equation (6.5) ($\sigma/y = M/I$) to give:

$$\frac{\sigma}{y} = \frac{M}{I} = \frac{E}{R} \tag{6.8}$$

These equalities summarise the basic theory of bending and should be remembered because they are used in a wide variety of structural analysis and design problems.
What do these equations suggest to you about the radius of curvature of a loaded beam?

26

> Since the value of bending moment will normally vary along the length of a beam, then the radius of curvature will also normally vary along the length of that beam.

Now can you remember the assumptions we made in developing the basic theory of bending?

27

> (1) The material is linearly elastic
> (2) E is the same in tension and compression.
> (3) The material is homogeneous (same physical properties throughout).
> (4) Plane sections at right angles to the longitudinal axis remain plane after bending.

These assumptions are unlikely to be fully satisfied by any construction material. The theory does nevertheless provide a valuable tool which is widely used in structural analysis and in design.

Now see of you can complete the relationship: $\dfrac{?}{?} = \dfrac{M}{I} = \dfrac{?}{?}$

28

$$\boxed{\dfrac{\sigma}{y} = \dfrac{M}{I} = \dfrac{E}{R}}$$

A steel beam with an I shaped cross-section as shown carries a single concentrated load of 30 kN at the mid point of a simply supported span of 3 m. We wish to determine the maximum bending stress and the curvature of the beam at the point of maximum bending moment. First we must calculate the maximum bending moment.
(i) Where will the bending moment be greatest?
(ii) What is the value of the maximum bending moment (neglecting the self-weight of the beam)?

29

$$\boxed{\begin{array}{c} \text{at mid span} \\ M = WL/4 = (30 \times 3)/4 = 22.5 \text{ kN m} \end{array}}$$

We must now locate the neutral axis and calculate the value of I_{XX}. Since the section is symmetrical we know that the centroid is at mid-depth, at 70 mm from the top and the bottom face. We then calculate I_{XX} using the expression $bd^3/12$ and the parallel axes theorem.

Part	$I_{CC} = bd^3/12$ $(\times 10^6)$	Area (A)	h	Ah^2 $(\times 10^6)$
top flange	0.07	$100 \times 20 = 2000$	60	7.2
bottom flange	0.07	$100 \times 20 = 2000$	60	7.2
web	1.25		0	0.0
totals	1.39			14.4

Thus for the complete section $I_{XX} = (1.39 + 14.4) \times 10^6 = 15.79 \times 10^6 \text{ mm}^4$.

We now use the equality $\sigma/y = M/I$ to find the maximum stress. Since the cross-section is symmetrical about the X–X axis, the maximum tensile and compressive stresses will be the same.

The maximum value of y is $140/2 = 70$ mm (distance from neutral axis to furthermost fibre), thus the maximum bending stress is

$$\sigma_{\text{max}} = \frac{My_{\text{max}}}{I} = \frac{(22.5 \times 10^6) \times 70}{(15.79 \times 10^6)}$$
$$= 99.75 \text{ N/mm}^2$$

Calculate the maximum bending stress in the web.

30

$$\boxed{71.25 \text{ N/mm}^2}$$

We obtain this answer by putting $y = 50$ mm in the above expression. Why? Because the outer layer of the web is 50 mm from the neutral axis.

Now calculate the radius of curvature of the loaded beam at mid span if the elastic modulus for steel is 200 kN/mm².

31

> 140.36 m

If you did not get this answer, check your working;

$$\frac{M}{I} = \frac{E}{R}$$

thus
$$R = (E \times I)/M$$

$$= \frac{(200 \times 10^3) \times (15.79 \times 10^6)}{22.5 \times 10^6}$$

$$= 140\,355 \text{ mm} = 140.36 \text{ metres}$$

Remember that this is the radius at the mid span of the beam only, because with this loading the value of bending moment varies at all other points along the beam. The radius will therefore vary along the length of the beam.

Now attempt the following problems.

32

> PROBLEMS

1. A rectangular timber beam 50 mm wide and 150 mm deep carries a bending moment of 1.12 kN m. Determine the maximum bending stress.

Ans. (± 5.97 N/mm^2)

2. A beam with a symmetrical cross-section, an I_{XX} value of 25×10^6 mm^4 and of depth 150 mm is simply supported over a span of 5 m. The beam supports a single concentrated load of 40 kN at the mid span. Determine the maximum bending stress at mid span and at a point 1 m from one end of the beam. Neglect the self-weight of the beam.

Ans. (± 150 N/mm^2, ± 60 N/mm^2)

3. A beam with cross-section as shown is simply supported and carries a bending moment of 7.0 kN m. Determine
 (i) the maximum tensile stress
 (ii) the maximum compressive stress.

   (*Hint*—the value of y for maximum tensile stress is not the same as the value of y for maximum compressive stress.)

Ans. (146.61 N/mm^2, 67.48 N/mm^2)

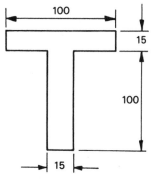

STRUCTURAL MECHANICS

33

In the problems just completed you have calculated the bending stresses and radius of curvature produced by specific external loadings. You will remember however that in earlier frames we did the reverse of this and determined the moment of resistance of a beam, which is the maximum bending moment a beam can carry such that the bending stresses do not exceed a prescribed maximum permissible value. To do that we used the equation $M = \sigma_{max}Z$.

What is Z in this equation?

34

Z is the elastic modulus of the section ($= I/y_{max}$)

If you are not sure of this, refer back to Frame 18 before proceeding. The equation $M = \sigma_{max}Z$ is used in structural design where we are usually only concerned with the maximum bending stress which occurs in the outer fibres at the top and bottom of the beam section. If however we wish to determine the stress at any layer other than the top or bottom surface of a beam, then we must use the original form of the equation given by:

$$\frac{\sigma}{y} = \frac{M}{I} \quad \text{OR} \quad M = \frac{\sigma I}{y}$$

35

STRENGTHENING BEAMS BY THE USE OF FLANGE PLATES

We saw in Frame 15 that, of the three shapes of section considered, the I shape provided the greatest moment of resistance. In general, the I section is most efficient in this respect and uses the material of the beam to the best advantage. A beam with an I shaped cross-section has a high strength to weight ratio. As a consequence of this fact, structural steel manufacturers have developed a large comprehensive range of I beams for constructional purposes. These beams may be used by themselves or may be strengthened, if necessary, by the addition of one or more steel flange plates fastened to the flanges of the original beam. These flange plates may extend along the entire length of the beam or may be used only at those places where bending moments are greatest. In the latter case the plates will be cut (curtailed) at points along the beam where the applied bending moment decreases to such a value that the unstrengthened beam section can resist it. At these points the bending moment has a value equal to the moment of resistance of the unstrengthened beam.

The beam shown below comprises a standard steel section $102 \times 64 \times 9.6$ kg/m with 2 No. 70 mm $\times$ 7 mm steel flange plates extending over the central 2 metre length. The standard beam has an I_{XX} value of 218 cm^4.

Given the maximum permissible bending stress is 165 N/mm^2 in tension or in compression, we wish to calculate the maximum central point load W that can be supported. We will neglect the self-weight of the beam and the flange plates.

First we must calculate I_{XX} for the composite section. By symmetry the centroid and hence the neutral axis is at mid depth.

Part	$I_{CC}(bd^3/12)$ $(\times 10^6)$	Area A	h	Ah^2 $(\times 10^6)$
top plate	0.002	490	54.5	1.455
standard beam	2.180	–	0	0
bottom plate	0.002	490	54.5	1.455
	2.184			2.910

Total $I_{XX} = (2.184 + 2.910) \times 10^6 = 5.09 \times 10^6$ mm^4

Then the modulus of section:
$$Z = \frac{I}{y_{max}} = \frac{5.09 \times 10^6}{(51 + 7)}$$
$$= 87.76 \times 10^3 \text{ mm}^3$$

and
$$M = \sigma_{max} \times Z = 165 \times 87.76 \times 10^3$$
$$= 14.48 \times 10^6 \text{ N mm}$$
$$= 14.48 \text{ kN m}$$

Now the maximum bending moment (at mid span) is given by:
$$M = WL/4 = W \times 6/4 \text{ kN m}$$

Thus
$$W = 4M/6 = 4 \times 14.48/6 = 9.65 \text{ kN}$$

Can we be sure that this is a safe load to put on the beam?

37

No not without checking the maximum stresses at the points D and E where the flange plates are curtailed.

At points D and E the strength of the beam depends only upon the strength of the unstrengthened section. The effective I value at these points (just beyond the end of the flange plates) is 218 cm^4, thus the maximum stress at D (and E) which occurs at the top and bottom faces is given by the following calculation:

$$\text{The section modulus:} \quad Z = \frac{I}{y_{\max}} = \frac{218 \times 10^4}{51}$$

$$= 42.75 \times 10^3 \text{ mm}^3$$

and $\qquad M = \sigma_{\max} \times Z \quad$ or $\quad \sigma_{\max} = \frac{M}{Z}$

Thus the maximum stress at D (and E) $\quad \sigma_{\max} = \dfrac{M}{42.75 \times 10^3}$

where M is the value of the bending moment at D.

If the load on the beam was 9.65 kN, what would be the bending moment at D?

38

9.65 kN m

If the central load on the beam is $W = 9.65$ kN

then the reaction at A $= V_A = W/2 = 4.83$kN

and the bending moment at D $= V_A \times$ distance AD

$$= 4.83 \times 2.0$$

$$= 9.65 \text{ kN m}$$

Since we know that the bending moment at D is 9.65 kN m we can now check the value of maximum stress at D since:

$$\sigma_{\max} = \frac{M}{42.75 \times 10^3} = \frac{(9.65 \times 10^6)}{42.75 \times 10^3} = 225.73\text{N/mm}^2$$

This is obviously very much greater than 165 N/mm^2, so we see that 9.65 kN is not a safe load to put on the beam. Point D is a critical point which determines the load-bearing capacity of the beam.

Now try to calculate the truly safe value for the load W.

39

> 7.05 kN

If you did not get 7.05 kN, check your working as follows:
The maximum possible value for the bending moment at D is given by:

$$M = \sigma_{max} \times Z = 165 \times 42.75 \times 10^3 = 7053 \times 10^3 \text{ N mm}$$
$$= 7.05\text{kN m}$$

But the bending moment at D is given by:

$$M = V_A \times 2 \qquad = (W/2) \times 2$$
$$= W\text{kN m}$$

Hence if the maximum allowable moment at D is 7.05 kN m (as above) if follows that:

$$M = 7.05 = W$$
$$\text{or} \qquad W = 7.05 \text{ kN}$$

This example emphasises the need for you to identify and check all possible critical points when analysing or designing any structure.
We have learnt that for a given weight of material, an effective shape of cross-section for a beam is the I section. Why is this?

40

> because the larger part of the material
> is concentrated at the outer layers

This is of course true if we use the beam so that it bends about the *X–X* axis as shown in figure (a). Suppose however we turn the beam through 90° and load it such that it bends about the *Y–Y* axis as shown in figure (b).

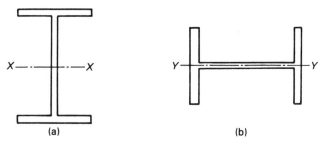

(a)   (b)

Could it resist the same bending moment as when the axis X–X is the axis of bending?

41

$$\boxed{\text{No}}$$

In figure (b) the greater part of the sectional area is concentrated near the neutral axis where it is not used efficiently in bending. Very little of the material is stressed to its full capacity.

PRINCIPAL AXES

For any shape of cross-section a value for I can be calculated about any axis which passes through the centroid. The value of I will normally be different for different axes depending upon the orientation of the axes, but for any section it can be shown that the maximum and minimum values of I occur about two axes at right angles to one another. These axes are the *principal axes* of the section. The maximum value of I occurs about the major *principal axis*, the minimum value of I occurs about the minor *principal axis*.

The simple theory of bending that we have developed is strictly only applicable if bending takes place about either the major or the minor principal axis. If this is not the case then the calculations are more complex and beyond the scope of this book. For most structural uses however, the bending of the beam sections does take place about a principal axis and the theory that we have investigated is valid.

The principal axes of a section can usually be readily identified because, if the section has a geometrical axis of symmetry then that axis is one of the principal axes and the other principal axis is at right angles to it. Both axes pass through the centroid of the section.

It is important for any beam, or any other structural member subject to bending, that it is positioned correctly to provide the greatest possible resistance to bending moments. This normally implies that the member should be positioned such that bending takes place about the major principal axis.

Identify the principal axes of the following sections:

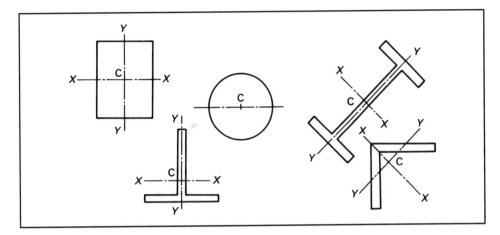

The principal axes are indicated on the above sketches. X–X is the major principal axis in each case. For the circle, the I value will be the same for any axis through the centroid, all of which are axes of symmetry.

The values of I and Z about the principal axes of standard structural sections, both symmetrical and asymmetrical, and the position of those axes may be obtained from standard tables.

43

TO REMEMBER

$$\frac{\sigma}{y} = \frac{M}{I} = \frac{E}{R}$$

For a solid rectangle: $I = bd^3/12$ (units $= \mathrm{mm}^4$)

and $Z = bd^2/6$ (units $= \mathrm{mm}^3$)

$$I_{XX} = I_{CC} + Ah^2$$

$$I_{XX} = bd^3/12 + bdh^2 \text{ (for a solid rectangular section)}$$

Moment of resistance $= \sigma_{\max} \times Z$

Maximum bending stress $= M/Z$

$$Z = I/y_{\max}$$

44

FURTHER PROBLEMS

1. A beam of rectangular cross-section 75 mm wide and 225 mm deep carries a uniformly distributed load (including the self-weight of the beam) of 1.4 kN/m over a simply supported span of 4 m. Determine the maximum bending stress in the beam.

Ans. (4.42 N/mm^2)

2. A steel I section joist has an I_{XX} value of 2294 cm^4 and a depth of 203.2 mm. If the maximum permissible bending stress is not to exceed 165 N/mm^2, determine the moment of resistance of the section.

Ans. (37.25 kN m)

3. If the beam of question 2 has a self-weight of 248 N per metre length, determine the maximum uniformly distributed load which could be safely applied to the entire simply supported span of 4 m.

   (*Hint:* the beam has to resist the bending moment produced both by its own weight and by the external load.)

Ans. (18.37 kN/m)

4. If the beam of question 3 is required to carry a single concentrated load at mid span instead of the uniformly distributed load, determine the maximum value of that concentrated load.

Ans. (36.75 kN)

5. A beam is fabricated by bending and welding a 6 mm thick plate to form a hollow rectangular box section 100 mm overall width and 160 mm overall depth. If the safe permissible bending stress is 165 N/mm^2, calculate the moment of resistance of the section.

   (*Hint:* to calculate the I of the section, calculate the I for a solid beam of the same overall dimensions and deduct the I of the material removed to give the hollow section.)

Ans. (21.36 kN m)

6. A hollow tube of 50 mm external diameter and 44 mm internal diameter is subjected to a bending moment of 0.75 kN m. Determine the maximum bending stress.

   (*Hint: I* for a circular area about a diameter is $\pi D^4/64$.)

Ans. (152.67 N/mm^2)

7. If the material of the beam in question 6 can be stressed to 165 N/mm^2, determine the value to which the external bending moment may be safely increased.

Ans. (0.81 kN m)

8. A beam has a cross-section as shown in figure Q8.
 (i) Identify the principal axes and calculate I_{XX} and I_{YY}.
(ii) If the beam is positioned to bend about the X–X axis and if the maximum permissible stress in tension or compression is 165 N/mm^2, calculate the moment of resistance of the section.
   (*Hint:* the maximum tensile stress is not the same as the maximum compressive stress. The moment of resistance corresponds to that moment which would produce the maximum permissible bending stress in either the tensile or compressive zones. Neither stress may exceed 165 N/mm^2 but the one may be less).

Ans. ($I_{XX} = 110.80 \times 10^6$ mm^4, $I_{YY} = 7.64 \times 10^6$ mm^4, (ii) 132.30 kN m)

Q8

9. A horizontal cantilever beam 2 m long has a T shaped cross-section as in figure Q9 below and carries a uniformly distributed load of 11 kN/m along the entire length. Calculate the greatest tensile and compressive bending stresses. Ignore the self-weight of the beam.

Ans. (Maximum tensile stress = 73.00 N/mm^2, compressive stress = 163.48 N/mm^2)

Q9

10. A compound beam consists of a standard I section Universal beam 381 mm deep and 152 mm wide with a steel plate 200 mm by 10 mm thick welded to each flange and extending for 1.5 metres both sides of the mid span. I_{XX} for the Universal beam is 212.76×10^6 mm^4.

If the beam is simply supported over a span of 6 m, determine the maximum uniformly distributed load which the beam can carry. Ignore the self-weight of the beam. The maximum permissible stress is 165 N/mm^2.

Ans. (327.61 kN)

11. A Universal beam (of I shaped cross-section) is 203 mm deep, 133 mm wide and has a steel plate 150 mm by 10 mm thick fastened to each flange. I_{XX} for the Universal beam is 28.80×10^6 mm^4.

(i) Determine the total uniformly distributed load which may be carried over a simply supported span of 6 metres if the permissible bending stress is 165 N/mm^2. Neglect the self-weight of the beam.

(ii) Determine the minimum central length over which the additional plates are required.

Ans. ((i) 124.02 kN, (ii) 4.23 m)

Programme 7

COMBINED
BENDING AND
DIRECT STRESS

On completion of this programme you will be able to:

- Calculate the state of stress in a structure subject to a combination of axial load and bending about a single axis
- State and apply the 'middle third rule'
- Calculate the state of stress in a structure subject to a combination of axial load and bending about two axes
- Identify and determine the 'core' of a section for no-tension conditions to apply
- Calculate the maximum permissible loading in a section with given limiting stress conditions
- Analyse wall structures subject to vertical and lateral load conditions
- Determine the state of stress under the foundations of simple column and wall structures

1

In Programmes 5 and 6 we learnt how to calculate uniform stress due to direct axial loads and bending stresses resulting from bending moments.

In this programme we will see how to determine the stresses at any point in a structural member which is subjected to a combination of both direct axial loading and bending moments.

There are many practical situations where *both* axial loads and bending moments act in combination on a structural member.

Can you think of any?

2

The list of situations is almost endless but you might have thought about columns with load applied eccentrically to the centroidal axis of the section; walls subjected to their own self-weight plus lateral loading from wind or retained earth, etc. We will look at some of these examples in detail so if you cannot immediately appreciate that these are circumstances where combined stresses arise, don't worry!

Now, by way of revision, consider a beam which is subjected to lateral loading which gives rise to a bending moment M at mid span:

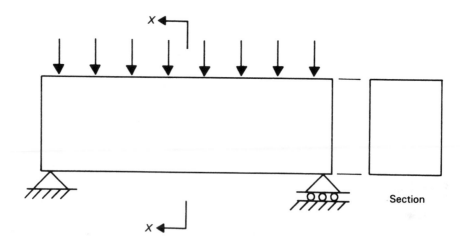

Write down the general expression which would be used to calculate the maximum bending stresses at the mid-span section X–X and sketch the shape of the distribution of bending stresses across the section.

3

$$\sigma = \pm \frac{M}{Z}$$

If this expression is unfamiliar, revise Programme 6 before proceeding. Note that in the above stress diagram *compressive stresses are shown as positive*, as in the type of problem that we are going to look at the stresses are predominantly compressive and the sign convention is chosen with this fact in mind.

Now consider the same beam alternatively subjected to a longitudinal axial force P which passes through the centroid of the beam's cross-section:

What is the expression for the axial stresses resulting from this force? Sketch the shape of the stress distribution across the beam's cross-section.

4

$$\sigma = + \frac{P}{A}$$

Again this expression should be quite familiar, but if in doubt turn back to and revise Programme 5.

If the beam is now subjected to *both* the longitudinal force P and the lateral loading which causes a mid-span moment M, the longitudinal stress distribution will obviously be different from the two cases considered above. As the beam is subjected to two loading effects, the resulting stresses will be a combination of the two individual stress distributions.

Write down the expressions that would give (a) the stress at the top of the beam at mid span and (b) the corresponding stress at the bottom of the beam.

5

$$\sigma_{\text{top}} = \frac{P}{A} + \frac{M}{Z} \qquad \sigma_{\text{bot}} = \frac{P}{A} - \frac{M}{Z}$$

The correct expressions are obtained by adding together the stresses for the two separate loading cases. **Signs** are important: at the top of the beam, we are adding together two compressive stresses; at the bottom, one of the stress components is compressive while the other is tensile. The stresses must be combined *algebraically*.

What we have done here is again to make use of the *Principle of Superposition* which we first saw in Programme 4. This is acceptable provided that the stresses and loads are linearly related. In such circumstances the stresses due to individual loads may be determined separately and the stresses resulting from a system of loads may be obtained by superimposing (or simply adding together) the separate stresses. This principle, as we have already seen, can be applied to other structural 'effects' such as bending moments and shear forces, and also to deflections, rotations, etc.

Before you continue to the next frame, sketch the shape of the stress distribution across the beam's cross-section.

6

You may have drawn either of these diagrams. Both could be correct! In the first figure the compressive stress due to axial load is large enough to negate the tensile bending stress ($P/A > M/Z$). In the second figure the tensile bending stress at the bottom of the beam is greater than that due to axial compression ($M/Z > P/A$).

What do you deduce from both diagrams about the position of the neutral axis?

7

> The neutral axis does not pass through the centroid of the section

You should have recalled from Programme 6 that the neutral axis normally passes through the centroid of the cross-section. From the diagrams in the last frame, this is obviously not the case. In the first figure there is no apparent neutral axis. In the second diagram the axis of zero strain lies within the section but lower down than the centroidal axis. The effect of applying an axial load to the section is to move the neutral axis away from the centroidal axis. Note, however, that the maximum stresses due to bending and axial loads still occur at the extreme fibres of the cross-section. Now try a problem:

A rectangular beam section has a sectional area of 180×10^3 mm² and an elastic modulus of 18×10^6 mm³. What are the stresses on the outer faces of the section when subjected to a compressive axial load of 100 kN and a moment of 80 kN m?

8

> $\sigma_{\text{top}} = 5 \text{ N/mm}^2$: $\sigma_{\text{bot}} = -3.89 \text{ N/mm}^2$

The solution to the above comes from a direct substitution of terms into the equations at the top of Frame 5.

A very common situation that can give rise to axial loading together with a moment is that of a column with a load applied *eccentrically* to the centroidal axis of the cross-section. In the figure the column cross-section is subjected to a thrust P which acts along the Y–Y axis and at an eccentricity e from the X–X axis which passes through the centroid (C) of the cross-section. If moments are taken about the centroidal axis, this is equivalent to an axial load P passing through the centroid together with a moment of Pe acting about the centroidal X–X axis.

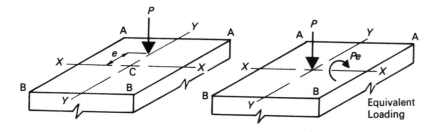

Equivalent Loading

Which is the most highly compressed face, A–A or B–B? Where is tension most likely to develop? Write down the general expressions that would give the maximum stresses in the outer faces A–A and B–B.

9

$$\sigma_{A-A} = \frac{P}{A} + \frac{Pe}{Z}; \quad \sigma_{B-B} = \frac{P}{A} - \frac{Pe}{Z}$$

A–A is the most highly compressed face.
Tension is likely to develop along the face B–B.

The expressions are of course similar to those given in Frame 5, but the moment M is replaced by the equivalent moment Pe.

In many practical situations we may be interested in the condition that no tensile stresses are allowed to develop in the cross-section for any eccentricity of loading. For example, masonry or brickwork cannot resist tensile stresses. According to our sign convention, this requirement means that all stresses must be positive. The face where tension is most likely to develop is face B–B. In this case, remembering that compressive stresses are positive, we may write the equation for the stress condition along the bottom face, B–B, as:

$$\sigma_{B-B} = \frac{P}{A} - \frac{Pe}{Z} \geq 0$$

$$\therefore \quad \frac{P}{A} \geq \frac{Pe}{Z}$$

$$\text{or} \quad e \leq \frac{Z}{A}$$

For a solid rectangular section of breadth b and depth d this can be expressed as:

$$e \leq \frac{\left(bd^2/6\right)}{bd} \quad \text{or} \quad e \leq \frac{d}{6}$$

Hence, provided that the load P acts at an eccentricity of no greater than $d/6$ above the centroidal axis, then no tensile stresses will occur along the outer bottom edge, B–B.

What is the maximum eccentricity at which this load can act below the centroidal axis such that tension will not occur along the top outer edge, A–A?

10

$d/6$ again

In other words provided that the load acts within a distance of $d/6$ either side of the centroidal axis, then no tension can develop in either of the two outer faces. Put another way – if the load acts within the middle third ($d/6 + d/6 = d/3$), no tension can occur anywhere in the section. This is often referred to as the *Middle Third Rule*. Note, however, that it only applies to solid rectangular sections. Now try some problems.

202

PROBLEMS

1. A force of 50 kN acts normally to the face of a column of square cross-section. It acts at an eccentricity of 15 mm from the centre of the column and along one of the axes of symmetry which passes through the centroid and is parallel to one side of the section. The sides of the section measure 350×350 mm. Determine the stress along each outer edge of the section.

Ans. (0.51, 0.30 N/mm²)

2. What is the maximum possible eccentricity of the force in question 1 if no tensile stresses are allowed to develop in the column section?

Ans. (58.33 mm)

3. A short steel column carries a 100 kN load as shown in plan view. It has a sectional area of 95 cm² and an elastic section modulus for bending about the *X–X* axis of 550 cm³. Calculate the stresses along the outer faces A–A and B–B. (*Hint:* you must convert cm units to mm units.)

Ans. (1.44, 19.62 N/mm²)

4. If the maximum compressive stress in the column in question 3 is not to exceed 160 N/mm², what is the maximum possible eccentricity of the 100 kN load measured from the *X–X* axis of the section?

Ans. (822 mm)

So far we have calculated the maximum stresses on the outer edges of the section using the elastic section modulus in our calculation. If we wished to calculate the stresses at any position at a distance *y* from the axis of bending we can make use of the general form of the equation of simple bending which incorporates the second moment of area of the section for the relevant axis of bending.

   Write down the general form of this equation for the column section given in Frame 8.

13

$$\sigma_Y = \frac{P}{A} \pm \frac{(Pe)y}{I}$$

The form of this equation is similar to that used previously except that the section modulus (Z) has been replaced by the second moment of area I together with the distance y measured from the centroidal axis. If you are not sure about this then revise Programme 6 before proceeding.

A force of 100 kN acts at B at an eccentricity of 50 mm along the Y–Y axis of the column section shown in plan view. Calculate the stresses at the two points marked A and B.

14

$$0.72 \text{ N/mm}^2, \ 1.19 \text{ N/mm}^2$$

For bending about the X–X axis the second moment of area, I, is given by:

$$I = \tfrac{1}{12}bd^3 = \tfrac{1}{12} \times 250 \times 400^3 = 1333 \times 10^6 \text{ mm}^4$$

The moment of the force will give rise to tension at A and hence the stress at A is given by:

$$
\begin{aligned}
\sigma_A &= \frac{P}{A} - \frac{(Pe)y}{I} \\
&= \frac{100 \times 10^3}{(400 \times 250)} - \frac{(100 \times 10^3 \times 50) \times 75}{1333 \times 10^6} \\
&= 1.00 - 0.28 \\
&= \underline{0.72 \text{ N/mm}^2}
\end{aligned}
$$

The moment of the force will give rise to compression at B and hence the stress at B is given by:

$$\sigma_B = \frac{P}{A} + \frac{(Pe)y}{I}$$

$$= \frac{100 \times 10^3}{(400 \times 250)} + \frac{(100 \times 10^3 \times 50) \times 50}{1333 \times 10^6}$$

$$= 1.00 + 0.19$$

$$= \underline{1.19 \text{ N/mm}^2}$$

If the load is now positioned on the X–X axis with an eccentricity of 50 mm in the direction of the point marked C, calculate the stresses at C and D.

15

$$\boxed{\sigma_C = 1.72 \text{ N/mm}^2 \qquad \sigma_D = 0.28 \text{ N/mm}^2}$$

Did you remember to calculate the second moment of area about the Y–Y axis ($I_Y = 400 \times 250^3/12 = 521 \times 10^6 \text{ mm}^4$) and use this in your calculations of the stresses? Now let's extend the theory developed so far to the case of a section that is subjected to axial loading combined with bending about both principal axes of the section.

Consider a section of (say) a column which is subjected to an axial compressive load P together with a moment M_X about the X–X axis and a moment M_Y about the Y–Y axis.

Assume that the moment M_X acts in a direction which will cause compression along the face AB and tension along the face CD. Likewise assume that the moment M_Y causes compression along the face BD and tension along the face AC.

Which corner of the section is subjected to the largest compressive stress and which is subjected to the least compressive stress?

16

largest stress at B: least stress at C

You should have predicted that the largest compressive stress occurs at B. You should have reasoned that, because of the application of the axial load P there is uniform compression over the whole section. However the moment M_X causes compression along the face A–B and the moment M_Y causes compression along the face B–D. Hence corner B is subjected to three compressive stress components and is therefore the most highly stressed corner. A similar argument should lead you to the conclusion that corner C is subjected to the least compressive stress.

How are these stresses calculated? Consider the stress component acting at B arising from the axial force and bending taking place separately about the two principal axes of the section:

(1) Due to the axial force
$$\sigma_B = \frac{P}{A}$$
(compression over the whole section)

(2) Due to the moment M_X acting about axis X–X
$$\sigma_B = \frac{M_X}{Z_X}$$
(compression along edge A–B)

(3) Due to the moment M_Y acting about axis Y–Y
$$\sigma_B = \frac{M_Y}{Z_Y}$$
(compression along edge B–D)

where compressive stresses are taken as positive and the subscripts denote the axis of bending.

Using the principle of superposition, the total stress at B is given as:
$$\sigma_B = \frac{P}{A} + \frac{M_X}{Z_X} + \frac{M_Y}{Z_Y}$$

Can you write down the corresponding stress expression for the total stress at C?

17

$\sigma_C = \dfrac{P}{A} - \dfrac{M_X}{Z_X} - \dfrac{M_Y}{Z_Y}$

You should have deduced that the stress components at C are given by:

(1) Due to the axial force
$$\sigma_C = \frac{P}{A}$$
(compression over the whole section)

(2) Due to the moment M_X acting about axis X–X
$$\sigma_C = -\frac{M_X}{Z_X}$$
(tension along edge C–D)

(3) Due to the moment M_Y acting about axis Y–Y
$$\sigma_C = -\frac{M_Y}{Z_Y}$$
(tension along edge A–C)

where the negative sign denotes tension. Using superposition or simple adding together the individual stress components will give the expression for σ_C.

Likewise the stresses at the corners A and D are given by:
$$\sigma_A = \frac{P}{A} + \frac{M_X}{Z_X} - \frac{M_Y}{Z_Y}$$
$$\sigma_D = \frac{P}{A} - \frac{M_X}{Z_X} + \frac{M_Y}{Z_Y}$$

The column section shown in the figure in Frame 15 carries an axial load of 100 kN together with moments about the X–X axis and Y–Y axis of 50 kN m and 70 kN m respectively. If the column section has an area of 160×10^3 mm^2 and a section modulus for both axes of bending of 10.67×10^6 mm^3, what are the stresses in the four corners of the section?

18

$$\sigma_A = -1.25 \text{ N/mm}^2: \quad \sigma_C = -10.62 \text{ N/mm}^2$$
$$\sigma_B = +11.87 \text{ N/mm}^2: \quad \sigma_D = +2.50 \text{ N/mm}^2$$

If you didn't obtain these answers, check your working against that given below:

Stress due to axial force $= \frac{P}{A} = \frac{100 \times 10^3}{160 \times 10^3} = +0.625 \text{ N/mm}^2$

Stress due to moment $M_X = \frac{M_X}{Z_X} = \frac{50 \times 10^6}{10.67 \times 10^6} = \pm 4.686 \text{ N/mm}^2$

Stress due to moment $M_Y = \frac{M_Y}{Z_Y} = \frac{70 \times 10^6}{10.67 \times 10^6} = \pm 6.560 \text{ N/mm}^2$

The stresses at the four corners are given by the algebraic addition of these three stress components using the four expressions given in Frames 16 and 17.

19

Previously we considered the case of a column section with a load applied eccentrically to one of the principal axes of the section. If the load is applied eccentrically to both axes of the section, this is a particular case of axial load combined with *biaxial bending*.

Consider the column shown in plan view in the figure with the loading applied at an eccentricity to both axes:

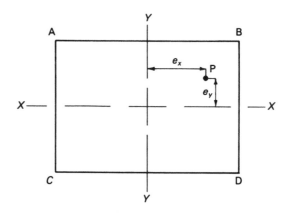

Write down the expressions for the equivalent moments acting about both the X–X and Y–Y axes

20

$$\boxed{M_X = +Pe_Y \qquad M_Y = +Pe_X}$$

Note the sign convention that we are using. We are assuming that the load is acting in the top right-hand quadrant of the section, causing maximum compression in the corner B. Hence to be consistent with our previous conventions, we must take the equivalent moment about the principal X–X axis as positive if it causes compressive stresses along the face AB. Likewise the equivalent moment about the Y–Y axis must be positive if it causes compression along the face BD.

If the load is moved to the bottom right-hand quadrant of the section what expressions would you write down for the equivalent moments about both principal axes? Think carefully about the signs.

21

$$M_X = -Pe_Y \qquad M_Y = +Pe_X$$

The sign of the moments obviously depends on which quadrant the load is acting. If the load is at the bottom left-hand quadrant the moments are given as:

$$M_X = -Pe_Y \qquad M_Y = -Pe_X$$

and if it is in the top left-hand quadrant they are given by:

$$M_X = +Pe_Y \qquad M_Y = -Pe_X$$

Obviously signs are important but with a little practice you should easily recognise whether a load is causing compression or tension at any particular point on the cross-section. All you then have to remember is the general convention used that compressive stresses are positive.

Now write down the complete expression that would give the maximum stress in the corner B of the section shown in Frame 19.

22

$$\sigma_B = \frac{P}{A} + \frac{(Pe_Y)}{Z_X} + \frac{(Pe_X)}{Z_Y}$$

And the stress at any other corner is given by the same expression with an appropriate adjustment to the signs of the relevant bending stress components.

The section in the figure of Frame 19 carries a compressive load of 100 kN acting at an eccentricity of 50 mm to the X–X axis and 75 mm to the Y–Y axis within the top right-hand quadrant of the section. If the section has an area of 160×10^3 mm² and a section modulus of bending for both axes of 10.67×10^6 mm³, what are the stresses in the four corners of the section?

23

$$\sigma_A = +0.39 \qquad \sigma_C = -0.55 \text{ N/mm}^2$$
$$\sigma_B = +1.80 \qquad \sigma_D = +0.86 \text{ N/mm}^2$$

If you didn't obtain these answers, check your working against that given in the next frame.

24

Moment about X–X axis: $M_X = Pe_Y = 100 \times 50 = 5000$ kN mm

Moment about Y–Y axis: $M_Y = Pe_X = 100 \times 75 = 7500$ kN mm

Stress due to axial force $= +\dfrac{P}{A} = \dfrac{100 \times 10^3}{160 \times 10^3} = 0.625$ N/mm^2

Maximum stress due to moment $M_X = \pm\dfrac{M_X}{Z_X} = \dfrac{5000 \times 10^3}{10.67 \times 10^6} = \pm0.468$ N/mm^2

Maximum stress due to moment $M_Y = \pm\dfrac{M_Y}{Z_Y} = \dfrac{7500 \times 10^3}{10.67 \times 10^6} = \pm0.703$ N/mm^2

By inspection, the moment of the force about the X–X axis caused compression at A and the moment of the force about the Y–Y axis causes tension at A. Hence the stress at corner A is given by:

$$\sigma_A = \frac{P}{A} + \frac{(Pe_Y)}{Z_X} - \frac{(Pe_Y)}{Z_Y}$$
$$= 0.625 + 0.468 - 0.703$$
$$= +0.39 \text{ N/mm}^2$$

The stresses at the other three corners are obtained in the same way.

25

The example that we have considered has been for a single load acting eccentrically on a cross-section. How do we deal with the case where more than one load is acting on the section? Again, we can use the Principle of Superposition to calculate the resultant forces and moments acting on the section, and use the same formulae to calculate the stresses within the section. Let's look at a typical example.

A column section carries two loads acting eccentrically as shown in plan view in the figure. We wish to determine the stresses at the corners B and C.

Cross-sectional area: $A = 400 \times 400 = 160 \times 10^3 \text{ mm}^2$

Section modulus: $Z = \frac{1}{6} \times 400 \times 400^2 = 10.67 \times 10^6 \text{ mm}^3$ for both axes

Total axial load: $P = 120 + 80 = 200 \text{ kN}$

Total moment X-axis: $M_X = (120 \times 40) + (80 \times (-40)) = 1600 \text{ kN mm}$

Total moment Y-axis: $M_Y = (120 \times 50) + (80 \times 20) = 7600 \text{ kN mm}$

(Note the signs used in the moment calculations which are consistent with our previously adopted sign conventions.)

Stress due to axial force $= +\dfrac{P}{A} = \dfrac{200 \times 10^3}{160 \times 10^3} = 1.250 \text{ N/mm}^2$

Maximum stress due to moment $M_X = \pm\dfrac{M_X}{Z_X} = \dfrac{1600 \times 10^3}{10.67 \times 10^6} = \pm 0.150 \text{ N/mm}^2$

Maximum stress due to moment $M_Y = \pm\dfrac{M_Y}{Z_Y} = \dfrac{7600 \times 10^3}{10.67 \times 10^6} = \pm 0.712 \text{ N/mm}^2$

Hence stress at corner B $= \dfrac{P}{A} + \dfrac{(P e_Y)}{Z_X} + \dfrac{(P e_X)}{Z_Y}$

$= 1.250 + 0.150 + 0.712$

$= \underline{2.11 \text{ N/mm}^2}$

And stress at corner C $= \dfrac{P}{A} - \dfrac{(P e_Y)}{Z_X} - \dfrac{(P e_X)}{Z_Y}$

$= 1.250 - 0.150 - 0.710$

$= \underline{0.39 \text{N/mm}^2}$

Complete this problem by working out the stresses at the other two corners.

$$\sigma_A = 0.69 \text{ N/mm}^2 \qquad \sigma_D = 1.81 \text{ N/mm}^2$$

28

PROBLEMS

1. A column, as shown in figure Q1, carries a load of 200 kN at an eccentricity of 30 mm to each of the principal axes. Calculate the stresses at each of the four corners of the section.

Ans. (3.33, 2.18, 1.02, −0.13 N/mm²)

Q1

2. If no tensile stresses are allowed to develop in the section in question 1, determine the maximum eccentricity of the load from the X–X axis if the eccentricity from the Y–Y axis is maintained at 30 mm.

Ans. (26.67 mm)

3. A steel column section carries three loads as shown. Calculate the stresses in the corners A and B. The sectional area is 100 cm² and the second moments of area about the X–X axis and Y–Y axis are $I_{XX} = 9000$ cm⁴ and $I_{YY} = 3000$ cm⁴ respectively. (*Hint:* the section properties are given in terms of second moments of area. As a start point in your calculations, determine the elastic section modulus for both axes of bending.)

Ans. (37.75, 15.25 N/mm²)

Q3

4. If the maximum compressive stress in the steel column in question 3 is not allowed to exceed 160 N/mm², to what value can the load of 100 kN be increased while maintaining the other two loads at their given values?

Ans. (606 kN)

In Frame 8 we considered the case of a column with a load acting eccentrically to one of the principal axes of the section. We developed the *middle third rule* to show that provided the load was applied within the middle third of the cross-section, then no tension could develop in the section no matter how large the load is.

Let's look at the extension of this theory to the case of a section where the load is applied eccentrically to both axes of the section.

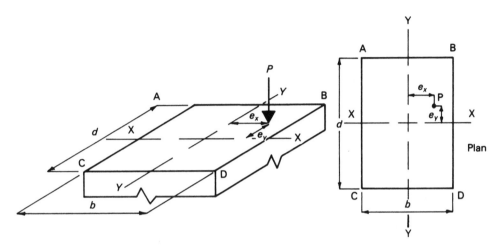

Which corner of the section is most likely to be in tension?

corner C

If you have followed the previous frames, this should now be self-evident and you should know that the stress at C is given by:

$$\sigma_C = \frac{P}{A} - \frac{(Pe_Y)}{Z_X} - \frac{(Pe_X)}{Z_Y}$$

If we are interested in the condition that no tensile stresses are allowed to develop in the section we can write this as:

$$\frac{P}{A} - \frac{(Pe_Y)}{Z_X} - \frac{(Pe_X)}{Z_Y} \geq 0$$

If the section is rectangular of breadth b and depth d, re-write this expression in terms of b and d.

31

$$1 - \frac{6e_Y}{d} - \frac{6e_X}{b} \geq 0$$

You should have obtained this expression as follows:

$$\text{Area } A \qquad\qquad = b \times d$$

$$\text{Section modulus } Z_X = \tfrac{1}{6}b \times d^2$$

$$\text{Section modulus } Z_Y = \tfrac{1}{6}d \times b^2$$

hence substituting into the final expression in Frame 30:

$$\frac{P}{bd} - \frac{6(Pe_Y)}{bd^2} - \frac{6(Pe_X)}{db^2} \geq 0$$

Rearranging the expression and cancelling out the terms in P will give the expression at the top of this frame. This is now independent of the load P and defines the combinations of eccentricity which will ensure that no tension can occur at the corner C.

If the load acts along the Y–Y axis (that is, $e_X = 0$), what is the limit of eccentricity that ensures that no tension can occur at C?

32

$$e_Y \leq \frac{d}{6}$$

This solution follows directly from the inequality expression at the top of Frame 31 with the substitution that $e_X = 0$. This is marked as point E in the figure shown at the top of the next page.

Similarly if the load acts along the X-axis ($e_Y = 0$), then the limit of eccentricity is such that for no tension at C:

$$e_X \leq \frac{b}{6} \qquad \text{(point F on the figure)}$$

The line joining E to F defines a boundary beyond which the load must not lie if tension is not to develop at C. Any combination of e_X and e_Y which lies inside this line will satisfy the inequality expression in Frame 31. Any combination of e_X and e_Y that lies just on the line will just satisfy the equality (that is, $\sigma_C = 0$ if the load lies on the line).

If you are not sure about this, try any combination of e_X and e_Y that lies on the line (such as $e_X = b/12$ and $e_Y = d/12$) to show that this line defines the limit of application of load which will ensure that tension cannot occur at C.

Sketch the lines that define the boundary of load application which will ensure that tension cannot occur at any corner of the section.

33

You should have realised that the condition for no tension at C was governed by the position of the load within the section quadrant furthermost from C. Similar reasoning and mathematical solution should lead you to the conclusion that tension at A is determined by the location of the load within the bottom right-hand quadrant which is furthermost from A, and likewise for B and D. Hence the shape of the parallelogram in the figure defines the *core of the section.* Provided that the load (*irrespective of its magnitude*) is applied anywhere within the core, tension cannot occur anywhere within the section.

34

The application of this proof of the existence of a *core* is to be found in a number of practical situations. For example, masonry is a material which has little tensile strength and in the design of a masonry column it is usually necessary to ensure that no tension develops anywhere in the column cross-section. Hence if the loading on the column is arranged in such a way that it always lies within the core, then the condition of no tension is assured irrespective of the magnitude of the load.

The core indicated in the figure in Frame 33 is *only for a solid rectangular section*. However, working from first principles the shape of the core for practically any shape of cross-section can be determined. Try the next problem for yourself – the proof is given in the next frame.

Show that for a circular section of radius R the core of the section is defined by a concentric circle of radius $\frac{1}{4}R$. (Hint: the second moment of area of a circle about a diameter is $\frac{1}{4}\pi R^4$ and no matter where a load is placed on the circle you can always draw a diameter passing through the load.)

35

The point of most likely tensile stress is at A which is the point on the same diameter as the load and on the opposite side of the perimeter. The diameter which is at right angles to this one is therefore the critical axis of bending.

$$\sigma_A = \frac{P}{A} - \frac{Pe}{Z} = \frac{P}{A} - \frac{(Pe)R}{I} = \frac{P}{\pi R^2} - \frac{(Pe)R}{\frac{1}{4}\pi R^4}$$

No tension will occur provided that:

$$\sigma_A = \frac{P}{\pi R^2} - \frac{(Pe)R}{\frac{1}{4}\pi R^4} \geq 0$$

Rearrange to give:

$$e \leq \tfrac{1}{4}R$$

As the load can lie anywhere on the section, the core is therefore a circle of radius $\frac{1}{4}R$.

WALLS AND FOUNDATIONS

One important application of the theory developed in this programme is to be found in the analysis of stresses in walls and under foundations which are subjected to combinations of vertical and lateral loading. The vertical loading can include the self-weight of the wall and that of any supported structure. The lateral loading can arise from lateral wind pressure or the pressure of earth or water retained behind the wall.

In designing a wall it is often necessary to calculate the stresses arising in the wall to see if they are within the limits of the stress capacity of the material of construction. In designing a foundation the stresses underneath the foundation are important, as excessively large stresses can result in movement of the foundation if the ground on which the foundation is standing is overstressed. In addition, it is often necessary to check that the ground is in compression over the whole area of contact with the foundation, as it is impossible to develop tensile stresses between the ground and the supported foundation.

Consider a wall carrying a vertical load V and a lateral loading H:

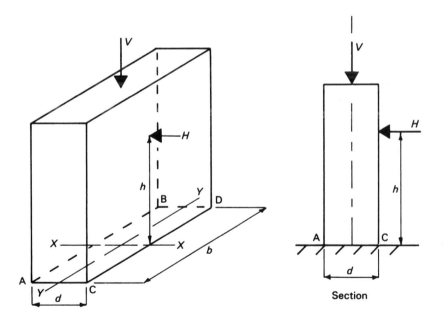

If stresses are to be calculated at the base of the wall, which is the axis of bending: X–X or Y–Y?

37

> the $Y-Y$ axis

If you consider the horizontal force H, it has a moment $H \times h$ about the $Y-Y$ axis, but does not have a moment about the $X-X$ axis. Bending must therefore be taking place about the $Y-Y$ axis.

Which is the most highly compressed face of the wall at the base section?

38

> the face AB

If you consider the direction of the moment of the force H about the $Y-Y$ axis, this should be fairly apparent. However, try to use intuition and envisage what is happening to this wall. If you can appreciate that the horizontal load is causing the wall to overturn about the base, it must result eventually in the back face of the wall (CD) tending to lift and transferring more of the load towards the front face (AB) which is hence subjected to higher compressive stresses.

The stresses at the base of the wall can be determined using the expression that have been developed previously:

$$\sigma = \frac{P}{A} \pm \frac{M}{Z}$$

or in terms of the symbols given in the figure in Frame 36:

$$\sigma_{AB} = \frac{V}{bd} + \frac{(Hh)}{bd^2/6}$$

and

$$\sigma_{CD} = \frac{V}{bd} - \frac{(Hh)}{bd^2/6}$$

A concrete wall is rectangular in section and has a length of 2 metres, a thickness of 1 metre and a height of 3 metres. It is subjected to a horizontal wind pressure of 0.75 kN/m² which can be considered to act as a uniformly distributed load acting over the full height of the wall. If the unit weight of concrete is 24 kN/m³, determine the maximum and minimum base stresses.

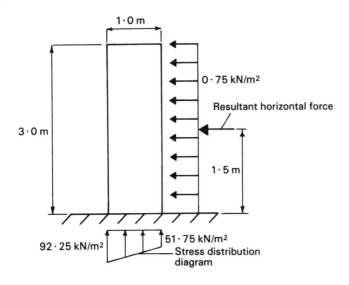

Check your working:

Vertical load V = weight of wall   = length × breadth × height × unit weight

= 2 × 1 × 3 × 24

= 144 kN

Horizontal load H from wind   = length × height × wind pressure

= 2 × 3 × 0.75

= 4.50 kN

Moment due to wind pressure   = $H \times h$

= 4.50 × 1.5

= 6.75 kN m

Base stresses given by $\sigma = \dfrac{V}{bd} \pm \dfrac{(Hh)}{bd^2/6}$

$= \dfrac{144}{2 \times 1} \pm \dfrac{6.75 \times 6}{2 \times 1^2}$

= $\underline{92.25 \text{ (max.) and } 51.75 \text{ (min.) kN/mm}^2}$

You could also calculate these stresses by considering the forces acting on a length of 1 metre of the wall.

40

The previous problem illustrates the application of the method to a simple problem using a simple set of loads. If the structure is subjected to a complex set of loads, the principle is the same: just reduce the loads to a set of resultant axial loads and moments acting about the axis of bending.

Before we finish this programme, let's look at the application of the *middle third rule* to problems of this nature.

A very common problem is that of determining the minimum width of a base such that no tension is allowed to develop along the contact surface between the base and the ground. Of course, a base size could be guessed and the general stress equations used to calculate the stresses under the base with the calculations repeated until a suitable section size is determined. However this is not necessary if the principle of the *middle third rule* is recalled and used.

Can you write down the 'middle third rule'?

41

| The line of action of the **resultant** force should lie within the middle third of a **solid rectangular** base if no tensile stresses are to develop. |

Let's consider a typical problem with horizontal and vertical loading as shown. The width (*d*) of the footing is unknown but we wish to determine its minimum dimension to ensure that no tension can occur at the back of the footing:

*Using simple principles of statics, can you determine where the **resultant** of the vertical and horizontal force meets the ground in relation to the centreline of the base?*

42

0.33 metres to the front of the centreline

Using a geometrical construction you should find that the resultant meets the base at a distance of one-third of a metre from its centreline. (You could have calculated this instead, using simple analytical techniques.)

Now, remembering the principle of the middle third rule, can you estimate how wide the base should be?

43

2 metres

If the resultant lies just on the edge of the middle third of the base, then it is just on the point of 'no tension'. The limiting distance of the resultant measured from the centreline of the base is therefore one-sixth of the base width. Hence, as the resultant in our example meets the base at 0.33 metres from the base centreline, the minimum base width is given by:

$$\text{minimum width} = 6 \times 0.33$$

that is
$$d = 2 \text{ metres}$$

44

TO REMEMBER

$$\sigma = \frac{P}{A} \pm \frac{M_X}{Z_X} \pm \frac{M_Y}{Z_Y}$$

The *Middle Third Rule*: the line of action of the resultant force should lie within the middle third of a **solid rectangular** section if no tensile stresses are to develop anywhere within the section.

45

FURTHER PROBLEMS

1. A 400×250 mm column support a vertical load of 150 kN as shown in figure Q1. Determine the maximum and minimum stresses in the section.

Ans. (2.40, 0.60 N/mm^2)

2. Determine the maximum eccentricity of the load in question 1 for no tension to develop anywhere in the section.

Ans. (66.67mm)

3. If the column in question 1 carries an additional load of 120 kN positioned along the Y–Y axis at an eccentricity of 25 mm, calculate the maximum and minimum compressive stresses in the cross-section.

Ans. (4.32, 1.08 N/mm^2)

4. A hollow box section is stressed by the application of a compressive force of 100 kN acting as shown in figure Q4. calculate the stresses at the outer faces of the section and also at the point marked A.

Ans. (2.82, 0.04, 0.87 N/mm^2)

5. In question 4, to what value can the force be increased if the maximum compressive stress in the section is limited to 25 N/mm²?

Ans. (888 kN)

6. If the force in question 4 is maintained at 100 kN, what is the maximum possible eccentricity of this force such that no tension is allowed to develop in the section?

Ans. (56.67 mm)

7. A bracket is formed from two steel T sections welded together and supported as shown in figure Q7. If the bracket carries a load of 10 kN, what are the maximum and minimum stresses in the section marked *X–X*?

Ans. (−45.46, 66.06 N/mm²)

Q7   Section *X–X*

8. A concrete wall is 5 m high, 0.8 m thick and 3 m long. It is subjected to a uniform lateral wind pressure of 0.5 kN/m². If the unit weight of concrete is 24 kN/m³, calculate the maximum and minimum stresses at the base of the wall.

Ans. (178.59, 61.41 kN/m²)

9. A masory pier is 0.75 m thick and has the section shown in figure Q9. If the unit weight of masonry is 20 kN/m³, what are the stresses at the base of the wall due to the self-weight of the masonry?

Ans. (48.89, 4.45 kN/m²)

Programme 8

SHEAR STRESS

On completion of this programme you will be able to:

- Define 'average shear stress'
- Calculate the average shear stress in bolts with simple bolted connections
- Design simple bolted connections with given limiting stress conditions
- Recognise the existence of and define 'complementary stress'
- Define 'shear strain'
- State the relationship between shear stress and shear strain
- Sketch the distribution of shear stress across a regular beam section
- State the relationship between average and maximum shear stress in a rectangular beam section
- State the formula for shear stress in beam sections
- Calculate the distribution of shear stress in rectangular and non-rectangular beam sections

1

So far we have shown how to calculate direct stresses arising from direct axial loading acting on a structural member, and bending stresses within structural elements which are subjected to bending moments. In Programme 7 we also saw how to calculate stresses arising from combinations of axial loading and bending moments.

However there are other stresses which can arise within a structural element and in this programme we are going to examine a further type of stress. To be precise we will study the development of shear stress and see how shear stresses can arise and how to calculate the magnitude and distribution of such stresses for a range of common problems. The development of *shear stress* is related to the existence of *shearing forces* and the magnitude of such stress depends upon the magnitude and distribution of the shearing forces. Hence it is important that, in the case of a beam, you can sketch the shape of the shearing force diagram. If you are not confident about this, revise Programme 4 before proceeding.

Now, let's see how and where shear stresses can arise.

2

As a simple example, consider a simple cantilevered beam of rectangular cross-section which supports a load W:

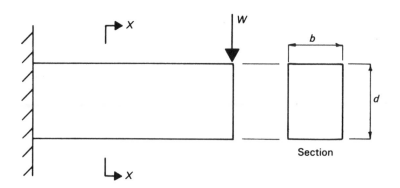

What is the magnitude of the shearing force at the section marked X–X?

3

$$\boxed{\text{shearing force at } X\text{–}X = W}$$

If you didn't get this answer straight away, you must revise Programme 4 before going any further. In particular, remind yourself of the definition of a shearing force.

Now let's define the *average* shear stress as:

$$\textit{average shear stress } = \frac{\text{force}}{\text{area over which force acts}}$$

The reason why we have used the word *average* to define the shear stress will become more apparent later in this programme. However, let's further consider the beam in Frame 2.

If the beam has a breadth b and depth d, write down the expression for the average shear stress at section X–X.

4

$$\boxed{\text{average shear stress } = \frac{W}{bd}}$$

You should have reasoned that the shearing force is vertical and acts on the vertical cross-section of the beam whose area is given by $b \times d$. Hence the expression given for the average shear stress. You should appreciate, however, that the nature of this stress is different from those that you have studied in previous programmes as, in this case, the force acts in the same plane as the beam's cross-section. The forces that cause axial and bending stress are normal to the sections on which they act.

Now let's try a problem:

A timber beam is simply supported at both ends over a span of 4 metres. It carries a uniformly distributed load (inclusive of self-weight) of total magnitude 50 kN and has a breadth of 100 mm and a depth of 200 mm. Where does the largest shearing force occur, what is its magnitude and what is the average shear stress resulting from this force at the critical section?

The answer and the worked solution are given in the next frame.

5

> 25 kN at either support
> 1.25 N/mm^2

If you didn't get the correct answer, check your working:

For a simply supported beam supporting a uniformly distributed load, the maximum shearing force occurs at either end. (If you don't realise that this is the case, draw the shearing force diagram. Revise Programme 4, if in doubt.)

The magnitude of the shearing force either end $= \frac{1}{2}$ the total load

$$= \frac{1}{2} \times 50$$

$$= 25 \text{ kN}$$

The average shear stress $= \dfrac{W}{bd}$ $\qquad = \dfrac{25 \times 10^3}{100 \times 200}$

$$= 1.25 \text{ N/mm}^2$$

6

Shear stresses can arise in structural elements other than beams. For example, consider two steel plates fastened together with a single bolt and subject to two equal and opposite tensile forces P. If the equilibrium of the upper plate and the upper half of the bolt is considered, it is apparent that there must be a shearing force acting across the cross-section of the bolt at the interface of the two plates to maintain equilibrium. This shearing force must be equal in magnitude to the applied force P.

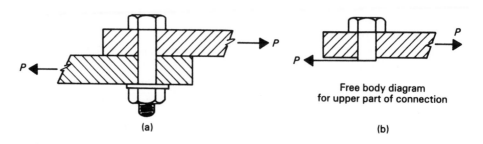

(a)

Free body diagram
for upper part of connection

(b)

If the bolt is circular in section with a diameter D, write down an expression for the average shear stress within the bolt.

7

$$\boxed{\text{average shear stress} = \frac{P}{\pi \frac{1}{4} D^2}}$$

The expression follows directly from the definition of average shear stress given in Frame 3 and can be used to design the bolts in simple connections in steelwork structures. Try the following example:

Calculate the strength of a single 20 mm diameter steel bolt based on an allowable average shear stress of 80 N/mm².

8

$$\boxed{25.14 \text{ kN}}$$

You should have obtained the answer as follows:

Allowable average shear stress $= \dfrac{P}{\pi \frac{1}{4} D^2} = \dfrac{4P}{\pi \times 20^2} = 80 \text{ N/mm}^2$

Hence
$$P = 80 \times \pi \times 20^2 \times \tfrac{1}{4}$$

$$= 25\,143 \text{ N}$$

$$= 25.14 \text{ kN}$$

The strength of a connection fabricated using a single bolt in this way depends on a number of factors other than just the bolt shear strength. The design of connections is outside the scope of this text but the example given illustrates one of the calculations associated with such a design.

Later in this programme we will see that the concept of an average shear stress is a simplification of what actually happens in many real structures which are subject to the action of shearing forces. We have in fact assumed that the shear stress is uniform over the shear surface which, as we will show later, is often not the case. However in the practical design of structural elements in, for example, steel or reinforced concrete, *average* shear stresses are usually calculated for comparison with allowable values as this provides a simple, convenient and generally acceptably accurate method of calculation.

9

Before we move on to some problems, let's consider one further illustration of the development of shear stress. Let's look again at the connection shown in Frame 6 (figure (b)). The figure below shows the same connection but this time also viewed in plan.

If the force P is large enough, it is possible for the joint to 'fail' by shearing of the surfaces marked aa, resulting in the shaded area of the upper plate to the left of the bolt being 'pushed out'.

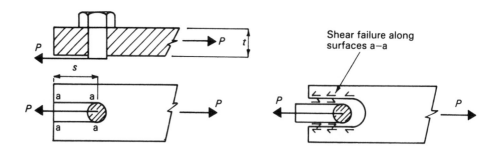

If the plate is of thickness t and the bolt is located at a distance s from the end of the plate, write down an expression for the average shear stress developed along the vertical faces of the shaded area to the left of the bolt.

10

$$\boxed{\text{average shear stress} = \frac{P}{2st}}$$

Again, the expression follows from the definition of average shear stress. In this case there are two shear surfaces resisting the force P and by considering the equilibrium of the shaded area you should have written down:

$$P = \text{average shear stress} \times \text{area on which shear stress acts}$$

$$\therefore \quad P = \text{average shear stress} \times 2(s \times t)$$

Hence
$$\text{Average shear stress} = \frac{P}{2st}$$

Let's look at an example:

A 20 mm diameter bolt connects two steel plates which are both 12 mm thick and which carry a tensile load of 40 kN. Calculate the shear stresses in both the bolt and the steel plates if the bolt is positioned with its centre 15 mm from the edge of each plate.

Average shear stress in the bolt $= \dfrac{P}{\pi\frac{1}{4}D^2} = \dfrac{40 \times 10^3}{\pi\frac{1}{4}20^2} = 127.32 \text{ N/mm}^2$

Average shear stress in plate $= \dfrac{P}{2st} = \dfrac{40 \times 10^3}{2 \times 15 \times 12} = 111.11 \text{ N/mm}^2$

As the shear stress in the bolt is higher than that in the plate it follows that, if the load was increased and both bolt and plate had the same limiting failure stress, then the bolt would fail in shear before the plate.

Now try some problems:

11

PROBLEMS

1. A concrete beam, 300 mm $\times$ 600 mm, is simply supported over a span of 5 metres and carries a point load of 100 kN at a distance of 1.5 metres from one support. If the unit weight of concrete is 24 kN/m³, determine the largest shearing force in the beam and the average shear stress resulting from this force.

Ans. (80.80 kN, 0.45 N/mm²)

2. Solid circular discs are punched out of 3 mm thick steel plates using a 50 mm diameter circular punch. If the ultimate shear strength of steel is 460 N/mm², determine the compressive force that must be developed by the punch.

Ans. (216.86 kN)

3. A hanger bar is fastened to a steel bracket made from 10 mm steel plate with a single 16 mm diameter steel bolt. If the shear stress in both the bracket and the bolt is not to exceed a maximum value of 80 N/mm², calculate the maximum load that can be carried by the hanger based on the strength of the bolt and the bracket.

Ans. (16.09 kN)

12

Now that we have looked at some simple examples involving shear stress, we can extend our study of shear to cover more complex situations. However, before we do it is necessary to introduce some new concepts.

Let's reconsider the cantilever beam shown in Frame 2 and look at a small element of the beam between the two sections A–A and B–B as indicated:

Neglecting the self-weight of the beam, what is the value of the shearing force acting on the vertical sections A–A and B–B?

13

W on both sections

If you draw the shearing force diagram for this beam, you should realise that the shearing force is in fact **W** at all sections along the beam. This implies that the shear *stresses* on the two vertical faces of the small element are equal in magnitude as the equal shearing forces act on equal cross-sectional areas. However, for vertical equilibrium the shear stresses (and forces) must act in opposite directions as indicated in the figure. Note that we have used the symbol τ to indicate shear stress.

Now let's consider the rotational equilibrium of this element. If moments are taken about the corner c of the forces acting on the element then it is apparent that there is a clockwise moment given by:

$$\text{Moment} = \text{force} \times \delta x$$

$$= \text{stress} \times \text{area on which stress acts} \times \delta x$$

$$= \tau \times (t \times \delta y) \times \delta x$$

where t is the thickness of the beam.

What force provides the anticlockwise moment that for rotational equilibrium must resist this clockwise moment?

15

a shearing force on the face a–b

You may have answered that there is no such force. There has to be if equilibrium of the element is to be maintained—which it has to be! The only possible answer is that there must be a shearing force acting on the horizontal face a–b which provides an anticlockwise moment equal to the moment given by the expression in Frame 14.

The existence of this shearing force implies the existence of shear stress acting on the horizontal face a–b. Such a shear stress is known as a complementary shear stress. In other words, the shear stresses on the vertical faces of the element must be accompanied by a set of shear stresses acting on the horizontal faces.

On a sketch of this element, can you draw the direction of the shear stresses acting on all four faces?

16

On to the next frame.

17

The stresses act in such a way that for horizontal and vertical equilibrium they provide equal and opposite forces on any two parallel faces of the element and such that the clockwise moments due to the shear stresses on the vertical planes are resisted by the anticlockwise moments due to the shear stresses on the horizontal planes.

Referring back to the diagram in Frame 16, can you think of a simple rule that will enable you to remember the relative directions of the shear stresses acting on any two intersecting perpendicular planes.

18

> The shearing stresses both act either towards or away from the line of intersection of the two planes.

This is a simple observation which is true for any situation in which shear stress and complementary shear stress act at a point within a structure.

The diagrams below show the direction of shear stress on one face of a small element of material. Complete each diagram by showing the direction of the stresses acting on the other three faces.

19

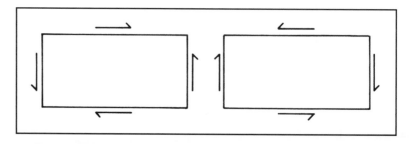

We have shown the existence of the complementary stresses and have identified their direction. What about the magnitude of the complementary stresses? The magnitude of the average stress on the vertical planes can be determined using the methods discussed earlier in previous frames. Let's see how to calculate the magnitude of the complementary stress:

In Frame 14 we wrote down an expression for the clockwise moment about corner c resulting from the shear stress acting on the vertical face b–d of the element shown in Frame 12.

Can you write down a corresponding expression for the anticlockwise moment about c resulting from the complementary shear stress acting on the horizontal face a–b? In this case, use the symbol τ' to indicate the complementary shear stress.

20

$$\boxed{\text{moment} = \tau' \times (t \times \delta x) \times \delta y}$$

Your reasoning should have been as follows:

If moments are taken about the corner c then there must be an anticlockwise moment given by:

$$\text{Moment} = \text{force} \times \delta y$$
$$= \text{stress} \times \text{area on which stress acts} \times \delta y$$
$$= \tau' \times (t \times \delta x) \times \delta y$$

where t is the thickness of the beam.

You know that for equilibrium the clockwise moment and the anticlockwise moment must be equal. *By equating the moment expressions in Frame 14 and this frame, what do you deduce about the relationship between the shear stress τ and the complementary shear stress τ'?*

21

$$\boxed{\tau = \tau'}$$

In other words the magnitude of the complementary shear stress on the horizontal planes is equal to the magnitude of the shear stress on the vertical planes. This is an important conclusion and should be remembered for use later. If you didn't arrive at this answer than check your solution in the next frame:

22

$$\text{Clockwise moment about c} = \tau \times (t \times \delta y) \times \delta x$$
$$\text{Anticlockwise moment about c} = \tau' \times (t \times \delta x) \times \delta y$$

Equating moments:

$$\tau \times (t \times \delta y) \times \delta x = \tau' \times (t \times \delta x) \times \delta y$$

Cancelling terms:

$$\tau = \tau'$$

So far we have been looking at a small finite element of material. If the sides of the element become very small ($\delta x \to 0$ and $\delta y \to 0$) then effectively the element will represent a *point* within the beam and the proof that we have developed will therefore apply to any point. It is important to realise and remember that if a shearing stress acts on any plane at a point within a structure, there must be a *complementary shearing stress* of *equal magnitude* acting on a plane at right angles to the first plane and passing through the same point.

In developing the above proof, we started out by considering the case of a cantilevered beam with a point load at the end, resulting in a constant shearing force and hence constant shear stresses at all sections throughout the structure. In most practical situations, the shear stresses will vary throughout the structure. However if the element under consideration represents a point within a structure, it is reasonable to assume that the shear stress at that point is uniform and constant. The conclusions that we have arrived at are still therefore quite valid.

23

SHEAR STRAINS AND MODULUS OF RIGIDITY

In Programme 5 you learnt that direct stresses acting on a material give rise to direct strains. In the same way *shear stresses* give rise to *shear strains*.

Consider the small element of material shown below which is subjected to shear stresses and complementary shear stresses on the vertical and horizontal faces.

Assuming that the element is made from an elastic material, can you sketch the distorted shape of the element due to the application of the applied shear stresses?

24

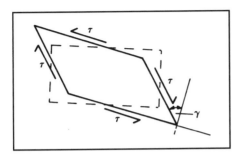

You should have drawn a diagram similar to the one shown. The initially rectangular element will be distorted into a diamond shape. The corners of the element, which were initially at right angles, will have rotated through a *small* angular distortion shown as γ in the diagram.

γ is known as the *shearing strain* and as γ is measured in radians it is dimensionless. τ, as we have already shown, is the shear stress, and both shear stress and shear strain are related by a physical property known as the *Shear Modulus (G)* sometimes referred to as the *Modulus of Rigidity*.

In Programme 5 you learnt that direct stress and strains are related by the material property known as Young's Modulus. With this relationship in mind, can you guess the relationship between shearing stress and strain and the Shear Modulus?

25

$$\boxed{\text{Shear Modulus } (G) = \frac{\text{Shear stress } (\tau)}{\text{Shear strain } (\gamma)}} \qquad (8.1)$$

The relationship between shear stress and strain is analogous to that for direct stress and strain but they are related by the Shear Modulus rather than Young's Modulus.

What are the units of Shear Modulus?

26

$$\boxed{\text{N/mm}^2}$$

Shear stress has units of N/mm^2; shear strain is an angular measurement and is dimensionless and therefore the Shear Modulus, which is the ratio of the shear stress and strain, must have units of stress, that is, N/mm^2 (or other similar stress units: kN/mm^2, GN/mm^2, etc.).

27

For most metals, Young's Modulus (E) is approximately 2.5 times as great as the Shear Modulus (G). Typically, steel has a Young's Modulus of 200 kN/mm^2 and a Shear Modulus of 80 kN/mm^2.

Can you find out typical Young's Modulus and Shear Modulus values for other materials? (No answers given for this question but, by consulting Materials textbooks, you should be able to determine some typical figures. Try to remember at least the typical orders of magnitude of these figures and the range of values between different materials.)

28

Now try a problem:

An aluminium plate is subjected to a shear stress of 50 N/mm^2 and has a Shear Modulus of 26 kN/mm^2. Determine the shear strain and hence calculate the total shear distortion (δ) shown in the figure:

29

$$\boxed{0.0019 \text{ radians: } 0.058 \text{ mm}}$$

Check your working:

$$\text{Shear strain } (\gamma) = \frac{\text{Shear stress } (\tau)}{\text{Shear modulus } (G)}$$

$$= \frac{50}{26 \times 10^3}$$

$$= 0.0019$$

$$\text{Shear distortion } (\delta) = \text{shear strain } (\gamma) \times \text{side length of element}$$

$$= 0.0019 \times 30$$

$$= 0.058 \text{ mm}$$

30

SHEAR STRESSES IN BEAMS

At the beginning of this programme we saw how to calculate the magnitude of average shear stresses in simple structural components. The use of the word *average* indicated that we were assuming that the shear stresses were distributed uniformly and evenly across the shear surface. However this is not always the case and in many practical situations the shear stresses are distributed in a non-uniform manner, and recognition of this non-uniformity can be important in the design of some structural elements. We are now going to investigate some typical situations where such non-uniform shear stress distributions occur and see how to calculate the shape of the stress distributions and the magnitude of such stresses. In doing this we will be making use of the facts that we have learnt about the existence of complementary shear stresses.

Let's look again at the cantilever beam of rectangular cross-section that we first considered in Frame 2. In this case, for convenience the beam is viewed with the support on the right:

Consider the section AC at a distance x from the free end of the cantilever and the section BD which is at a distance δx from AC.

Can you write down expressions for the bending moments at these two sections?

31

$$\boxed{\begin{aligned} \text{bending moment at AC} &= W \times x \\ \text{bending moment at BD} &= W \times (x + \delta x) \end{aligned}}$$

If you didn't get this answer, revise Programme 4 before Proceeding.

Now can you remember the general expression that you learnt in Programme 6 which will give the variation of bending stress at section AC? Write the expression down.

32

$$\sigma = \frac{My}{I} = \frac{(W \times x)y}{I}$$

This should be a familiar expression. I is the second moment of area of the section about the neutral axis and y is the distance from the neutral axis at which the stress is being calculated. You should also know that the bending stress varies linearly from zero stress at the neutral axis to a maximum value at the outer fibres of the section.

Now let's look at the equilibrium of that segment of the beam which is bounded by the two sections AC and BD and by the plane EF which is at a distance z from the neutral axis:

Section A–C

The figure shows the distribution of tensile bending stress acting on the face AE of the section under consideration. The tensile force due to the tensile stress acting on the small hatched area of thickness δy shown on the cross-section is given by:

$$\text{Force} = \sigma_y \times b \times \delta y$$

where σ_y is the bending stress at a distance y from the neutral axis.

Hence the total tensile force acting on the face AE is given by the summation of the above expression over the whole face:

$$\text{Total force acting on face AE} = \int_z^{d/2} \sigma_y b \, dy$$

and substituting the expression at the top of this frame:

$$\text{Total force acting on face AE} = \int_z^{d/2} \frac{(W \times x)y}{I} b \, dy$$

Complete the integration to give an expression for the total force on AE.

$$\text{Total force on AE} = \frac{Wxb}{2I}\left[\frac{d^2}{4} - z^2\right]$$

Similarly you should be able to show that the force acting on the opposite face (BF) of this segment of beam is given by a similar expression where the moment is given by the second equation at the top of Frame 31. This expression is given as:

$$\text{Total force on BF} = \frac{W(x + \delta x)b}{2I}\left[\frac{d^2}{4} - z^2\right]$$

$$= \frac{Wxb}{2I}\left[\frac{d^2}{4} - z^2\right] + \frac{W\delta xb}{2I}\left[\frac{d^2}{4} - z^2\right]$$

$$= \text{Total force on AE} + \frac{W\delta xb}{2I}\left[\frac{d^2}{4} - z^2\right]$$

This expression implies that this segment of beam is subject to an out-of-balance tensile force acting to the right and given by:

$$\text{Out of balance force} = \frac{W\delta xb}{2I}\left[\frac{d^2}{4} - z^2\right]$$

For equilibrium of this beam segment, there must be a force resisting this out-of-balance force. Can you deduce where this resisting force comes from and indicate its position and direction on a sketch of the beam segment?

You should have reasoned that there must be a shearing force acting on the horizontal face EF as shown in the figure and which must equate to the out-of-balance force.

Why is there no shear force acting on the face AB?

STRUCTURAL MECHANICS

35

Because for there to be a shearing force on the face AB there has to be a shear stress. However AB is a free surface and there can be no shear stress along a free surface as there is no adjacent reactive surface against which shear stresses can develop.

The final equation in Frame 33 gives the out-of-balance force acting on the beam segment, which as we have deduced must equal the total shearing force along the face EF. What we are interested in is the magnitude of the shear stress along this face. As the segment is small it is reasonable to suggest that the shear stress is uniform over the shear surface EF. Knowing the total shear force and the area over which it acts, the shear stress can be calculated:

$$\text{Shear stress} = \frac{\text{Total shear force}}{\text{Area}}$$

$$= \frac{W\delta x b}{2I}\left[\frac{d^2}{4} - z^2\right] \times \frac{1}{(\delta x \times b)}$$

$$\frac{W}{2I}\left[\frac{d^2}{4} - z^2\right] \tag{8.2}$$

The above expression gives the shear stress on the horizontal face EF which is at a distance z from the neutral axis.

What is the value of the shear stress on a vertical plane at a distance z from the neutral axis?

36

$$\boxed{\frac{W}{2I}\left[\frac{d^2}{4} - z^2\right]}$$

This should be easy! You should remember that the stress on any horizontal plane must be accompanied by a complementary shear stress on the vertical plane passing through the same point. The expression above therefore gives the shear stress on the vertical cross-section (AE) of the beam at a level at a distance z measured from the neutral axis:

For a rectangular section, the second moment of area (I) about the neutral axis is given by:

$$I = \frac{b \times d^3}{12}$$

Substituting this into equation (8.2), we therefore obtain the following equation:

242

$$\text{Shear stress} = \frac{6W}{bd^3}\left[\frac{d^2}{4} - z^2\right]$$

$$= \frac{6W}{bd}\left[\frac{1}{4} - \left(\frac{z}{d}\right)^2\right] \tag{8.3}$$

What is the value of the shear stress at the level of the neutral axis? (Hint: at the level of the neutral axis z must be equal to zero.)

37

$$\frac{3}{2}\frac{W}{bd}$$

You should have obtained this solution by substituting $z = 0$ into equation (8.3). If you think about the work that we did at the beginning of this programme this might seem surprising. You will remember that we started out in this programme looking at average shear stress which we defined as the shear force divided by the area of the surface resisting shear.

What do you deduce to be the relationship between the average shear stress and the shear stress at the level of the neutral axis for a rectangular beam section?

38

shear stress at neutral axis = 1.5 × average shear stress

The average shear stress is given by W/bd and hence the shear stress at the level of the neutral axis is 1.5 times this value. This value is in fact the *maximum* shear stress that will occur in the cross-section of a rectangular beam. The fact that the *maximum* is 1.5 times the *average* is an important result and is used in the design of, for example, timber beams where maximum shear stresses are usually calculated for comparison with allowable or permissible values.

Equation (8.3) defines the distribution of shear stress on the cross-section of a rectangular beam at any position at a distance z from the neutral axis.

Either by inserting suitable values into the equation or by examining the form of the equation, sketch the shape of the distribution of shear stress across the full depth of the beam section.

39

The equation is in fact the equation of a parabolic curve and, either by recognising this fact or by simply inserting values into the equation, you should have plotted a stress distribution diagram as shown above. As we have already stated, the maximum shear stress is found to occur at the neutral axis and at the outer fibres of the section the shear stress is zero.

We have used the simple example of a cantilever with a point load at the end to arrive at the expressions for shear stress distribution. In this case the shearing forces are constant throughout the beam and the shear stress distribution will be the same at any cross-section. In the more general case of a beam subject to any loading or support conditions, the theory is equally valid provided that the appropriate shear force at the section under consideration is used in the formula for shear stress.

40

PROBLEMS

1. A bridge is supported by rubber bearing pads of dimensions as shown in figure Q1. The bridge deck exerts a shearing force on each pad of 8 kN. If the shear modulus (G) of rubber is 0.0012 kN/mm^2, determine (a) the average shear stress, (b) the average shear strain and (c) the total horizontal deformation, δ, in the pad.

Ans. (0.267 N/mm^2, 0.222, 5.56 mm)

2. To test the shear strength of brickwork, a simple test is devised using three bricks mortared together and loaded in a test rig as shown. If at some point in the test a load of 5 kN is applied, determine the average shear stress along the surface between the mortar and the bricks.

Ans. (0.125 N/mm²)

3. A timber beam, 250 mm wide × 400 mm deep, is simply supported over a span of 4 metres and carries a uniformly distributed load (inclusive of its own weight) of 5 kN/m. Determine (a) the maximum shear force in the beam, (b) the average shear stress at the section of largest shear force and (c) the maximum shear stress at this section.

Ans. (10 kN, 0.10 N/mm², 0.15 N/mm²)

4. The beam in question 3 now supports an additional point load, *W*, at a distance of 1 metre from one support. If the maximum shear stress in the beam is limited to a value of 0.8 N/mm², calculate the maximum value of this additional load *W*. (Note that the answer given is based on considerations of shear only. Other considerations such as bending stresses and deflections will normally have to be considered.)

Ans. (57.78 kN)

5. The beam in question 3 is loaded with the UDL of 5 kN/m and a point load of 40 kN at a distance of 1 metre from the left-hand support. Consider a section at a distance of 1.5 metres from this support. (a) Calculate the maximum bending stress and the maximum shear stress at this section. (b) Sketch the shape of the bending stress diagram and the shear stress diagram. (c) At this section, where does the maximum bending stress occur and where does the maximum shear stress occur?

Ans. (5.16 N/mm², 0.11 N/mm². The maximum bending stress of 5.16 N/mm² occurs at the outer fibres of the section. The maximum shear stress of 0.11 N/mm² occurs at the level of the neutral axis)

41

SHEAR STRESS IN NON-RECTANGULAR BEAMS

So far, we have looked at the distribution of shear stress in beams of rectangular cross-section. The theory developed can now be extended to enable us to determine the distribution of shear stress in non-rectangular sections, for example in I section beams which are a common type of structural element.

We will confine our further study of this problem to beam sections having at least one axis of symmetry. In this case, bending of the beam will be taking place about a principal axis. The proof that will follow is merely an extension of the proof for the rectangular section, but the equations that we will develop can be more generally applied.

Consider the same cantilever beam that we examined in Frame 30, but this time let us assume that it has a more general shape of cross-section as shown below. The vertical Y-axis is taken to be the axis of symmetry.

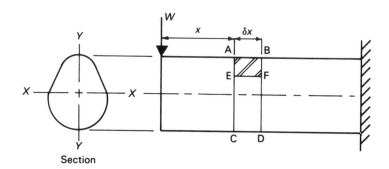

Section

As before, the bending moment at a distance x along the beam is given by:

$$\text{Moment at section AC} = W \times x$$

and the bending stress at a distance y from the neutral axis is given as:

$$\sigma = \frac{My}{I} = \frac{(W \times x)y}{I}$$

where the symbols have the usual meaning.

The second moment of area, I, is taken about the X–X axis which is also the neutral axis of bending.

Now, as before, let's look at the equilibrium of that part of the beam which is bounded by the two section AC and BD and by the surface, EF, which we will take to be a curved surface to emphasise the generality of the proof that will follow. EF can of course represent a flat horizontal plane, which in many problems where this theory is applied will often be taken as the case.

The figure on the next page shows the face AE of the part of the beam under consideration. Let's look at the force developed on the small element of area δA at a distance y from the neutral axis.

Can you write down an expression for the force developed over this area δA?

$$\text{force} = \sigma_y \times \delta A$$

This is similar to the expression developed in Frame 32 where previously we were taking an elemental area of width b and thickness δy.

The *total* force acting on the face AE (hatched in the above figure) is obtained from the summation of this expression over the whole face. If we assume that the area of this face is A, then this total force can be expressed as:

$$\text{Total force on face AE (hatched in figure)} = \int_A \sigma_y \, dA = \int_A \frac{(W \times x)y}{I} \, dA$$

$$= \frac{(W \times x)}{I} \int_A y \, dA$$

Similarly the force on the opposite face (BF) is given by a similar expression with the moment given by $W(x + \delta x)$. Hence as the area of face BF is the same as face AE, the force on the face BF is given by:

$$\text{Total force on face BF} = \int_A \sigma_y \, dy = \int_A \frac{(W \times \{x + \delta x\})y}{I} \, dA$$

$$= \frac{(W \times \{x + \delta x\})}{I} \int_A y \, dA$$

$$= \frac{(W \times x)}{I} \int_A y \, dA + \frac{(W \times \delta x)}{I} \int_A y \, dA$$

$$= \text{Total force on AE} + \frac{(W \times \delta x)}{I} \int_A y \, dA$$

What is the expression for the shear force acting over the curved surface EF?

43

$$\frac{(W \times \delta x)}{I} \int_A y \, dA$$

As in frame 34, you should have reasoned that this element of beam is subjected to an out-of-balance force due to the effects of the bending stresses, and this can only be resisted by a shearing force acting along the surface EF. If you are not sure about this, turn back to and Revise Frames 32 to 35.

Does the integral expression look familiar? It is in fact the first moment of area of the small elemental area δA about the neutral axis summed over the whole hatched area A of the face of the beam section. Let's assume that the centroid of this area A is at a distance $\bar{y}$ from the neutral axis.

Can you think of another way of writing the above expression for the shear force in terms of the area A and the centroidal distance $\bar{y}$. (Hint: remember how you calculate the centroidal position of an irregular shape.)

44

$$\frac{(W \times \delta x)}{I} A \bar{y}$$

You should know that when calculating the centroidal position of an irregular shape we make use of the general expression:

$$A \bar{y} = \int_A y \, dA$$

Hence the above expression for the shear force. If the length of the side of the curved shear surface (EGE) is b, then the average shear stress acting over the curved shear surface EF is given by:

$$\text{Average shear stress} = \frac{\text{Total shear force}}{\text{Area of surface}}$$

$$= \frac{(W \times \delta x)}{I} A \bar{y} \times \frac{1}{(b \times \delta x)}$$

$$= \frac{W A \bar{y}}{Ib}$$

If the curved surface EF is in fact a horizontal plane, what would be the expression for the shear stress on the vertical face AE at the level of EF?

$$\tau = \frac{WA\bar{y}}{Ib}$$

This expression follows again from considerations of complementary shear stresses. The vertical shear stress at any point on the face AE must be complementary to the horizontal shear stress on the plane EF at the same distance from the neutral axis. Hence the expression derived can be used to determine the shear stress acting on the vertical cross-sections of beams of almost any shape section.

The general form of this expression is

$$\tau = \frac{QA\bar{y}}{Ib} \qquad (8.4)$$

where: $\tau =$ the shear stress
$Q =$ the shear force acting at the section
$A =$ the area of the section to the outside of the level where shear stress is being calculated
$\bar{y} =$ the distance to the centroid of the area A measured from the neutral axis
$I =$ the second moment of area of the *whole cross-section* about the neutral axis
$b =$ the width of the cross-section at the level where shear stress is being calculated.

Make use of this general expression to show that for a rectangular beam section of width b and depth d, the maximum shear stress at the level of the neutral axis is 1.5 times the average shear stress. The solution is in the next frame.

Area above the neutral axis $= b \times 0.5d$
Distance to centroid of this are $= 0.25d$
Second moment of area about neutral axis $= \frac{1}{12}bd^3$
Width of section $= b$

Hence shear stress at level of neutral axis

$$= \frac{QA\bar{y}}{Ib}$$

$$= \frac{Q(b \times 0.5d)0.25d}{(b \times d^3/12)b}$$

$$= \frac{3}{2}\frac{Q}{bd}$$

$$= 1.5 \times \text{average shear stress}$$

47

Now let's look at the application of the theory to a very common structural problem: an I beam section. Steel I section beams are widely used in building construction and must be designed to resist the shear stresses resulting from the applied loading system. As an illustration of the theory, we will look at the I section shown in the figure and examine the distribution of shear stress under the application of a 50 kN shearing load.

Firstly let's consider the derivation of the shear stress expression at a level which is at a distance z from the neutral axis and within the web of the section.

The first moment of area about the centroidal axis ($A\bar{y}$) of that part of the section above the level being considered can be taken as the sum of the first moment of area of the flange plus the first moment of area of that part of the web that lies above the level being considered. Hence:

$$A\bar{y} = \left[(100 \times 10) \times 155 + \{6 \times (150 - z)\} \times \frac{(150 + z)}{2}\right]$$

$$= 155\,000 + 3(150^2 - z^2) \text{ mm}^3$$

Second moment of area, I (for the whole section)

$$= 2\left[(\tfrac{1}{12} \times 100 \times 10^3) + (100 \times 10 \times 155^2)\right] + \tfrac{1}{12} \times 6 \times 300^3$$

$$= 61.57 \times 10^6 \text{ mm}^4$$

Breadth of section, $b = 6$ mm

Shear force, $Q = 50$ kN

Shear stress, $\tau = \dfrac{QA\bar{y}}{Ib}$

$$= \frac{(50 \times 10^3) \times [155\,000 + 3(150^2 - z^2)]}{(61.57 \times 10^6) \times 6}$$

$$= 135.35 \times 10^{-6}[155\,000 + 3(150^2 - z^2)] \text{ N/mm}^2$$

Note that this equation is the equation of a parabola. Hence the shear stress in the web of the beam must vary parabolically.

Use this equation to calculate the shear stress at the level of the neutral axis and the level of the intersection between the web and flange (positions 1 and 2 in the diagram).

48

$$30.12 \text{ and } 20.98 \text{ N/mm}^2$$

These stresses are obtained by substituting $z = 0$ and $z = 150$ mm respectively into the shear stress expression.

Now determine an expression for the shear stress at any level within the flange of the beam, and use the expression developed to calculate the stresses at the interface of the web and flange and at the top of the section (positions 3 and 4). Use the diagram below and check your solution against the working in the next frame.

49

$$1.26 \text{ and } 0.00 \text{ N/mm}^2$$

$$A\bar{y} = 100 \times (160 - z) \times \frac{(160 + z)}{2} = 50 \times (160^2 - z^2) \text{ mm}^3$$

$$I = 61.57 \times 10^6 \text{ mm}^4 \text{ (as before)}$$

$$b = 100 \text{ mm}$$

$$\text{Shear stress, } \tau = \frac{QA\bar{y}}{Ib} = \frac{50 \times 10^3 \times \left[50 \times (160^2 - z^2) \right]}{61.57 \times 10^6 \times 100}$$

$$= 406.04 \times 10^{-6} \times (160^2 - z^2) \text{ N/mm}^2$$

At position 3: $z = 150$: $\tau = 406.04 \times 10^{-6} \times (160^2 - 150^2) = 1.26 \text{ N/mm}^2$

At position 4: $z = 160$: $\tau = 406.04 \times 10^{-6} \times (160^2 - 160^2) = 0.00 \text{ N/mm}^2$

Can you sketch the shear stress distribution throughout the full beam depth?

50

The shape of the stress distribution diagram follows from the general expressions which we developed in the preceding frames for the shear stress distribution in both the web and flange. Both of the expressions were parabolic: hence the parabolic shape of the above diagram. Note that the maximum shear stress occurs at the level of the neutral axis.

Another point of interest is the abrupt change of stress at the level of the intersection of the web and flange. The underside of the flange is a free surface and therefore the shear stress along this surface must be zero. The above diagram, however, shows a shear stress at this level of 1.26 N/mm^2, not zero. This would indicate that the simple theory that we have developed does not give accurate results within the flange. In fact, at the web–flange junction the localised stress distribution is much more complex than we have indicated, resulting in large localised stress gradients which cannot be determined by such a simple analysis. However, the methods that we have examined are nevertheless accurate enough for determining the shear stresses within the web of the beam.

You will note from the above diagram that the largest part of the stress diagram is acting on the web of the section. By multiplying the area of this part of the stress diagram by the thickness of the web of the beam, we can calculate what proportion of the shear *force* acting on the section is resisted by shear stresses which are developed in the web. This will lead to an interesting conclusion.

Can you calculate the total shear force carried by the web of the beam and hence determine what percentage of the total applied shearing force is resisted by shearing stresses within the web. The answer and worked solution are in the next frame. (You will need to remember the formula for the area enclosed by a parabola.)

$$\boxed{48.73 \text{ kN} \qquad 97\%}$$

Shear force in web $=$ thickness of web $\times$ area of stress diagram

$$= 6 \times [(300 \times 20.98) + (300 \times \tfrac{2}{3}\{30.12 - 20.98\})] \times \frac{1}{10^3}$$

$$= 48.73 \text{ kN}$$

Note that in the above calculation, in order to work out the area of the stress diagram, we have to split it into a rectangular segment and a parabolic segment.

The total shear force acting on the cross-section is 50 kN.

$$\text{Percentage of shear force carried by web } = \frac{48.73}{50.00} \times 100$$

$$= 97\%$$

This figure indicates that the web of this I beam resists nearly all the applied shear force. In fact, this is the principal structural function of the web. Of course we have only looked at one example, but the finding is generally true   for most standard section I beams the web resists between 90% to 98% of the total shear force.

For this reason it is common in the design of structural steel I beams to assume for simplicity that the web of the beam resists *all* the shear force.

*If we make this assumption for the beam section that we have been considering, what is the **average** shear stress in the web?*

$$\boxed{27.78 \text{ N/mm}^2}$$

$$\text{Average shear stress} = \frac{\text{applied shear force}}{\text{web breadth} \times \text{web depth}}$$

$$= \frac{50 \times 10^3}{6 \times 300}$$

$$= 27.78 \text{ N/mm}^2$$

What is the percentage difference between this average shear stress and the maximum shear stress of 30.12 N/mm² which we have previously determined?

53

$$\boxed{7.78\%}$$

Although there is a difference between the average and the maximum shear stress, the difference is small and, again, although this figure only applies to this particular problem, for most I section beams the average shear stress is within 10% of the maximum shear stress.

For this reason it is common practice in the design of steel I section beams not only to assume that all the shear force acting at a section is carried by the web but to base design calculations on the *average* shear stress which will be compared with an allowable average shear stress value.

For further simplicity in design, it is also common practice to take the depth of the beam as equal to the overall depth of the beam section. The error introduced by this assumption is small but leads to much simplification in practical design.

54

$$\boxed{\text{TO REMEMBER}}$$

$$Average \text{ shear stress } = \frac{\text{Shear force}}{\text{Area over which shear force acts}} \ (\text{N/mm}^2)$$

Shear stresses are always accompanied by *complementary* shear stresses.

$$\text{Shear Modulus } (G) = \frac{\text{Shear stress } (\tau)}{\text{Shear strain } (\gamma)} \ (\text{N/mm}^2)$$

For a rectangular section: maximum shear stress $= 1.5 \times$ average shear stress

$$\text{Shear stress } \tau = \frac{QA\bar{y}}{Ib} \ (\text{N/mm}^2)$$

(see Frame 45 for definition of terms)

In commonly occurring structural sections such as rectangular or I-shaped beams, the maximum shear stress usually occurs at the level of the neutral axis.

The web of an I beam resists nearly all the applied shear force at any cross-section.

55

FURTHER PROBLEMS

1. Figure Q1 shows a simple test arrangement to determine the shear strength of timber. The timber specimen is 50 mm wide. At an applied load of 10 kN, calculate the average shear stress along the plane $X-X$.

Ans. (2.0 N/mm²)

2. A steel bracket, as shown in figure Q2, supports a load of 60 kN. if the allowable shear stress in each bolt is 80 N/mm², calculate the minimum diameter of bolt required and select a bolt size from available diameters of 16, 20, 24 and 30 mm.

Ans. (21.84 mm, 24 mm)

3. Figure Q3 shows a cross-section through a reinforced concrete footing which supports a centrally placed column. In plan, the footing measures 2.5×2.5 metres and, because of the loading transmitted down the column, the pressure under the footing is 200 kN/m². Calculate the average shear stress at section $X-X$ and, if the allowable shear stress is 0.5 N/mm², state whether the average shear stress is acceptable.

Ans. (0.30 N/mm², acceptable)

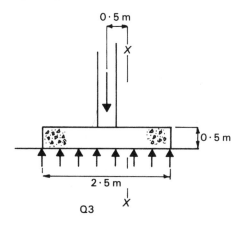

Q3

4. A crane hook is connected to a shackle by a single bolt as shown in figure Q4. If the hook is to carry a load of 80 kN and the safe shearing stress in the bolt is 80 N/mm², determine the minimum diameter of bolt required. (*Hint*: the bolt is in double shear.) *Ans.* (*25.23 mm*)

Q4

5. A T shaped beam is fabricated from two sections of timber nailed together as shown in figure Q5. If the section is subjected to a shearing force of 5 kN, calculate the shear stress at the level of the neutral axis and at the level just below the junction of the two pieces of timber. (*Hint*: first calculate the position of the centroid of the section.) *Ans.* (*0.58, 0.53 N/mm²*)

6. If each nail in the beam in figure Q5 can carry a shear force of 0.95 kN, what is the maximum spacing of the nails along the length of the beam? (*Hint*: you may have to think a little harder about this problem! In Q5 you calculated the shear stress at the level of the junction of the two pieces of timber. This will also be the complementary shear stress along the horizontal interface between the two timbers. What will be the shear *force* along, say, one metre length of the beam? If this shear force is resisted solely by the shearing action of the nails, how many nails are then required?) *Ans.* (*36.13 mm*)

7. A Universal beam is strengthened by welding a 200 mm × 20 mm steel plate to the top flange as shown in figure Q7. If the allowable maximum shear stress is 70 N/mm², calculate the maximum shearing force that this section can sustain. *Ans.* (*119.02 kN*)

8. A cantilevered beam carries a point load of '*W*' at its free end. If the beam has a circular cross-section of diameter *D*, what are the average and maximum shear stresses developed in the section? (*Hint*: to work out the first moment of area term in the general shear stress expression, you will have to employ numerical integration methods. You will also have to recall the expression for the second moment of area of a circle about a diameter.)

Ans. $(1.27W/D^2, 1.70W/D^2)$

9. If the beam in Q8 is replaced by a hexagonal cross-section of side length *s* and is orientated with one diagonal horizontal, what would now be the maximum shear stress in the section?

Ans. $(0.462W/s^2)$

10. A timber beam is simply supported at both ends and carries a central point load of 50 kN. The beam is fabricated from two 250 mm × 250 mm timbers which are bolted together as shown in figure Q10. The timbers are also keyed together by wooden keys which are spaced 700 mm apart. The maximum allowable average shear stress in the keys is 2 N/mm^2 and it is assumed that the keys resist all the longitudinal shear stress between the main timbers. Calculate the least required thickness of the keys, shown as dimension *t* on the drawing. (*Hint*: this problem is similar to Q5 and Q6. Do not attempt this question until you have done these two previous questions.)

Ans. (105.0 mm)

Q10

Programme 9

TORSIONAL
STRESS

On completion of this programme you will be able to:

- Define 'Torsional Moment'
- Define and calculate the Polar Second Moment of Area of solid and hollow circular sections
- Recognise the stress variation across a circular section
- State the equations of torsional stress
- Recognise and explain the terms within the torsional stress equations
- Calculate the state of stress within a circular shaft
- Calculate the angular deformation of a circular shaft
- Define the term 'Power'
- State the units of power
- Calculate the power transmitted in a rotating circular shaft
- Carry out simple design procedures for circular shafts in both static and rotational situations
- Calculate the state of stress in non-prismatic circular sections

1

Our study of stress would not be complete without examining the stresses that can occur within a structural element due to torsional or twisting effects. In this programme we are going to determine the governing equations that will define the behaviour and response of an element which is subjected to torsion, and see how these equations can be applied to the solution of a number of common problems. A complete treatment of torsional stress for all shapes of structural sections can be quite complex and is beyond the scope of this text. We will however look at the analysis of circular sections whereby the fundamental analytical expressions can be developed and the basic principles established.

Hollow circular or solid circular sections subject to torsional or twisting effects are quite common in everyday life. Can you think of an example?

2

The drive shaft of a car connecting the engine to the rear axle is a very common example. The rotation of the shaft will cause twisting, resulting in the development of torsional stresses. Other examples could include the propeller shaft of a ship or aircraft. On a more simple level the application of a spanner to tighten up a steel bolt would result in torsional stresses which, if excessive, could lead to fracture of the bolt, as anybody who has tried to release a rusty bolt well knows.

In order to calculate torsional stress it is necessary first to calculate the magnitude of the torsional moment acting on the structural cross-section. Hence it is important to have a sound grasp of the fundamental principles of structural analysis which were examined in Programmes 1 and 2.

For example, consider the cranked cantilevered beam shown below.

Can you show on a sketch the magnitude and direction of all the reactive forces and moments that occur at the fixed support, A?

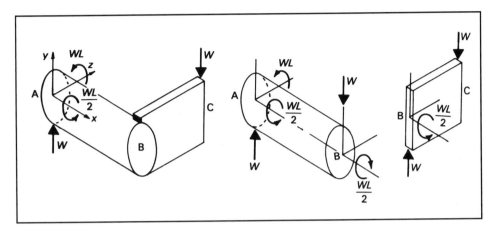

The first figure shows the free body diagram for the beam ABC with a moment about the Z-axis of WL kN m (W kN × L metre), a moment about the X-axis of $0.5WL$ kN m (W kN × $0.5L$ metre), and a vertical reaction of W kN. The moment about the Z-axis is a bending moment giving rise to bending stresses within the section, but the moment about the longitudinal X-axis is a *torsional moment* (or *torque*) which will cause twisting along the length of the beam AB.

The second diagram on the right shows more fully the reactive components acting on each section of the beam, AB and BC, from which it is more apparent that AB is subject to torsion, with twisting taking place along the longitudinal X-axis.

The length of the beam, AB, is in fact subject to a complex stress situation, as within the section there will be shear stresses, bending stresses and torsional stresses. We have already seen how to calculate the shear and bending stresses, so let us concentrate on the torsional stresses by considering the beam AB to be in a state of pure torsion, that is, subject only to a torsional moment which in general terms we will give the symbol T. We will also assume that the beam section is a solid circular section and that the end A is fixed.

If under the action of the torque, T, the free end, B, rotates through an angle θ, can you sketch the beam showing the distorted shape of the small surface element, abcd? What do you deduce about the stresses acting on the faces of this element?

4

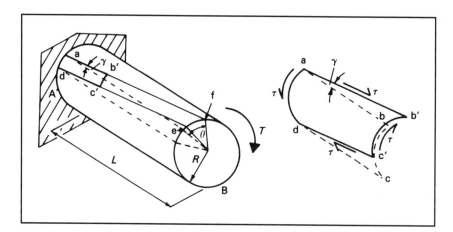

We assume that the deformations are small and that all cross-sections rotate about the longitudinal axis with radii remaining straight and cross-sections remaining plane and circular.

Hence the small element, abcd, will distort to the shape shown, with the sides remaining straight but the corners rotating through a small angle γ. The shape of this element should lead you to the conclusion that it is in the state of pure shear with shear stresses (τ) and complementary shear stresses acting on the faces of the element as indicated.

What is the significance of the angle γ?

5

γ is the shear strain

Look back at Programme 8, Frame 24, if you don't recall this.

If the beam is made up of a material having a shear modulus G, what is the relationship between τ, γ and G?

6

$$G = \frac{\tau}{\gamma}$$

Again, revise Programme 8, Frame 25, if in doubt. Now, on to the next frame, but keep refering to the diagram in Frame 4 above.

7

Because of the rotation, point e on the circumference of the end section will move to point f through the arc length ef. If the radius of the circular section is R, the arc length ef is given by:

$$ef = R \times \theta$$

and if the geometry of the surface segment is considered, then the arc length ef can also be written as:

$$ef = L \times \gamma$$

Hence combining the above two equations:

$$R \times \theta = L \times \gamma$$

or

$$\gamma = \frac{R \times \theta}{L}$$

But recalling the relationship for shear stress and shear strain at the top of Frame 6, we can rewrite this expression in terms of the shear stress, τ:

$$\frac{\tau}{G} = \frac{R \times \theta}{L}$$

or

$$\frac{\tau}{R} = \frac{G\theta}{L} \tag{9.1}$$

If this expression is to be used to calculate the shear stress, τ, in units of N/mm^2, what are the appropriate units for the other terms in the expression?

8

$$\boxed{R\text{---mm: } G\text{---}N/mm^2\text{: } L\text{---mm: } \theta\text{---radians}}$$

Most of these units should be obvious. However, note that θ is expressed in radians to justify the expression $ef = R \times \theta$ in Frame 7.

Now try an example:

A solid circular aluminium shaft has a diameter of 150 mm and is 1.5 metres long. An applied torque causes an angular twist of 2 degrees. What is the torsional shear stress at (1) the centre of the shaft and (2) the outside surface? Take $G = 26\ kN/mm^2$. Check your answer against the solution given in the next frame.

9

$$\boxed{(1) \text{ zero} \qquad (2)\ 45.50\ \text{N/mm}^2}$$

Solution to part (2):

Angle of twist $= 2° = \dfrac{\pi}{180°} \times 2 = 0.035$ radians

Shear stress, τ

$$= \frac{RG\theta}{L} \text{ (from equation 9.1)}$$

$$= \frac{75 \times (26 \times 10^3) \times 0.035}{1500}$$

$$= 45.50\text{N/mm}^2$$

The solution to part (1) should follow from intuition. The central longitudinal axis of a circular shaft is analogous to the neutral axis of bending of a beam. When the shaft is twisted it will rotate about the central longitudinal axis. All points along this axis will remain unstressed.

In fact, inspection of equation (9.1) shows that the shear stress τ increases linearly with increasing radius. Although we developed the expression by examining a small element on the outer surface of a beam, we could have derived the same expression by considering any small element at any radius, r, measured from the central longitudinal axis. The equation shows that shear stress due to torsion in a solid circular section increases linearly from zero at the centre to a maximum at the outer surface of the section. Hence to calculate the stress at any radius r, equation (9.1) may be written as:

$$\frac{\tau}{r} = \frac{G\theta}{L} \tag{9.2}$$

Now let's look at a small segment of cross-section (bcb'c') which we will assume has a thickness δr, and side length δs. Assume that the centre of the segment is at a radius r from the longitudinal axis.

Cross-section

If the shear stress along the face of the segment is τ, what is the total force acting along the face and hence what is the moment of this force about the longitudinal axis?

10

$$\text{force} = \tau \times \delta r \times \delta s: \qquad \text{moment} = (\tau \times \delta r \times \delta s) \times r$$

This moment is in fact the torque exerted by the shear stress acting on the face of the small segment about the longitudinal axis. The torque exerted by all the segments at a radius r is given by the summation of this term over the complete circumference of the circle, that is:

$$\text{Torque exerted} = \Sigma \tau \times \delta r \times \delta s \times r$$
$$= \tau \times \delta r \times r \times \Sigma \delta s$$
$$= \tau \times \delta r \times r \times (2\pi r)$$
$$= \tau 2\pi r^2 \delta r$$

However the solid section consists of concentric rings of elements, all of which will exert a torque about the central longitudinal axis. Hence the total torque exerted will be the summation of the above expression over the whole surface area of the cross-section.

Write down the integral expression which will express the total torque over the complete surface of the solid circular section.

11

$$\int_0^R \tau 2\pi r^2 \, dr$$

The equation

$$\int_0^R \tau 2\pi r^2 \, dr$$

gives the torque exerted by the internal stress system about the longitudinal axis. If the externally applied torsional moment is T, this must also be an expression for T because for equilibrium the external torsional moment must equal the internal torsional resistance.

Hence T may be expressed as:

$$T = \int_0^R \tau 2\pi r^2 \, dr$$

Previously we derived an expression for τ in terms of the shear modulus, G. Use this expression (Frame 9—equation 9.2) to eliminate τ from the above equation and write the torsional moment T in terms of G, θ and L.

12

$$T = \frac{G\theta}{L} \int_0^R 2\pi r^3 \, \mathrm{d}r$$

The integral expression is known as the *Polar Second Moment of Area* and in general terms is given the symbol J. Hence the above expression can be written as:

$$T = \frac{G\theta J}{L}$$

or
$$\frac{T}{J} = \frac{G\theta}{L} \tag{9.3}$$

If equation (9.3) and equation (9.2) are combined, a complete set of equations for the torsional analysis of a circular section can be written down:

$$\frac{\tau}{r} = \frac{T}{J} = \frac{G\theta}{L} \tag{9.4}$$

where τ = shear stress
$\qquad$ r = radius at which shear stress is being calculated
$\qquad$ T = applied Torsional Moment
$\qquad$ J = Polar Second Moment of Area
$\qquad$ G = Shear Modulus
$\qquad$ θ = angle of rotation
$\qquad$ L = length of the member.

At first sight this might seem a difficult set of equations to remember. However in Programme 6 we developed a similar looking set of equations for the analysis of a beam in bending.
Call you recall and write down these equations?

13

$$\frac{\sigma}{y} = \frac{M}{I} = \frac{E}{R}$$

The form of the equations is identical. Hence if you can recall the equations for the analysis of a beam in bending, you shouldn't have too much trouble in remembering the analogous equations for a section in torsion.

Now move to the next frame to look at some examples of the application of these equations.

Many common applications of torsional theory are concerned with the torsion of solid circular sections.

If the radius of a typical solid circular section is R, can you use the integral expression at the top of Frame 12 to derive a general expression for the Polar Second Moment of area of a solid circular section? The solution is in the next frame.

$$J = \frac{\pi R^4}{2}$$

The solution to this is simple:

From the expression at the top of Frame 12:

$$J = \int_0^R 2\pi r^3 \, dr$$

$$= \left[\frac{2\pi r^4}{4}\right]_0^R$$

$$= \frac{\pi R^4}{2}$$

This is a useful expression to commit to memory, to use in the solution of problems involving solid circular sections.

If such a solid circular section has a radius of 10 mm, is 4 metres long and is subject to a torque of 0.10 kN m, can you determine (1) the angle of twist of one end of the shaft relative to the other and (2) the maximum shearing stress within the section. Take G = 75 kN/mm².

$$0.34 \text{ rads} \qquad 63.69 \text{ N/mm}^2$$

If you didn't obtain these answers, check your solution against that given in Frame 17 on the next page.

17

When tackling torsion problems you should first write down the basic expression given in equation (9.4):

$$\frac{\tau}{r} = \frac{T}{J} = \frac{G\theta}{L}$$

You should then identify which part of the equation should be used to give the answer that is being sought. In this case we know the torque (T), the shear modulus (G), the length of the section (L) and the radius, from which we can immediately calculate the Polar Second Moment of Area (J). By inspection, you should be able to see that to calculate the angle of twist (θ) it is necessary to use the middle and third group of terms in the above expresson, in which the only unknown is θ. Similarly to calculate the shear stress (τ), the first and middle group of terms can be used as the only unknown in these terms is the shear stress τ.

The polar second moment of area (J) is given by:

$$J = \frac{\pi R^4}{2}$$

$$= \frac{\pi 10^4}{2}$$

$$= 1.57 \times 10^4 \text{ mm}^4$$

Hence the angle of twist (θ) is given by:

$$\frac{T}{J} = \frac{G\theta}{L}$$

or
$$\theta = \frac{TL}{JG} = \frac{(0.1 \times 10^6) \times 4000}{(1.57 \times 10^4) \times 75\,000}$$

$$= 0.34 \text{ radians}$$

Likewise, the maximum shear stress (τ), which will occur at the outer surface of the section, is given by:

$$\frac{\tau}{r} = \frac{T}{J}$$

or
$$\tau = \frac{Tr}{J} = \frac{(0.1 \times 10^6) \times 10}{1.57 \times 10^4}$$

$$= 63.69 \text{ N/mm}^2$$

Now try some problems:

18

PROBLEMS

1. A hollow circular section has an outer radius of R and an inner radius of r. Determine an expression for the Polar Second Moment of Area of this section.

Ans. $(\pi\{R^4 - r^4\}/2)$

2. A hollow circular steel drive shaft has an outer radius of 20 mm and an inner radius of 12 mm. If the maximum allowable shear stress in the shaft is 75 N/mm², what is the maximum permitted torsional moment that can be transmitted by the shaft?

Ans. (0.82 kN m)

3. The hollow shaft in question 2 is to be replaced by a solid circular shaft of identical material which is required to transmit the same torsional moment. Calculate the radius of this shaft if the maximum allowable shear stress remains the same.

Ans. (19.1 mm)

4. If the shafts referred to in questions 2 and 3 are 2 metres long and are rigidly fixed at one end, calculate the rotation of the free end for both the hollow and solid shafts. (Take $G = 75$ kN/mm².)

Ans. (0.1, 0.105 radians)

5. If the 'efficiency' of the shafts is measured in terms of the amount of material used in their construction, which is the most efficient the hollow or solid section? What general conclusion do you draw about the advantages of a hollow shaft compared with that of a solid shaft when made from the same material and used to transmit identical torsional moments?

Ans. (The solid shaft is 42.5% heavier than the hollow one and hence less efficient. Generally a hollow section is more efficient that an equivalent solid one)

6. A steel shaft is to be designed as a hollow circular section to carry a torsional moment of 1.5 kN m. If the maximum shear stress is limited to 100 N/mm², determine the minimum required dimensions of a hollow section with an outside diameter 1.5 times the inside diameter.

Ans. (15.22 and 22.84 mm)

19

CIRCULAR SECTIONS USED AS DRIVE SHAFTS TO TRANSMIT POWER

A common torsional problem is to be found in the case of a drive shaft used to transmit power from, say, a car engine to the drive wheels or a ship's engine to the propeller. We can use the theory that we have developed so far to look at the design or analysis of drive shafts which are required to rotate at a given constant speed to transmit a given torque. The combination of the torque and speed of revolution is a measure of power transmitted.

Before we proceed further, let's remind ourselves what we mean by power.
Can you write down the definition of power and state its units?

20

> Power is the rate of doing work.
> Units = Watts = G Joule/second

Within the above definition we have included the term 'work'. You will no doubt remember that if a force moves through a displacement, the work done is equal to the product of the force and the displacement, that is:

$$\text{Work} = \text{Force} \times \text{Displacement}$$

In the case of problems involving torsion, the definition of work will be different.
Can you write down an expression for the work done by an applied torque (T) acting on a shaft which it rotates through an angle θ? State the units of work.

21

> Work = $T \times \theta$
> Units = Joules = G N m

Note that the torque should be expressed in N m and the rotation in radians.

Now try the following problem – answers and solution are given in the next frame:
A drive shaft to an engine transmits a torque of 0.5 kN m. It rotates at a constant speed of 50 revolutions per minute. Determine (1) the work done during one revolution of the drive shaft and (2) the power transmitted.

| 3141.59 Joules | 2616.95 Watts |

Check your solution:

$$1 \text{ revolution} = 2\pi \text{ radians}$$

$$\text{Work done per revolution} = \text{Torque} \times \theta$$

$$= (0.5 \times 1000) \times 2\pi$$

$$= 3141.59 \text{ Joules}$$

$$\text{Number of revs per second} = \frac{\text{revs per minute}}{60}$$

$$= 50/60$$

$$= 0.833$$

$$\text{Power} = \text{Work done per second}$$

$$= \text{Work per rev} \times \text{number revs per second}$$

$$= 3141.59 \times 0.833$$

$$= 2616.95 \text{ Watts}$$

Note that in the above calculations the torque or torsional moment transmitted by the shaft appears in the calculations of both work and power. In the theory that we have developed previously in this programme, the torsional moment is one of the main parameters in our stress equations. Hence to design or analyse drive shafts we need only combine the general relationship between power and torque with the previous relationships.

From the equations used above, a general relationship between power and torque can be written as:

$$P = T \times 2\pi \times \frac{n}{60}$$

or

$$T = \frac{30P}{\pi n}$$

where:

$$T = \text{torque (N m)}$$

$$P = \text{power (Watts)}$$

$$n = \text{number of revs per minute}$$

Try the following problem. If you have understood what we have done above and you have completed the previous set of problems, this shouldn't prove too difficult. The complete solution is given in the next frame.

A solid circular shaft is to transmit 2×10^6 Watts at 150 revs per minute. What is (1) the torque transmitted, (2) the minimum required radius of the section if the maximum allowable shear stress is 100 N/mm²?

23

| (1) 0.127×10^6 N m   (2) 93.14 mm |

The torque transmitted is given by:

$$T = \frac{30P}{\pi n}$$

$$= \frac{30 \times (2 \times 10^6)}{\pi \times 150}$$

$$= 0.127 \times 10^6 \text{ N m}$$

If the minimum required radius of the section is R, then the polar second moment of area is given by:

$$J = \frac{\pi R^4}{2}$$

and the relationship between the torque and the allowable shear stress is given by:

$$\frac{\tau}{r} = \frac{T}{J}$$

or:

$$\frac{J}{r} = \frac{T}{\tau}$$

Hence:

$$\frac{\pi R^4}{2R} = \frac{0.127 \times 10^6 \times 1000}{100}$$

Therefore

$$R^3 = \frac{2 \times 0.127 \times 10^6 \times 1000}{\pi \times 100}$$

Which solves to give

$$R = 93.14 \text{ mm}$$

Calculate the angular twist (in degrees) over a 5 metre length of this shaft if a radius of 95 mm is provided, Take G = 80 kN/mm².

24

| 3.56 degrees |

If you didn't get this solution, look back at the first part of the problem in Frame 17—it's practically the same.

25

PROBLEMS

1. A solid circular propeller shaft has a radius of 75 mm. If the shearing stress is limited to 100 N/mm², what is the maximum power that can be transmitted at a speed of 200 revs per minutes?

Ans. (1.39 × 10⁶ Watts)

2. If the shaft in question 1 is 3 metres long and the angular twist must not exceed 2 degrees, what is the maximum power that can be transmitted based on this limiting condition? Take $G = 80$ kN/mm².

Ans. (0.969 × 10⁶ Watts)

3. A hollow circular drive shaft is to transmit 100 kW at a constant speed of 150 revs per minute. If the outside diameter of the shaft is to be twice the inside diameter, determine the dimensions of the shaft if the shear stress is not to exceed 50 N/mm².

Ans. (outside diameter 88.44 mm)

26

OTHER TORSIONAL PROBLEMS

The problems that we have dealt with so far have been fairly straightforward, but using a combination of the torsion equations and simple statics it is possible to analyse a wide range of similar problems to those that we have already considered.

However there is a range of more complex problems where it is necessary to extend the theory further. It is not possible to examine every single type of problem but by looking at some typical situations the general principles may be established.

As an example, consider a circular shaft which has both ends rigidly fixed and a torque of 100 kN m applied at its centre point as shown.

Can you say what is the torque resisted by the left-hand segment AB and what is the torque resisted by the right-hand segment BC?

27

| 50 kN m on both segments |

You should have arrives at this answer by considerations of symmetry. As both halves of the shaft are identical, there is geometrical symmetry and symmetry in loading and therefore the torque will be resisted equally by both segments.

Now let's consider the same shaft with a torque T applied at some point along the shaft which is not at the centre. Let's also assume for generality that the two segments of the shaft either side of the point of loading are of different radii and therefore have different Polar Second Moments of Area and are made of different material having different shear moduli.

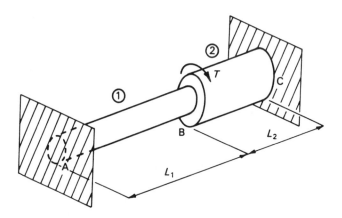

Can you say what proportion of the torque is resisted by the left-hand segment AB and what proportion is resisted by the right-hand segment BC?

28

| The answer should be that you can't! |

The reason that you can't is that the problem that we are looking at is *statically indeterminate*. This is because there are too many support reactions to permit analysis using only the simple principles of static equilibrium, and it is therefore necessary to supplement the equations of equilibrium by additional relationships which will lead to the complete analysis of the structure.

If we assume that the effect of applied torque, T, is to induce a torque T_1 in the segment AB and a torque T_2 in the segment BC, write down the relationship between T, T_1 and T_2.

$$\boxed{T = T_1 + T_2}$$

This is our *equation of equilibrium* for the externally applied and internally induced torsional moments. It can't of course be solved as it contains two unknowns: T_1 and T_2. The figure below shows the free body diagram for both segments of shaft, indicating the direction of the torsional moments.

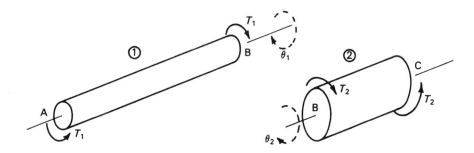

Under the action of the torsional moments both sections of shaft will obviously rotate. Let's say that AB rotates by an amount θ_1 and BC by an amount θ_2. As both A and C are fixed supports, this rotation will in both cases represent the amount of twisting that takes place at B.

Can you write down a relationship between θ_1 and θ_2?

$$\boxed{\theta_1 = \theta_2}$$

This is common sense. As both segments of shaft are connected rigidly at B, it follows that they must rotate by the same amount. This is in fact an expression of *compatibility of displacements* and effectively provides the additional equation that is necessary to solve for T_1 and T_2. The concept of displacement compatibility is common and is widely used in structural analysis. You have already come across this concept in Programme 3 and you will come across it again many times in your further studies of structural analysis.

Now move to the next frame to see how the equation of equilibrium and the equation of displacement compatibility can be combined to solve for the unknown Torsional Moments: T_1 and T_2.

31

So far we have produced two equations:

(1) Equilibrium: $\qquad\qquad\qquad\qquad\qquad\qquad\qquad T = T_1 + T_2$

(2) Displacement Compatibility: $\qquad\qquad\qquad\qquad\theta_1 = \theta_2$

We are attempting to find a solution for T_1 and T_2 but the second equation is expressed in terms of θ_1 and θ_2. It is therefore necessary to eliminate θ_1 and θ_2 from this second equation, replacing both terms by expressions in T_1 and T_2 and hence leading to two equations which can be solved directly for the torsional moments in each segment of the shaft.

Can you write down a relationship between θ_1 and T_1, and a corresponding relationship between T_2 and θ_2? (Hint: think back to the work that we have been dealing with previously in this programme.)

32

$$\theta_1 = \frac{T_1 L_1}{G_1 J_1} \qquad \theta_2 = \frac{T_2 L_2}{G_2 J_2}$$

These relationships follow directly from the general torsion equations (see Frame 12—equation 9.4, if in doubt). It follows therefore that the two fundamental equations describing the behaviour of this shaft can be written as:

(1) Equilibrium: $\qquad\qquad\qquad\qquad\qquad\qquad T = T_1 + T_2$

(2) Displacement Compatibility: $\qquad\qquad\qquad\dfrac{T_1 L_1}{G_1 J_1} = \dfrac{T_2 L_2}{G_2 J_2}$

Rearrange the above two equations to give the expressions for T_1 and T_2.

33

$$T_1 = \frac{T(G_1 J_1 L_2)}{(G_1 J_1 L_2 + G_2 J_2 L_1)} \qquad T_2 = \frac{T(G_2 J_2 L_1)}{(G_1 J_1 L_2 + G_2 J_2 L_1)}$$

These final two equations give the torsional moments in each segment of the shaft. If the stress in each segment is required or the angular rotation is to be determined, the general torsional stress equations (equation 9.4) can be applied to either segment using the appropriate torsional moment calculated from the equations in Frame 33.

The equations in Frame 33 need not necessarily be remembered. If you understand the principles behind their derivation, you should be able to tackle any similar problem working from first principles of *equilibrium* and *displacement compatibility*.

Now try a problem:

A solid circular shaft, AC, has a total length of 1.5 metres and is 50 mm in diameter. A gear wheel positioned at B, 0.5 metres from the left-hand end of the shaft (A), exerts a torque of 0.45 kN m. If the ends A and C are instantaneously locked in position by brakes just before the torque is applied, determine the torsional moment induced in both segments of the shaft.

35

$$T_1(\text{AB}) = 300 \text{ N m} \qquad T_2(\text{BC}) = 150 \text{ N m}$$

The solution follows directly from the equations in Frame 33, although for practice you should develop the answers from first principles. However if the equations are used, they can be considerably simplified as both segments of the shaft are identical and therefore the Shear Modulus terms (G) and the Polar Second Moment of Area terms (J) cancel.

Hence:

For the length AB: $\quad T_1 = \dfrac{T \times L_2}{(L_1 + L_2)} = \dfrac{450 \times 1.0}{(0.5 + 1.0)} = 300 \text{ N m}$

For the length BC: $\quad T_2 = \dfrac{T \times L_1}{(L_1 + L_2)} = \dfrac{450 \times 0.5}{(0.5 + 1.0)} = 150 \text{ N m}$

Note that for a continuous uniform shaft, the distribution of a torque applied at some point along its length is merely proportional to the lengths into which the shaft is divided by the point of application of the torque.

Using the result that you have just obtained, calculate the maximum shearing stresses in both segments of the shaft and the angular twist of the shaft after the torque is applied. Take G = 80 kN/mm².

36

$$\boxed{\text{(AB) } 12.22 \text{ N/mm}^2 \qquad \text{(BC) } 6.11 \text{ N/mm}^2 \qquad 0.0031 \text{ radians}}$$

Check your solution:

The polar second moment of area (J) is given by:

$$J = \frac{\pi R^4}{2}$$

$$= \frac{\pi 25^4}{2}$$

$$= 61.38 \times 10^4 \text{ mm}^4$$

Hence the maximum shear stress (τ) in the length AB which will occur at the outer surface of the section is given by:

$$\tau = \frac{TR}{J}$$
$$\frac{300 \times 1000 \times 25}{61.38 \times 10^4}$$
$$= 12.22 \text{N/mm}^2$$

The torsional moment in the length BC is half that in the length AB and hence the maximum shear stress in this length will be half as great, that is, 6.11 N/mm^2.

To calculate the angular twist in the shaft, either AB or BC can be considered as the result must be the same in both cases. Hence considering the length AB:

The angle of twist (θ) is given by:

$$\theta = \frac{TL}{JG}$$

$$= \frac{300 \times 1000 \times 500}{61.38 \times 10^4 \times 80\,000}$$

$$= 0.0031 \text{ radians}$$

You should also calculate the angle of twist for the length BC to convince yourself that either calculation will give the same answer.

This is just one example of this type of problem. There are more examples of a similar nature for you to practice on at the end of this programme.

Now on to the next frame:

NON-PRISMATIC SECTIONS

So far we have looked mainly at problems involving prismatic sections, that is, those having constant cross-sectional area. To complete our study of torsional stresses we will look briefly in some more detail at the case of non-prismatic sections where the cross-sectional area varies along the length of the member. The figure below illustrates a simple example of a non-prismatic shaft, in this case fabricated from two separate prismatic sections. For simplicity we will assume that the torsional moment is constant along the length of the shaft.

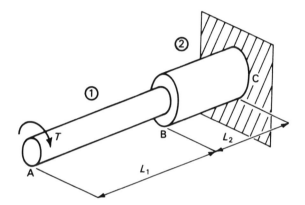

Generally there are two sets of calculations associated with the design or analysis of such a shaft:

(1) Calculation of sectional stress
The general formula to use is the one that by now you should be quite familiar with:

$$\frac{\tau}{r} = \frac{T}{J}$$

In order to calculate the stress at any particular cross-section it is only necessary to calculate and use the polar second moment of area based on the radius appropriate to the section under consideration.

(2) Calculation of the total angle of twist along the shaft
The formula that we would use for prismatic sections is given by:

$$\frac{T}{J} = \frac{G\theta}{L}$$

or

$$\theta = \frac{TL}{GJ}$$

In the case of the shaft shown in the diagram above, can you write down an expression for the total angular rotation along the shaft?

38

$$\boxed{\theta = \theta_1 + \theta_2 = \frac{TL_1}{GJ_1} + \frac{TL_2}{GJ_2}}$$

You should have reasoned that the total angular rotation along the length of the shaft is the rotation of the length AB plus that of the length BC.

If the shaft consists of several lengths of differing sectional properties, the total angular rotation can be expressed in general terms as:

$$\theta = \sum \frac{TL_i}{GJ_i}$$

The shaft shown in the figure in Frame 37 consists of two lengths each 500 mm long, but the polar second moments of area of AB and BC are 1×10^6 mm^4 and 2×10^6 mm^4 respectively. If a clockwise torque of 0.4 kN m is applied to the free end, A, what is the total angular rotation of the shaft? Take $G = 80$ kN/mm^2.

39

$$\boxed{0.00375 \text{ radians}}$$

The solution is as follows:

$$\theta = \frac{TL_1}{GJ_1} + \frac{TL_2}{GJ_2}$$

$$= \frac{0.4 \times 10^6 \times 500}{80\,000 \times 1 \times 10^6} + \frac{0.4 \times 10^6 \times 500}{80\,000 \times 2 \times 10^6}$$

$$= \frac{0.4 \times 10^6 \times 500}{80\,000} \left[\frac{1}{1 \times 10^6} + \frac{1}{2 \times 10^6} \right]$$

$$= 0.003\,75 \text{ radians}$$

We have of course assumed that the torsional moment is constant along the shaft. This may not always be the case.

If the shaft in Frame 37 now has a clockwise torque of 0.4 kN m applied at A and an additional clockwise torque of 0.3 kN m acting at B, what is the torsional moment acting on the lengths AB and BC?

| 0.4 kN m | 0.7 kN m |

The figure below shows the free body diagrams for the two segments of the shaft.

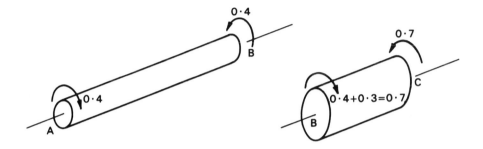

What would the total torsional rotation of the shaft be under the action of these two applied torques? (Hint: this is practically the same as the last problem, only the values of the moments have changed.)

41

| 0.00469 radians |

The solution is as follows:

$$\theta = \frac{T_1 L_1}{GJ_1} + \frac{T_2 L_2}{GJ_2}$$

$$= \frac{0.4 \times 10^6 \times 500}{80\,000 \times 1 \times 10^6} + \frac{0.7 \times 10^6 \times 500}{80\,000 \times 2 \times 10^6}$$

$$= \frac{10^6 \times 500}{80\,000} \left[\frac{0.4}{1 \times 10^6} + \frac{0.7}{2 \times 10^6} \right]$$

$$= 0.004\,69 \text{ radians}$$

Now let's look at a more complex case of torsion in non-prismatic sections.

42

Consider the case of a circular shaft with a continuously varying cross-section as shown. As the radius varies so does the polar second moment of area, and as the torsion equations have been derived for a shaft of constant sectional profile they can no longer be used directly to determine the rotation of this shaft.

However let's approximate the profile to a series of short prismatic sections, as shown in the second figure, the length of each section being very small and equal to δx.

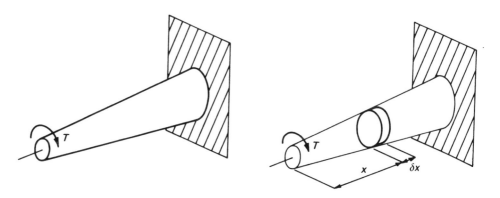

If each segment is assumed to be prismatic, the small rotation $\delta\theta$ of a typical segment at a distance x from the free end can be calculated using the usual formula with the length of the segment taken as δx:

$$\delta\theta = \frac{T\delta x}{GJ_x}$$

where J_x is the polar second moment of area of the segment which will vary with the distance x along the shaft.

Using this formula, can you write down an expression for the total angular rotation along the shaft?

43

$$\boxed{\theta = \int_0^L \frac{T\,\mathrm{d}x}{GJ_x}}$$

This expression follows from the application of the normal principles of calculus and can be used to determine the rotation of any non-prismatic section. However as the polar second moment of area varies with distance x, the integration of this expression can be quite complex and may necessitate the use of numerical integration methods.

Try the following problem – you may need to revise your integral calculus first!

A shaft of circular cross-section has a polar second moment of area that varies according to the formula $J_x = J_0(1 + \{x/L\})^4$, where x is measured from the end where the polar second moment of area is J_0. If the length of the shaft is L and it is subject to a torsional moment T, determine an expression for the total rotation of the shaft.

44

$$\frac{7}{24}\frac{TL}{GJ_0}$$

Using the formula in Frame 43, the rotation of the shaft is given by:

$$\theta = \int_0^L \frac{T\,dx}{GJ_x}$$

$$= \frac{T}{GJ_0}\int_0^L \frac{dx}{(1+\{x/L\})^4}$$

$$= \frac{T}{GJ_0}\left[-\frac{L}{3(1+\{x/L\})^3}\right]_0^L$$

$$= \frac{TL}{3GJ_0}\left[-\frac{1}{8}+1\right]$$

$$= \frac{7}{24}\frac{TL}{GJ_0}$$

We have looked only at shafts where the cross-sectional geometry varies and shown how to calculate the rotation due to a constant applied torque. In more complex problems, both the torque, T, and the shear modulus, G, could vary along the length of the shaft. Such problems can be tackled using the methods employed above, provided that the variation of all the parameters can be defined in terms of a variable distance x measured along the shaft enabling appropriate integration methods to be employed.

45

TORSION OF NON-CIRCULAR SECTIONS

In this programme we have confined our studies to the torsion of circular sections and have derived the relationship that:

$$\frac{T}{J} = \frac{G\theta}{L}$$

or

$$T = \frac{GJ\theta}{L}$$

For circular sections, J is the polar second moment of area. For sections that are not circular, the above equation can still be used provided that J is taken as the *torsional constant* for the section, which in general will not be equal to the polar second moment of area. The calculation of the torsional constant and the treatment of non-circular sections is outside the scope of this book.

46

$$\boxed{\text{TO REMEMBER}}$$

$$\frac{\tau}{r} = \frac{T}{J} = \frac{G\theta}{L} \quad \text{(See Frame 12 for definition of terms)}$$

For a solid circular section:
$$J = \frac{\pi R^4}{2}$$

For a non-prismatic section:
$$\theta = \int_0^L \frac{T \mathrm{d}x}{GJ_x}$$

47

$$\boxed{\text{FURTHER PROBLEMS}}$$

Note: in all these problems, take the shear modulus, G to be 80 kN/mm^2.

1. A solid circular drive shaft has a diameter of 50 mm. If it is subjected to a torque of 2 kN m, calculate (1) the maximum shear stress developed and (2) the angular twist per unit length of shaft.

Ans. (81.49 N/mm^2, 0.041 radians/metre)

2. The shear stresses in a solid circular shaft are limited to a maximum value of 75 N/mm^2 and the angular twist permitted is no greater that 0.5 degrees over a length of 1 metre. If both these limiting conditions are to be satisfied simultaneously, determine the least radius of the section and the maximum permissible torque.

Ans. (107.43 mm, 146.07 kN m)

3. A hollow circular shaft has an external radius of 50 mm and an internal radius of 25 mm. If the allowable maximum shear stress is 80 N/mm^2, what is the allowable torque and the twist of the shaft per metre length when this torque is applied?

Ans. (14.73 kN m, 0.02 radians/metre)

4. The shear stresses in a hollow circular shaft are limited to a maximum value of 50 N/mm^2 and the angular twist permitted is no greater than 0.5 degrees over a length of 1 metre. If both these limiting conditions are to be met simultaneously, when a torque of 25 kN m is applied determine the required internal and external diameters of the shaft.

Ans. (143.24 mm, 86.58mm)

5. A solid circular shaft rotates at 100 revs per minute and is to be designed to transmit 200 kW. If the maximum shear stress is not to exceed 50 N/mm^2, what is the least radius of the shaft?

Ans. (62.42 mm)

6. A propeller shaft of a ship is fabricated from a hollow circular section with external and internal diameters of 500 mm and 300 mm respectively. If it rotates at a constant speed of 100 revs per minute, what power can be transmitted if the maximum shear stress is not to exceed 40 N/mm^2?

Ans. (8.95 × 10^6 Watts)

7. Determine the maximum torque that can be applied at the junction of the two solid circular shafts shown in figure Q7. Both shafts are firmly restrained at their ends as shown and the maximum shear stress is limited to 75 N/mm^2.

Ans. (1.04 kN m)

Q7

8. The shaft in question 7 is to be redesigned to carry a 10% increase in the calculated torque. Determine the necessary diameters of both sections of the shaft if the ratio of the shaft diameters remains constant.

Ans. (41.28, 20.64 mm)

9. An alternative design for the shaft in question 7 is for the length BC to be fabricated from a hollow tubular section as shown in figure Q9. Calculate the torque that can be applied to this shaft if the maximum shear stress is still limited to 75 N/mm².

Ans. (981.75 N m)

10. The shaft in figure Q10 is fabricated from solid circular section and is stepped as shown. If torques are applied as shown, determine (1) the total angular twist of the section and (2) the maximum shear stress in each length of the shaft.

Ans. (0.0763 radians, 244.46, 71.30, 39.23 N/mm²)

Q9   Q10

11. A solid circular shaft has a diameter of d at one end and a diameter of $1.2d$ at the other end with a uniform taper between. The length of the shaft is L and it is subject to a torque T. Determine an expression for the total angular twist over the length of the shaft. (Hint: look at the equation in Frame 43 and the example in Frame 44.)

Ans. ($22.47TL/\pi Gd^4$)

12. The shaft in question 11 is 2 metres long, the diameter $d = 100$ mm and it rotates at 100 revs per minute. If the total angular twist over the length of the shaft is limited to 1 degree, what is the maximum power that can be transmitted based on this limiting condition?

Ans. (0.102×10^6 Watts)

Programme 10

STRESS
TRANSFORMATIONS
AND MOHR'S CIRCLE
OF STRESS

On completion of this programme you will be able to:

- Define 'Principal Stress'
- Define 'Principal Plane'
- Draw a Mohr's Circle of Stress for a range of complex stress situations
- Interpret a Mohr's Circle of Stress
- Determine the magnitude of the principal stresses
- Identify the orientation of the principal planes
- Determine the magnitude and orientation of the planes of maximum shear stress
- Identify the orientation of a plane of given stress condition
- Determine the state of stress on a plane of given orientation

1

In Programmes 5 to 9 we learnt how to calculate the stress within structural members when subjected to direct axial loading, bending, shear and torsion. The theories that we have developed are applicable to a wide range of problems and the formulae that we have used are widely utilised in the design of many types of structural component using different materials of construction.

In Programme 7 we made use of the principle of superposition to calculate the stresses due to a combination of direct and bending stresses and saw that in such situations we could calculate the components of stress separately and simply add the components together to give the final set of stresses. However, many structural components are subject to much more complex stress situations than those we have already considered. For example, we have already seen that the cranked cantilever beam in Frame 2 of Programme 9 is stressed from a combination of bending, shear and torsion, and indeed there are many practical situations where such complex stress combinations exist.

One of the most important combination of stresses that can occur in practical design is that of direct stresses acting on a body in combination with shearing stresses. The direct stresses may themselves be a combination of individual stresses acting in different directions. The stresses resulting from such a combination can be quite complex and in this programme we are going to look at both numerical and graphical techniques whereby detailed stress analysis may be performed.

The theory we will develop is widely used in practical engineering design. For example, in the design of building foundations the strength of the soil on which the building will be built is determined from a standard laboratory test where the results are interpreted using the graphical techniques that we are going to develop.

Before you proceed you should acquire a supply of graph paper, a compass and a protractor. When you have done so, move to the next frame.

2

Let's start by looking at a simple problem. The bar shown in the figure (top of next page) is stressed by the application of an axial load P. Let's assume that the bar has a uniform rectangular cross-sectional area A and we are interested in the stress condition at the cross-section B–B.

State what sort of stress condition exists at this cross-section and write down the equation for this stress in terms of P and A.

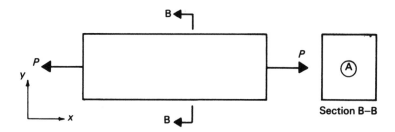

Section B–B

3

$$\boxed{\text{Uniform direct stress normal to the section} : \quad \sigma_x = \frac{P}{A}}$$

Now let's look at the same bar but this time take a section B–B which is inclined at an angle θ to the vertical. If we look at the free body diagram of that part of the bar to the right of section B–B it is obvious that, to ensure horizontal equilibrium, there must be a horizontal force P acting at the section. This force will have components acting normal to the section (N) and along the section (S). We will assume that the directions of these components are as shown. We will say more about sign conventions later

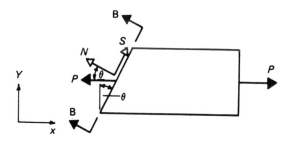

Can you write down two equation relating N and S to P and θ?

4

$$\boxed{N = P\cos\theta \qquad S = -P\sin\theta}$$

This follows from a straightforward resolution of the force P acting at the section into its components along and perpendicular to the section.
    What do you deduce about the stresses acting on the face B–B?

5

> There are both normal and shear stresses acting on this face.

As the face B–B has a normal force N acting on it, it follows that there must be normal *stresses* acting on this face. Likewise if there is a force S acting along the face B–B, there must be shearing stresses acting along this face.

This is an important conclusion. So far we have looked at members in tension or compression and determined the stresses in these members based on a simple calculation of axial force divided by the cross-sectional area of the member. The cross-sectional area was in fact the area *normal* to the direction of the force.

What we have now concluded is that if we take a section which is not normal to the direction of the force, the state of stress on this section is more complex and consists of a combination of normal and shear stresses. This is an important observation – make sure that you fully appreciate it.

Remembering that the basic definition of stress is force divided by the area on which the force acts, derive equations for the normal (σ) and shear (τ) stresses acting on the inclined face B–B expressed in terms of the stress $\sigma_x(= P/A)$.

6

> $$\sigma = \sigma_x \cos^2 \theta \qquad \tau = -\sigma_x \sin \theta \cos \theta$$

These equations follow directly from the general definition of stress and from the geometry of the section that we are considering:

The area (A') of the face B–B is given by:

$$A' = \frac{A}{\cos \theta} \quad \text{(simple geometry)}$$

Hence the normal stress is given by:

$$\sigma = \frac{N}{A'} = \frac{P \cos \theta}{A / \cos \theta} = \frac{P}{A} \cos^2 \theta = \sigma_x \cos^2 \theta$$

and the shear stress is given by:

$$\tau = \frac{S}{A'} = \frac{P \sin \theta}{A / \cos \theta} = -\frac{P}{A} \sin \theta \cos \theta = -\sigma_x \sin \theta \cos \theta$$

These stress equations can be more conveniently written as:

$$\sigma = \frac{\sigma_x}{2}(1 + \cos(2\theta)) \tag{10.1}$$

and

$$\tau = -\frac{\sigma_x}{2}\sin(2\theta) \tag{10.2}$$

which follow from the double angle relationships that:

$$\cos^2\theta = \frac{(1 + \cos(2\theta))}{2} \quad \text{and} \quad \sin\theta\cos\theta = \frac{\sin(2\theta)}{2}$$

Now try the following exercise.

A horizontal bar of rectangular cross-sectional area 100 mm² is subject to an axial tensile force of 10 kN. Prepare a table of values of normal and shearing stresses on planes inclined at θ to the vertical where θ varies from 0° to 180° in steps of 15°. Plot a graph of shearing stress (τ – as y coordinate) against normal stress (σ as x coordinate), and answer the following question:
(1) What is the shape of the graph?
(2) What is
  (a) the maximum normal stress
  (b) the minimum normal stress
  (c) the maximum shear stress and
  (d) the angle to the vertical of the planes on which these stresses occur?

7

> (1) Circular
> (2) (a) 100 N/mm² (b) 0 N/mm² (c) ±50 N/mm² (d) 0°, 90°, 45° and 135°

Check your solution:

$\theta°$	σ (N/mm²)	τ (N/mm²)
0	100	0
15	93.3	−25
30	75	−43.3
45	50	−50
60	25	−43.3
75	6.7	−25
90	0	0
105	6.7	25
120	25	43.3
135	50	50
150	75	43.3
165	93.3	25
180	100	0

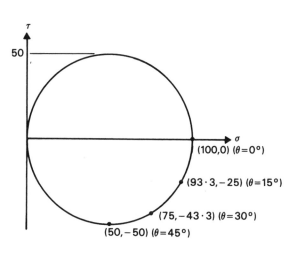

291

8

The diagram that you have just drawn is known as a *Mohr's Circle*. We will look more closely at Mohr's circles later in this programme but essentially this form of construction enables us to describe in graphical form the variation of stresses acting on any inclined plane.

We drew this circle by calculating and plotting a number of points. However once it is recognised that the plot is circular and the properties of the circle are identified, then it is possible to draw and use such a circle to carry out a stress analysis of any such similar problem without the necessity to calculate and plot every point.

Let's look again at the circle that we have just plotted. You have already deduced that the maximum normal stress was 100 N/mm^2 and occurred on the plane inclined at 0°. This is in fact the stress σ_x and corresponds to the point marked C in the diagram. The minimum normal stress was 0 N/mm^2 acting on a plane at 90° to the vertical which corresponds to the point A on the diagram.

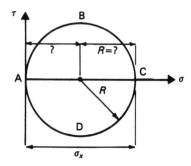

If σ_x is a known stress, what would be he radius of the circle and which point defines its centre?

9

$$\text{radius} = \frac{\sigma_x}{2} \qquad \text{centre at } \sigma = \frac{\sigma_x}{2}, \tau = 0$$

Hence for any structural element subject to a direct tensile (or compressive) stress, σ_x, acting on a vertical section then the Mohr's Circle of Stress can be drawn simply by locating the centre at coordinates of $\sigma_x/2$ and zero and drawing a circle of radius $\sigma_x/2$. This will now define fully the state of stress (σ and τ) on all inclined planes. How then do we use the circle to determine the stresses on any given plane?

Can you recall the expressions for σ and τ in terms of $\cos(2\theta)$ and $\sin(2\theta)$?

$$\sigma = \frac{\sigma_x}{2}(1 + \cos(2\theta)) \qquad \tau = -\frac{\sigma_x}{2}\sin(2\theta)$$

It follows from these two equations that if a point (E) is taken on the circle which is at an angular distance of 2θ from C, then this point must represent the stresses σ and τ on the plane inclined at an angle θ to the vertical. The angle 2θ is the angle subtended at the centre of the circle by the arc CE and is measured in a *clockwise* direction from C. The construction below shows that E must represent the two equations above.

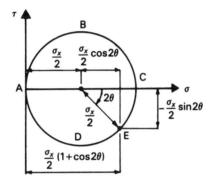

Refer to the Mohr's Circle of Stress that you drew in response to the problem in Frame 6 and use the circle to determine the stress on two perpendicular planes inclined at 20° and 110° to the vertical.

11

$\theta = 20°$	$\sigma = 88.3 \text{ N/mm}^2$	$\tau = -32.1 \text{ N/mm}^2$
$\theta = 110°$	$\sigma = 11.7 \text{ N/mm}^2$	$\tau = +32.1 \text{ N/mm}^2$

You should obtain these answers by turning off angles of 40° and 220° in a clockwise direction from C and reading off the appropriate stresses. You could have calculated these values using the equations, and it is probably a good idea to check your answers using the equations to convince yourself that the circle is merely a graphical interpretation of the equations.

The two planes that you have just considered are at right angles. What do you notice about the shear stress on these two planes and the corresponding points you have drawn on the diagram?

293

12

> The shear stresses have the same numerical value and
> the two points are on opposite ends of a diameter.

The fact that the shear stresses are of the same magnitude on two perpendicular planes should not surprise you – remember the concept of *complementary shear stresses*? The fact that the two points representing these two planes are on the opposite ends of a diameter is a useful result to remember, as the form of construction is such that any pair of points at the opposite ends of the diameter *always* represent the stresses on a pair of mutually perpendicular planes.

Now let's look at some other features of the Mohr's Circle of Stress which will be useful in our further studies of this subject.

Refer to the diagram in Frame 10 and write down what you think is significant about the stresses and the inclination of the planes represented by the points marked C, A, B and D.

13

Did you draw the following conclusions?

Point C. Point C represents a plane on which the shear stress is zero. It also represents the plane on which the normal stress is a maximum. A plane subject to zero shear stress is in fact called a *Principal Plane* and the normal stress on this plane is a Principal Stress – in this case a *maximum* principal stress.

Point A. Point A also represents a plane on which the shear stress is zero and on which the normal stress is a minimum. By the above definition it is also a Principal Plane and the normal stress (which in this instance happens to be zero) is a *minimum* principal stress.

Points A and C are also on the opposite ends of the diameter, thus indicating that the principal planes are at right angles to each other.

Points B and D. These two points represent the planes on which the shear stress is a maximum. Note that the magnitude of the maximum shear stress is equal to the radius of the circle which in this instance is one half of the maximum principal stress. The angular distance of D from C is $90°$ ($2\theta = 90°$). C represents a principal plane and hence the plane of maximum shear represented by D is inclined at $45°$ to the principal plane ($\theta = 45°$). As the other plane of maximum shear represented by B is on the opposite end of the diameter to D, it also follows that both planes of maximum shear are at right angles to each other and at $45°$ to the principal planes.

Make sure that you fully understand these conclusions as they are generally applicable to the work that we will do later in this programme. Now try some problems.

14

PROBLEMS

1. Define the terms *Principal Stress* and *Principal Planes*.
Ans. *(see Frame 13)*

2. A bar of circular cross-section of radius 20 mm carries an axial tensile load of 100 kN. Draw the Mohr's Circle of Stress and determine the values of the principal stresses and the maximum shear stresses.
Ans. *(79.6, 0, ±39.8 N/mm²)*

3. What is the inclination to the longitudinal axis of the bar of the planes on which the normal stress is 50 N/mm²?
Ans. *(52.5°, 127.5°)*

4. If the bar in question 2 has a shear strength of 50 N/mm², to what value can the axial load be safely increased?
Ans. *(125.66 kN)*

5. If the load on the bar is increased to the value calculated in question 4, what will be the values of the maximum and minimum principal stresses?
Ans. *(100 N/m², 0 N/mm²)*

15

The theory that we have developed so far has been based on an examination of the state of stress in a uniformly loaded bar where the stresses acting on the face of any inclined section are uniform throughout the section under consideration. In many other types of problem this is often not the case and it is more usual to describe the state of stress at a *point* in a structure. In the case of the bar that we originally considered in Frame 2, this point can be represented by the small rectangular element indicated in the figure below.

Although this element is infinitesimally small, in practice we draw it with a finite size so that we can clearly show the stresses acting at the point. The sides of this element represent the faces of the plane inclined at an angle θ and the plane which is perpendicular to this first plane. The stresses acting *at this point* on the two perpendicular planes can be indicated in magnitude and direction on this element.

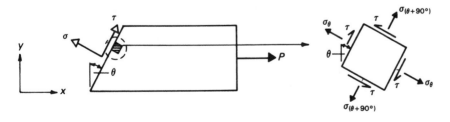

16

At this stage we will now define more clearly the sign convention that we have been using and that we will continue to use.

(1) The angle θ which defines the inclination of the face of an element is measured in a clockwise direction from the positive direction of the y-axis.

(2) A normal stress is positive if it causes tension in the element.

(3) Shear stresses are positive if the stresses on the two opposite faces on an element form a couple which would cause a *clockwise* rotation of the element.

Because shear stresses on two parallel faces must be accompanied by equal complementary shear stresses on the two perpendicular faces, it follows that one pair of faces will always be acted on by positive shear stresses and the other will be acted on by negative shear stresses.

The figure below shows the stresses acting on a typical element according to this convention. In the derivation of our original stress equations we in fact took the shear stresses on the face inclined at θ to the positive direction of the y-axis as positive. We will continue to do this.

In the problem in Frame 11 you calculated that the stresses on a plane inclined at 20° were $\sigma = 88.3 \ N/mm^2$ and $\tau = -32.1 \ N/mm^2$. The corresponding stresses on the perpendicular plane were $\sigma = 11.7 \ N/mm^2$ and $\tau = +32.1 \ N/mm^2$. Show this set of stresses on a sketch of a small element.

17

Now on the next frame.

18

We have looked at a relatively simple case of stress analysis and have developed some important concepts by examining the case of a bar in uniaxial tension. Let's extend these concepts by examining the state of stress at a point in a structure which is subject to *biaxial* stress; that is, subject to direct stresses in the direction of both the x and y axes as shown in the figure below. There are many practical situations where such a state of stress can exist including, for example, shell structures with thin walls used as pressure vessels to contain gas or liquids.

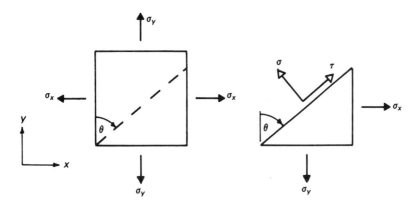

If we are interested in the state of stress on a plane inclined at an angle θ to the positive direction of the y-axis, we can take a section through our small element as shown in the second diagram above. We will assume that the stresses on the inclined face of the element consist of both normal stresses, σ, and shear stresses, τ.

By considering the equilibrium of this wedge shaped element, can you derive expressions for σ and τ in terms of σ_x, σ_y and θ? (Hint: take the area of the inclined face as A and remember that force = stress × area.)

19

$$\sigma = \sigma_x \cos^2 \theta + \sigma_y \sin^2 \theta \qquad \tau = -(\sigma_x - \sigma_y) \sin \theta \cos \theta$$

If you got this answer, can you express the equations in terms of 2θ by using the double angle relationship that:

$$\sin(2\theta) = 2 \sin \theta \cos \theta$$

and

$$\cos(2\theta) = 2 \cos^2 \theta - 1 = 1 - 2 \sin^2 \theta$$

The complete solution is in the next frame.

20

$$\sigma = \frac{(\sigma_x + \sigma_y)}{2} + \frac{(\sigma_x - \sigma_y)}{2}\cos(2\theta) \qquad \tau = \frac{-(\sigma_x - \sigma_y)}{2}\sin(2\theta)$$

The proof of these equations follows directly from considerations of the equilibrium of the small wedge shaped element which for clarity is reproduced below:

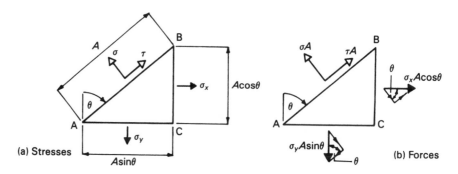

(a) Stresses

(b) Forces

The area of face AB $= A$

Hence:
the area of face BC $= A\cos\theta$

the area of face AC $= A\sin\theta$

The normal force on the face AB $= \sigma \times A$

The shear force on the face AB $= \tau \times A$

The force on the face BC $= \sigma_x \times (A\cos\theta)$

The force on the face AC $= \sigma_y \times (A\sin\theta)$

Resolving forces perpendicular to the plane AB (see figure (b)):

$$\sigma \times A = (\sigma_x \times A\cos\theta) \times \cos\theta + (\sigma_y \times A\sin\theta) \times \sin\theta$$

Hence:
$$\sigma = \sigma_x \cos^2\theta + \sigma_y \sin^2\theta$$

$$= \sigma_x \frac{(\cos(2\theta) + 1)}{2} + \sigma_y \frac{(1 - \cos(2\theta))}{2}$$

that is
$$\sigma = \frac{(\sigma_x + \sigma_y)}{2} + \frac{(\sigma_x - \sigma_y)}{2}\cos(2\theta) \qquad (10.3)$$

Likewise resolving along the face AB:

$$\tau \times A = -(\sigma_x \times A\cos\theta) \times \sin\theta + (\sigma_y \times A\sin\theta) \times \cos\theta$$

hence:
$$\tau = -(\sigma_x - \sigma_y)\sin\theta\cos\theta$$

$$= -\frac{(\sigma_x - \sigma_y)}{2}\sin(2\theta) \qquad (10.4)$$

If equations (10.3) and (10.4) are used to plot the variation of σ and τ for different values of θ, what shape of graph would you expect?

21

A circle – a *Mohr's Circle*

If equations (10.3) and (10.4) are compared with equations (10.1) and (10.2), you will see that they are almost identical. In fact if we put $\sigma_y = 0$ in both equations (10.3) and (10.4), we would obtain equations (10.1) and (10.2). We saw that equations (10.1) and (10.2) were in fact the equations of a circle, and indeed this second pair of equations also describes a circle.

If the equations are plotted as shown below, with tensile normal stresses taken as positive to the right of the origin, it can be seen that the circle is practically the same as the one that we considered for uniaxial stress. However, it no longer passes through the origin but is displaced so that its centre is at $(\sigma_x + \sigma_y)/2$ and its radius is equal to $(\sigma_x - \sigma_y)/2$. In all other respects the interpretation of the stresses from the circle is the same as before.

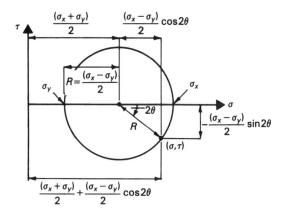

Note that as both σ_x and σ_y occur on planes on which the shear stress is zero, by our previous definitions σ_x and σ_y are Principal Stresses and the planes on which they occur are Principal Planes.

Sketch the Mohr's Circle of Stress for a point in a material where the Principal Stresses (σ_x and σ_y) are given by (1) 60 & 30 N/mm² (2) 60 & −30 N/mm², (3) −60 & −30 N/mm². In all cases, indicate the value of the maximum shear stress. Remember that tensile stresses are plotted as positive.

22

23

Now let's look at another stress situation: in this case a small element of a material which is subjected to shear stresses only. Note that in the diagram the shear stresses on the vertical faces (τ') have been drawn in a direction which would cause clockwise rotation of the element (positive shear according to our sign convention). Likewise the complementary shear stresses have been drawn as negative shear stresses according to our sign convention.

Again, if we are interested in the stress situation on a plane inclined at an angle θ to the positive direction of the y-axis, we can take a section through the small element and consider the equilibrium of the small wedge shaped element shown in the second figure.

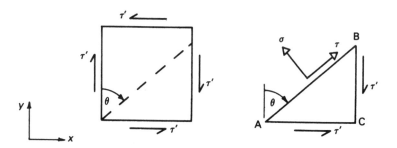

Can you derive expressions for σ and τ in terms of τ' and θ. Express your answers in terms of 2θ, making use of the double angle relationships.

24

$$\sigma = \tau' \sin(2\theta) \qquad \tau = \tau' \cos(2\theta)$$

This derivation is very similar to that given in Frame 20. Check your working:

$$\text{The area of face AB} = A$$

Hence:
$$\text{the area of face BC} = A\cos\theta$$

$$\text{the area of face AC} = A\sin\theta$$

$$\text{The normal force on the face AB} = \sigma \times A$$
$$\text{The shear force on the face AB} = \tau \times A$$
$$\text{The force on the face BC} = \tau' \times (A\cos\theta)$$
$$\text{The force on the face AC} = \tau'(A\sin\theta)$$

Resolving forces perpendicular to the plane AB:

$$\sigma \times A = (\tau' \times A \cos\theta) \times \sin\theta + (\tau' \times A \sin\theta) \times \cos\theta$$

Hence: $$\sigma = \tau' 2 \sin\theta \cos\theta$$

$$\therefore \qquad \sigma = \tau' \sin(2\theta) \tag{10.5}$$

Likewise resolving along the face AB:

$$\tau \times A = (\tau' \times A \cos\theta)\cos\theta - (\tau' \times A \sin\theta)\sin\theta$$

Hence $$\tau = \tau'(\cos^2\theta - \sin^2\theta)$$

$$\therefore \qquad \tau = \tau' \cos(2\theta) \tag{10.6}$$

These equations define the variation of σ and τ for different values of θ. By now you should be anticipating that a plot of these equations will be circular.

Can you sketch the Mohr's Circle given by these two equations, indicating the direction of the angle 2θ which would give the stresses on any plane inclined at an angle θ?

25

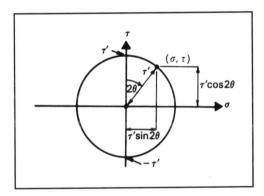

You will note that in the case of an element subjected to pure shear the Mohr's Circle is centred on the origin and has a radius equal to the value of the shear stress, τ'.

Expressed in terms of τ', what are the values of the principal stresses and the inclination of the planes on which they occur?

26

$+\tau'$ on a plane at $45°$ $(2\theta = 90°)$ and $-\tau$ on a plane at $135°$ $(2\theta = 270°)$

Now try some further problems to give you more practice at transforming stresses and drawing and interpreting Mohr's Circles of Stress.

27

1. Figure Q1 shows an element in a state of biaxial stress. If $\sigma_x = 70$ N/mm^2 and $\sigma_y = 30$ N/mm^2 determine, by working from first principles (a) the stresses acting on planes inclines at $20°$ and $110°$ to the positive direction of the y-axis and (b) the magnitude of the maximum shear stress.

Ans. ((a) 65.32, −12.86 & 34.68, 12.86 N/mm^2, (b) 20 N/mm^2)

2. Repeat question 1(a) but this time plot a Mohr's Circle of Stress and obtain the answers from the diagram. Show the magnitude and direction of the stresses on a sketch of a small rectangular element orientated at the correct angle.

Ans. (See diagrams at the bottom of the page)

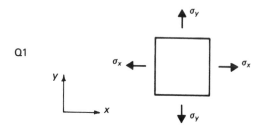

3. Repeat questions 1 and 2 for stresses of $\sigma_x = 70$ N/mm^2 and $\sigma_y = -20$ N/mm^2

Ans. ((a) 59.47, −28.92 & −9.47, 28.92 N/mm^2, (b) 45 N/mm^2)

4. Using the Mohr's Circle drawn in question 3, determine the inclinations of the planes which are subjected to only shear stresses. What is the value of shear stress on these planes?

Ans. (61.9° and 118.1° measured from the positive direction of the y-axis, ±37.5 N/mm^2)

5. Use your Mohr's Circle diagram from question 3 to determine the inclinations of the planes on which the normal stresses are 35 N/mm^2?

Ans. (38.6° and 141.4° measured from the positive direction of the y-axis)

(Q2 Ans.)   (Q3 Ans.)

So far we have considered particular cases of stress analysis: either uniaxial stress, biaxial stress or pure shear. The more general application of stress analysis is to the case of problems where a combination of direct and shear stresses acting on two perpendicular planes at a point in a structure are known. Typically the principal stresses and maximum shear stresses and the orientation of the planes on which these stresses act are to be determined. Often the failure of a stressed structural component will be governed by a limiting maximum principal stress or maximum shear stress being reached and it is these stresses which must be calculated from known stress components acting in known directions.

Let's now develop the theory further so that we can produce a general theory applicable to all types of problems. Consider the element shown which is now subjected to direct stresses combined with shear stresses. The problem again is to determine equations for the normal and shear stresses on a plane inclined at an angle θ to the positive directions of the y-axis.

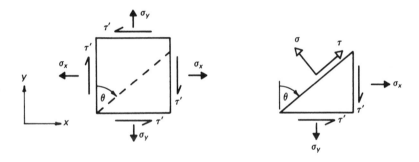

Based on the work that we have already done, can you write down the equations for the normal stress σ and the shear stress τ?

29

$$\sigma = \frac{(\sigma_x + \sigma_y)}{2} + \frac{(\sigma_x - \sigma_y)}{2}\cos(2\theta) + \tau'\sin(2\theta) \qquad (10.7)$$

$$\tau = -\frac{(\sigma_x - \sigma_y)}{2}\sin(2\theta) + \tau'\cos(2\theta) \qquad (10.8)$$

The figure above is merely a combination of the cases that we have already considered. The above equations are obtained simply by combining equations (10.3) with (10.5) and (10.4) with (10.6). In other words we have made use, yet again, of the principle of superposition. Let's look at the Mohr's Circle for this case.

30

It is not obvious that the equations given for σ and τ describe a circle. However they have been obtained by combining other equations that we have already seen are the equations of a circle and you should therefore anticipate that the two equations will describe a circle. The figure below gives the Mohr's Circle interpretation of these equations. As an exercise to show that this circle is in fact a correct interpretation, you should assign some numerical values to σ_x, σ_y and τ'. For differing values of θ, calculate and plot pairs of values of θ and τ using equations (10.7) and (10.8). (You did a similar exercise in Frame 7.)

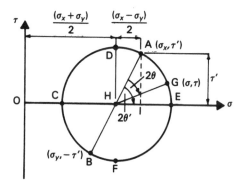

Note that provided σ_x, σ_y and τ' are known, the points A and B can be plotted and joined together. This describes a diameter of the circle, the intersection of which with the σ-axis will give the centre of the circle. Once the centre is found, the circle can be drawn. An angle 2θ turned in a *clockwise* direction from the line HA will define a point G which will give the stresses σ and τ acting on a plane inclined at an angle θ to the positive direction of the y-axis.

As in all the previous cases, the values of the principal stresses will be given by the points marked C and E, and the maximum shear stresses by the points marked D and F. The inclination of the planes on which these stresses act is given by the angle 2θ when G in coincident with whichever of these points is being considered.

For the element shown below, plot the Mohr's Circle and determine (1) the maximum and minimum principal stresses, (2) the maximum shear stresses.

(1) 44.1, 15.9 (2) 14.1 N/mm²

Referring back to the Mohr's Circle diagram in Frame 30, some useful and important relationships can be derived:

(1) The centre is at $((\sigma_x + \sigma_y)/2$. As the centre is in a fixed position it follows that for any given stress system at a point in a structure, the sum of the normal stresses on any two perpendicular planes must be a constant no matter to what angle the planes are rotated.

(2) From the geometry of the triangle HAI and using Pythagoras' Theorem, the radius of the circle is given by:

$$\text{Radius} = [\{0.5(\sigma_x - \sigma_y)\}^2 + \tau'^2]^{\frac{1}{2}}$$

(3) The maximum shear stresses equal the radius of the circle. Therefore:

$$\tau_{max} = \pm[\{0.5(\sigma_x - \sigma_y)\}^2 + \tau'^2]^{\frac{1}{2}} \tag{10.9}$$

(4) The maximum principal stress is represented by the point E such that:

$$
\begin{aligned}
\sigma_{max} &= OH + HE \\
&= OH + \text{Radius of circle} \\
&= 0.5(\sigma_x + \sigma_y) + [\{0.5(\sigma_x - \sigma_y)\}^2 + \tau'^2]^{\frac{1}{2}} \tag{10.10}
\end{aligned}
$$

(5) Similarly the minimum principal stress is represented by point C such that:

$$
\begin{aligned}
\sigma_{min} &= OH - HC \\
&= OH - \text{Radius of circle} \\
&= 0.5(\sigma_x + \sigma_y) - [\{0.5(\sigma_x - \sigma_y)\}^2 + \tau'^2]^{\frac{1}{2}} \tag{10.11}
\end{aligned}
$$

(6) The angle of inclination of the plane on which the maximum principal stress acts is given in the diagram by the angle $2\theta'$ where, from the geometry of the triangle HAI:

$$\tan(2\theta') = \frac{\tau'}{(\sigma_x - \sigma_y)/2} = \frac{2\tau'}{(\sigma_x - \sigma_y)} \tag{10.12}$$

The minimum principal stress acts on a plane at 90° to the plane of maximum principal stress and follows directly once θ' is known.

Use these equations to check your answers to the problem at the end of Frame 30 which you have already solved graphically.

32

You have already got the answers to this problem but you may wish to check your working:

$$\text{Radius} = [\{0.5(\sigma_x - \sigma_y)\}^2 + \tau'^2]^{\frac{1}{2}}$$
$$= [\{0.5(40 - 20)\}^2 + 10^2]^{\frac{1}{2}}$$
$$= 14.1 \text{ N/mm}^2$$

$$\sigma_{\max} = 0.5(\sigma_x + \sigma_y) + \text{Radius}$$
$$= 0.5(40 + 20) + 14.1$$
$$= 44.1 \text{ N/mm}^2$$

$$\sigma_{\min} = 0.5(\sigma_x + \sigma_y) - \text{Radius}$$
$$= 0.5(40 + 20) - 14.1$$
$$= 15.9 \text{ N/mm}^2$$

$$\tau_{\max} = \text{Radius of circle}$$
$$= 14.1 \text{ N/mm}^2$$

Now use the circle that you plotted for the last problem in Frame 30 to read off the stresses on planes inclined at $\theta = 30°$ and $60°$. Check your answers by calculation using the equations for σ and τ (10.7 and 10.8 in Frame 29).

33

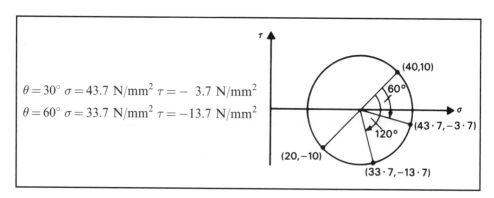

$\theta = 30° \quad \sigma = 43.7 \text{ N/mm}^2 \quad \tau = -\ 3.7 \text{ N/mm}^2$

$\theta = 60° \quad \sigma = 33.7 \text{ N/mm}^2 \quad \tau = -13.7 \text{ N/mm}^2$

The problem that you have been working on was for a particular set of stresses. Both the normal stresses σ_x and σ_y were positive (tensile), the shear stresses on the side faces of the element were positive (clockwise), and those on the top and bottom faces were negative (anticlockwise). The circle was drawn by plotting the two points $(\sigma_x, +\tau')$ and $(\sigma_y, -\tau')$ resulting in the construction that we have already examined.

If the signs of the stresses are different, the circle can still be drawn provided that when plotting the two points the signs of the stresses are taken into account.

For the elements shown, sketch the Mohr's Circles of Stress indicating the two points that are used to construct the circles.

34

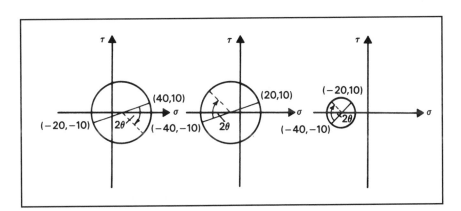

The most important points on all these Mohr's Circle diagrams are the points representing the principal stresses and the points representing the maximum shear stresses. As we have seen, these stresses can be calculated using equations but the use of the Mohr's Circle gives a complete visual picture of the state of stress at a point in a structure and is often easier to use and interpret than a set of equations.

Reading the values of stresses from a Mohr's Circle is relatively easy provided that the circles are drawn correctly, taking account of the signs of the known stresses. What can be a little more difficult is the determination of the inclination of the planes on which these stresses act, particularly when the circle relates to stresses which are not positive according to our chosen sign convention.

Let's look at a further geometrical construction that will enable us to determine the orientation of planes of stress using a previously constructed Mohr's Circle:

35

This geometric construction is known as the *Pole Method* and will enable us to determine the inclination of the principal planes (and indeed the planes for any other set of stresses) which can be plotted on the Mohr's Circle. This construction makes use of the fact that the angle subtended at the centre of a circle by a chord is twice that subtended by the same chord at the circumference.

The figure below reproduces the Mohr's Circle first introduced in Frame 30. Assuming, as we have done so far, that the element for which the stresses are known is orientated with its side faces vertical and its top and bottom faces horizontal, then the following geometric construction is shown and can be used in all cases:

(1) From A which represents the stresses σ_x, τ' on the *vertical* faces of the element, draw a vertical line to meet the circle at P. P is known as the *Pole Point*.

(2) The pole point P could have been alternatively located by drawing a *horizontal* line through B which represents the stresses σ_y, $-\tau'$ on the *horizontal* faces of the element.

(3) From P draw a line to connect P to G which represents the stresses σ_θ, τ on a plane inclined at an angle θ to the positive direction of the y-axis.

(4) The angle APG is equal to θ as this is the angle subtended by the chord AG on the circumference which is half the angle AHG (equal to 2θ) subtended at the centre. Hence the line PG is inclined at the same angle as the face of the element on which the stresses σ_θ and τ act. The line PG' is the inclination of the perpendicular plane on which stresses represented by the point G' act.

(5) If the pole point P is connected to E and C, this gives the inclinations of the principal planes.

(6) If P is connected to D and F, this will give the inclination of the planes of maximum shear stress.

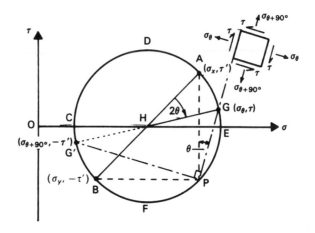

This form of construction can be used to either:

   (a) determine the inclination of the planes on which known stresses act
or (b) determine the stresses acting on planes of known inclination.

   Use this method of construction to determine the orientation of the principal planes and the planes of maximum shear stress for the problem we considered in Frames 30 and 31. The graphical solution is given in the next frame.

36

the principal planes are inclined at 22.5° and 112.5° to the positive direction of the *y*-axis. The planes on which the maximum shear stresses act are inclined at 67.5° and 157.5°.

Now use the same methods to determine the magnitude of the principal stresses and the orientation of the planes on which they occur for the stress system indicated on the element below.

37

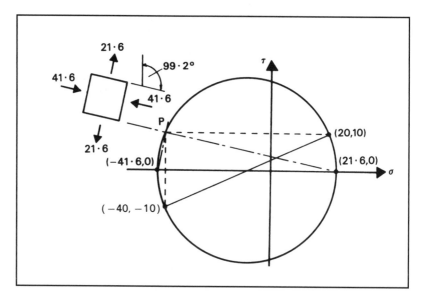

It is worth noting that, if the stresses shown in the figure at the end of Frame 36 represent a set of applied stresses at a point in a structure, then the principal stresses at the point are numerically greater than the applied direct stresses and the maximum shear stress at the point is greater than the applied shear stress. For example, the applied direct compressive stress is 40.0 N/mm^2 but the principal compressive stress is 41.6 N/mm^2. Similarly the applied shear stress is 10 N/mm^2 while the maximum shear stress is 31.6 N/mm^2 (the radius of the circle). You will find that this observation is true for most of the problems we have looked at in this programme, and is in fact generally true. It is important to recognise that this is the case as, under the action of complex stress systems material failure of a structural element may occur at apparently much lower applied stresses than if a single stress is applied.

*Now look at the element of material subjected to the direct **compressive** stresses and shear stresses shown. If the material will fail at a **tensile** stress of 5 N/mm^2, do you think that material failure is likely? Sketch the Mohr's Circle of Stress to help you answer this question.*

38

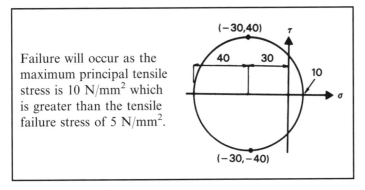

Failure will occur as the maximum principal tensile stress is 10 N/mm² which is greater than the tensile failure stress of 5 N/mm².

This example illustrates that, although a point in a structure may be subjected to shear stresses together with direct applied stresses of one type (in this case compressive stresses), one of the principal stresses induced at the point may be of opposite sign to the applied direct stresses. It is important to recognise the fact that, although a point in a structure may be subjected to one type of stress system, stresses of an opposite sense may be induced at the point on planes inclined at different angles to those on which the applied stresses act.

Finally it should be pointed out that, in all the problems in this programme, we have drawn Mohr's Circle of Stress for known stresses acting on small elements which for convenience were orientated with the side faces vertical and the top and bottom faces horizontal. The known stresses acting at a point in a structure may not necessarily be acting on horizontal and vertical planes and may in fact be acting on planes which are inclined. In this case all that is required is to orientate the x and y reference axis parallel to the side faces of the inclined element. The Mohr's Circle can then be drawn and interpreted in the usual way with the angle θ still taken as positive when measured clockwise from the positive direction of the y-axis.

39

┌─────────────────┐
│   TO REMEMBER   │
└─────────────────┘

This programme is more concerned with techniques rather than specific formulae to remember. However you should know how to construct and interpret a Mohr's Circle and should know that:

The principal stresses are the maximum and minimum normal stresses that occur at a point within a structure and they act on the principal planes which are orientated at 90° to each other.

There are no shear stresses acting on the principal planes.

The planes on which the maximum shear stresses act are at 45° to the direction of the principal planes.

40

(Note: the Mohr's Circle of Stress for the problems below are given at the end of the frame)

1. A short bar is loaded in axial compression to give uniform uniaxial stress of 80 N/mm^2. Working from first principles, determine the normal and tangential stresses on planes inclined to the longitudinal axis of the bar at (i) 35° (ii) 45° and (iii) 60°.

Ans. ((i) −26.32, 37.58 (ii) −40, 40 (iii) −60.0, 34.64 N/mm^2)

2. Repeat question 1 but use equations (10.7) and (10.8) in Frame 29.

3. Check your answers to question 1 by drawing the Mohr's Circle of Stress and reading the appropriate stresses from the circle.

4. A drive shaft of circular cross-section is designed to carry a uniform tensile stress of 15 N/mm^2 and a shearing stress of 20 N/mm^2 at a point on the outer surface of the shaft. Taking the shear stresses as positive, determine by calculation (i) the principal stresses at the point and the angle of the planes on which these occur and (ii) the maximum shear stresses at the point and the angles of the planes on which these occur.

Ans. ((i) 28.86, −13.86 N/mm^2 at 34.72° & 124.72° (ii) ±21.36 N/mm^2 at 79.72° & 169.72°—angles measured relative to an axis perpendicular to the longitudinal axis of the shaft)

5. Repeat question 4 but solve graphically using a Mohr's Circle of Stress.

6. Draw the Mohr's Circle of Stress for the element shown in figure Q6 and use the circle to determine the principal stresses and the orientation of the planes on which they act. Show the principal stresses on a drawing of a correctly orientated element as in the example in Frame 36.

Ans. (20.0, 95.0 N/mm^2 at 63.4° and 153.4°—angles measured relative to the given y-axis)

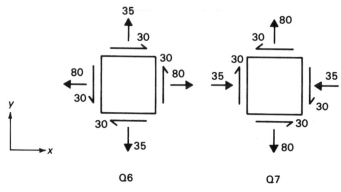

Q6   Q7

STRESS TRANSFORMATIONS AND MOHR'S CIRCLE OF STRESS

7. Draw the Mohr's Circle of Stress for the element shown in figure Q7 and use the circle to determine the principal stresses and the orientation of the planes on which they act. Show the principal stresses on a drawing of a correctly orientated element as in the example in Frame 36.

Ans. (87.4, −42.4 N/mm² at 76.2° and 166.2°—angles measured relative to the given y-axis)

8. Figure Q8 shows a point in a test specimen which is stressed with a constant stress of 50 N/mm² in the x direction and with an increasing stress σ_y in the y direction. If the specimen fails at a shearing stress of 35 N/mm², what is the value of σ_y at failure?

Ans. (20 N/mm²)

9. A glued joint between two lengths of material is shown in figure Q9. The glue used is one of a range of modern adhesives which have extremely high shear strength. The joint is therefore best orientated in a plane which has shear stresses only acting on it. Determine by drawing a Mohr's Circle the optimum value of the angle θ, and the magnitude of the corresponding shear stresses.

Ans. (57.7° or 122.3°, 9.5 N/mm²)

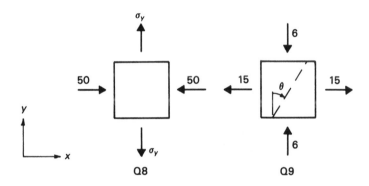

Q8   Q9

10. Figure Q10 (see overleaf) shows a cantilevered beam which carries a point load of 20 kN. Calculate the bending stresses and shear stresses at A, B and C and use these stresses to calculate for each of these points (i) the magnitude of the principal stresses and (ii) the values of the maximum shear stresses. In each case determine the orientation of the planes on which these stresses act.

(Hint: you may need to refer to Programme 8 to remind yourself how to calculate shear stresses at a cross-section.)

Ans.((i) A: 4.5, 0 N/mm² at 0° & 90° B: 2.72, −0.47 N/mm² at 157.5° & 67.5° C: 1.5, −1.5 N/mm² at 135° & 45° (ii) A: ±2.25 N/mm² at 135° & 45° B: ±1.59 N/mm² at 112.5° & 22.5° C: ±1.5 N/mm² at 90° & 0°—all angles measured relative to an axis at right angles to the longitudinal axis of the beam)

Q10

11. (a) A drive shaft is to be designed as a solid circular section to transmit a torsional moment of 2 kN m. If the maximum shear stresses are limited to 60 N/mm², what is the least diameter of section necessary?

(b) To allow for the possible existence of an axial force in the shaft which could give rise to the development of axial tensile stresses, the drive shaft is slightly overdesigned and is fabricated out of a solid 60 mm diameter circular section. If the allowable shear stresses are still 60 N/mm², what is the largest axial thrust that can be tolerated together with the torque of 2 kN m?

Ans. (55.37 mm, 209.69 kN)

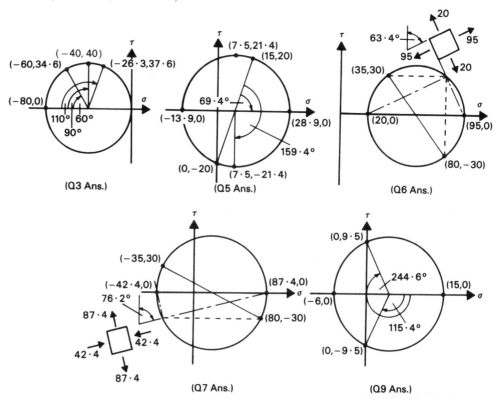

Programme 11

COMPOSITE
SECTIONS

On completion of this programme you will be able to:

- Recognise different forms of composite construction
- Calculate the stress conditions in axially loaded composite sections
- Define 'modular ratio'
- Explain the concept of a 'transformed section'
- Calculate the magnitude and state of stress within composite sections acting in flexure
- Explain the terms 'singly reinforced' and 'doubly reinforced' concrete sections
- Analyse a reinforced concrete beam to determine the magnitude and state of stress in both concrete and steel
- Analyse the effect of temperature variation on axially loaded composite sections

1

In all of the programmes so far that have been concerned with stress analysis we have looked at problems with structural elements fabricated from a single type of material. There are, however, a number of structural situations where a component may be fabricated from more than one type of material, each of which will have its own set of physical properties, such as Young's Modulus of Elasticity, and hence the response of a composite structure will be governed by the composite action of the materials out of which the structure is fabricated.

Typical composite structures are shown in the figures below and in this programme we will look at some aspects of stress analysis in such structures. Figure (a) shows a very common example of composite construction: the cross-section of a reinforced concrete beam where the flexural strength is derived from the composite action of reinforcing steel which has high tensile strength together with concrete which has high compressive strength. Figure (b) shows a composite slab construction with Universal steel beams acting together with a reinforced concrete decking.

Other examples of composites that you might think about could include the simple bi-metallic strip used in temperature control switches whose action depends on the different thermal expansion of two metals which are firmly bonded together. In this programme, in addition to considering the types of stress studied in previous programmes, we will also look at the way in which changes in temperature can induce stresses in composite structural components.

(a)

(b)

2

AXIALLY LOADED MEMBERS

Let's start by looking at the simplest type of problem: the analysis of stress in axially loaded members consisting of two or more materials acting together. Let's look at a bar made of two separate materials which are firmly glued together and subjected to an axial tensile force P. The force P could alternatively be a compressive force; the following theory will be equally applicable except that the resulting stresses will be compressive and not tensile. It should be noted that for any form of composite action it is important that the individual materials are firmly bonded to each other, otherwise they act as separate elements and not as composite whole.

If we call the two materials in the bar material 1 and material 2, then the load P will induce uniform axial stresses of σ_1 and σ_2 in 1 and 2 respectively. We will assume that the cross-sectional areas of the two materials are A_1 and A_2.

Can you write down an expression relating the axial load P to the internal stresses σ_1 and σ_2?

3

$$P = \sigma_1 A_1 + \sigma_2 A_2$$

This is an *equilibrium expression* which follows from the normal condition of equilibrium that the external force (P) must equal the internal force within the composite bar, which in this case must be the sum of the two internal forces ($\sigma_1 \times A_1$) and ($\sigma_2 \times A_2$).

As we are assuming that, as in all previous programmes, these two materials are linearly elastic and obey Hooke's Law, then the elastic strains experienced by both parts of the bar will be given by:

$$\varepsilon_1 = \frac{\sigma_1}{E_1}$$

and

$$\varepsilon_2 = \frac{\sigma_2}{E_2}$$

where E_1 and E_2 are the Young's Moduli for the two separate materials.

What do you think is the relationship between ε_1 and ε_2? State the reasons for your answer.

4

$$\boxed{\varepsilon_1 = \varepsilon_2}$$

We said originally that the two materials are firmly bonded together, It follows therefore that under load they must elongate by the same amount, and as they were originally of the same length their strains must be equal.

The above equation is an expression of *displacement compatibility*. The concept of compatibility of displacements is fundamental to all forms of structural analysis and the compatibility equation is simply an expression that relates the possible displacements or displacement components of a structure in a way that is geometrically feasible, taking into account the overall geometry of the structure.

Hence for this problem we have identified one equilibrium and one compatibility expression:

(a) *Equilibrium* $\qquad\qquad\qquad\qquad P = \sigma_1 A_1 + \sigma_2 A_2$

and

(b) *Compatibility* $\qquad\qquad\qquad\qquad \varepsilon_1 = \varepsilon_2$

where:

$$\varepsilon_1 = \frac{\sigma_1}{E_1} \quad \text{and} \quad \varepsilon_2 = \frac{\sigma_2}{E_2}$$

Re-arrange the above equations to eliminate the strains ε_1 and ε_2 and to produce two equations for σ_1 and σ_2 respectively.

5

$$\boxed{\sigma_1 = \frac{PE_1}{A_1E_1 + A_2E_2} \qquad \sigma_2 = \frac{PE_2}{A_1E_1 + A_2E_2}} \qquad \text{(11.1 \& 11.2)}$$

These two equations are a straightforward re-arrangement of the expressions given in the previous frame and now give the internal direct axial stresses for an applied axial load P. Make sure that you obtain these answers before proceeding.

If the section consisted of more than two materials then the procedure is exactly the same – we write down the expressions of equilibrium and compatibility and rearrange to give the stress in each one of the component elements of the bar.

In the next frame we will look at a typical example, and then you can try some problems for yourself.

A reinforced concrete column having the cross-section as shown in the figure supports a centrally placed compressive load of 2000 kN, The steel reinforcing bars which are shown in cross-section are each 25 mm diameter. If the Elastic Moduli of steel and concrete are 200 and 25 kN/mm^2 respectively, determine the stress in both steel and the concrete.

The area of the steel reinforcement, A_s, is given by:

$$A_s = 4(\pi d^2/4) = 4(\pi 25^2/4) = 1964 \text{ mm}^2$$

The area of the concrete, A_c, making allowance for that part of the gross section which is occupied by the steel, is given by:

$$A_c = 300^2 - A_s = 300^2 - 1964 = 88\,036 \text{ mm}^2$$

Hence, taking the steel as material 1 and the concrete as material 2 and using equation (11.1) in Frame 5:

$$\sigma_s = \frac{PE_s}{A_s E_s + A_c E_c}$$

$$= \frac{2000 \times 200 \times 10^3}{(1964 \times 200 + 88\,036 \times 25)}$$

$$= 154.22 \text{ N/mm}^2$$

Use equation (11.2) to calculate the stress in the concrete.

7

$$\boxed{\sigma_c = 19.28 \text{ N/mm}^2 \text{ (compression)}}$$

Check: $\sigma_c = \dfrac{PE_c}{A_s E_s + A_c E_c} = \dfrac{2000 \times 25 \times 10^3}{(1964 \times 200 + 88\,036 \times 25)} = 19.28 \text{ N/mm}^2$

Alternatively as: $\varepsilon_c = \varepsilon_s$

Then: $\dfrac{\sigma_c}{E_c} = \dfrac{\sigma_s}{E_s}$

or: $\sigma_c = \sigma_s \times (E_c/E_s) = 154.22 \times 25/200 = 19.28 \text{ N/mm}^2$

8

1. A reinforced concrete column has a rectangular cross-sectional shape of width 400 mm and depth 300 mm. If the column is to be designed to carry a compressive axial load of 2200 kN, determine what area of steel reinforcement is required in the column if the steel and concrete stresses are limited to 345 N/mm^2 and 15 N/mm^2 respectively. Both these stresses are assumed to be reached simultaneously.

Ans. (1212 mm²)

2. If the column which has been designed in question 1 is now subjected to an axial load of 1500 kN, determine the stress in both the steel and concrete if the Elastic Moduli of steel and concrete are 200 and 21 kN/mm^2 respectively.

Ans. (109.61, 11.51 N/mm²)

3. If the column in question 2 is 3 metres high, what is its change of length under the load of 1500 kN?

Ans. (1.64 mm)

9

FLEXURAL MEMBERS

The calculation of stress in composite axially loaded structural elements is fairly straightforward. The analysis of a composite section in bending is a little more complicated. The intention of such an analysis is to determine the bending stress in the component parts of a composite beam when the beam section is subjected to a bending moment. Consider the cross-section of a beam shown below which consists of two materials (material 1 and material 2) firmly bonded together. Under load, the beam will deform with both materials acting together to resist the applied bending moment.

What do you think is the relationship between the radius of curvature (R_1) of the left-hand side of the beam section and the radius of curvature (R_2) of the right-hand side?

> they are equal: $R_1 = R_2 = R$

As the two parts are firmly bonded together they must deflect together, and hence at any cross-section the radius of curvature, R, will be common to both parts of the beam. This is an expression of *displacement compatibility* which is valid only if the two materials are firmly bonded or fastened together.

Can you write down the Moment of Resistance of the left-hand side of the beam in terms of the Radius of Curvature, R_1, Young's Modulus, E_1, and the Second Moment of Area, I_1, about the neutral axis of bending?

$$M_1 = \frac{E_1 I_1}{R_1}$$

This should be quite familiar to you from your studies in Programme 6. It follows that the Moment of Resistance of the right-hand side of the beam (material 2) is given by

$$M_2 = \frac{E_2 I_2}{R_2}$$

If the externally applied bending moment at this section is M, what is the relationship between M, M_1 and M_2?

$$M = M_1 + M_2$$

This is our *equilibrium expression* which states that the external bending moment must equal the sum of the two internal moments of resistance. Hence our equilibrium and compatibility expressions are:

(a) *Equilibrium* $M = M_1 + M_2$ and   (b) *Compatibility* $R_1 = R_2 = R$

where

$$M_1 = \frac{E_1 I_1}{R_1} \quad and \quad M_2 = \frac{E_2 I_2}{R_2}$$

Re-arrange the above equations to give an equation for M but eliminating M_1 and M_2.

13

$$M = \frac{E_1 I_1 + E_2 I_2}{R}$$

This follows from a simple combination of the equilibrium expression and the compatibility expression in the previous frame. Inspection of the above equation shows that it is of the familiar form that you wrote down in Frame 11:

$$M = \frac{E_{\text{comp}} I_{\text{comp}}}{R}$$

where Young's Modulus of Elasticity, E_{comp}, and the second moment of area of the composite section, I_{comp}, are given by:

$$E_{\text{comp}} I_{\text{comp}} = E_1 I_1 + E_2 I_2$$

This shows that the equation of the simple theory of bending can be applied to the composite section provided that the beam stiffness, $E_{\text{comp}} I_{\text{comp}}$, is taken as the combination of terms for the two separate materials given in this last equation.

However a simplification can be made as follows to facilitate analysis:

$$E_{\text{comp}} I_{\text{comp}} = E_1 I_1 + E_2 I_2$$
$$= E_1 \left[I_1 + \frac{E_2}{E_1} I_2 \right]$$
$$= E_1 I_{1(\text{equiv})}$$

where the composite section can now be treated as an equivalent section made only of material 1 provided that an equivalent second moment of area, $I_{1(\text{equiv})}$, is taken, given by:

$$I_{1(\text{equiv})} = \left[I_1 + \frac{E_2}{E_1} I_2 \right]$$

where the ratio E_2/E_1 is known as the *modular ratio*.

If we do this we will have created a *transformed section* where effectively we have replaced the beam made of two materials by an equivalent beam made of one material *which would have the same response to loading as the original beam*.

For the rectangular beam that we are considering, the second moment of area of both sections of beam about the neutral axis, I_1 and I_2, are given by the equations:

$$I_1 = \frac{t_1 d^3}{12} \qquad \text{and} \qquad I_2 = \frac{t_2 d^3}{12}$$

So the expression for the equivalent second moment of area, $I_{1(\text{equiv})}$ can be written as:

$$I_{1(\text{equiv})} = \left[t_1 + \frac{E_2}{E_1} t_2 \right] \frac{d^3}{12} = t_{1(\text{equiv})} \frac{d^3}{12} \qquad (11.3)$$

In other words, the equivalent transformed section can be obtained by multiplying the width t_2 by the modular ratio (E_2/E_1) and treating this new beam as if it were a beam made of one material. This is illustrated in figures (a) and (b) below.

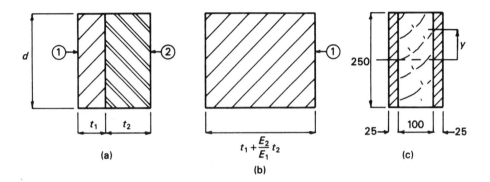

(a)

$t_1 + \dfrac{E_2}{E_1}t_2$

(b)

(c)

Figure (c) above shows a timber beam section reinforced with a steel plate on either side. Transform the beam into an equivalent timber section and calculate the second moment of area of this transformed section about the neutral axis. Take the elastic moduli of steel and timber as 200 and 10 kN/mm² respectively.

14

$$I_{1(\text{equiv})} = 1432 \times 10^6 \text{ mm}^4$$

The solution follows from equation (11.3) (taking the timber as material 1):

$$I_{1(\text{equiv})} = \left[t_1 + \frac{E_2}{E_1} t_2 \right] \frac{d^3}{12}$$

$$= \left[100 + \frac{200}{10} \times 50 \right] \frac{250^3}{12}$$

$$= 1100 \times \frac{250^3}{12}$$

$$= 1432 \times 10^6 \text{ mm}^4$$

Can you repeat the calculation but this time transform the composite beam into an equivalent steel section?

15

$$I_{1(\text{equiv})} = 71.6 \times 10^6 \text{ mm}^4$$

The solution follows from equation (11.3) (taking the steel as material 1):

$$I_{1(\text{equiv})} = \left[t_1 + \frac{E_2}{E_1} t_2 \right] \frac{d^3}{12}$$

$$= \left[50 + \frac{10}{200} \times 100 \right] \frac{250^3}{12}$$

$$= 55 \times \frac{250^3}{12}$$

$$= 71.6 \times 10^6 \text{ mm}^4$$

The last calculations show that we can transform the composite beam into either a timber beam 1100 mm wide or a steel beam 55 mm wide. Both beams and the original composite beam will respond identically to identical loading situations and the application of the equation of the simple theory of bending will produce identical answers. The choice of transformation, either to steel or timber, does not affect the end result.

You may be puzzled by the fact that the second moments of area of the two transformed sections are different, and hence it might appear unlikely that the two sections will respond identically to the same loading situation. However you should recall (see beginning of Frame 13) that the important factor relating moment to beam curvature is the *stiffness* of the beam which is the product of the elastic modulus, E_{comp}, and the second moment of area, I_{comp}. You should also recall that $E_{\text{comp}} I_{\text{comp}}$ is equal to $E_1 I_{1(\text{equiv})}$.

What is the stiffness of the composite timber and steel beam that we have just been considering?

16

$$14.32 \times 10^3 \text{ kN m}^2$$

This can be obtained from the transformed timber section:

$$E_{\text{timber}} I_{\text{timber}} = 10 \times (1432 \times 10^6) \times 10^{-6} = 14.32 \times 10^3 \text{ kN m}^2$$

Alternatively the beam stiffness can be obtained from the transformed steel section:

$$E_{steel}I_{steel} = 200 \times (71.6 \times 10^6) \times 10^{-6} = 14.32 \times 10^3 \text{ kN m}^2$$

which shows that both transformed sections are effectively the same.

So far we have shown how to transform a composite beam into an equivalent section. It remains now to determine how to use the transformed section to calculate the stresses in the materials of the original beam.

Look back to diagram (c) in Frame 13 which shows a level at a distance y from the neutral axis of the beam. Because of the application of a moment to the beam, the fibres of both materials will be stresses and strained.

What do you think is the relationship between the strain in material 1 (ε_1) and the strain in material 2 (ε_2) at this level?

17

$$\boxed{\varepsilon_1 = \varepsilon_2 = \varepsilon}$$

As the two materials are firmly bonded, the beam is behaving as a composite whole with a single axis of bending. We are assuming that the normal assumptions of the simple theory of bending are applicable and that the strains will vary linearly from zero at the neutral axis to a maximum at the outer fibres. As the two materials are bonded together, at any level in the beam the strains in the two materials must be the same.

Can you make use of this fact to find a relationship between the stress in material 2 (σ_2) and the stress in material 1 (σ_1) at the same level in the beam's cross-section?

18

$$\boxed{\sigma_2 = \frac{E_2}{E_1}\sigma_1}$$

The relationship follows from the fact that stress and strain are related by Young's Modulus of Elasticity, E.

Hence as the strains are equal: $\qquad \varepsilon_1 = \varepsilon_2$

or $\qquad\qquad\qquad\qquad\qquad \dfrac{\sigma_1}{E_1} = \dfrac{\sigma_2}{E_2}$

which re-arranges to give: $\qquad \sigma_2 = \dfrac{E_2}{E_1}\sigma_1 \qquad\qquad\qquad (11.4)$

On to the next frame:

19

The equations that we have established provide us with the method of analysing a composite beam. The steps in such as analysis can be summarised as:

(a) Transform the section into an equivalent section in one material (say material 1).
(b) Calculate the second moment of area of the transformed section (equation 11.3).
(c) Analyse the transformed section to determine the stress at any level using the equation of the simple theory of bending ($\sigma = yM/I$). This will give the stress in material 1.
(d) To obtain the stress in material 2 at a level where the stress in material 1 is known, use equation (11.4).

To illustrate this, let's calculate the maximum bending stresses in the steel reinforced timber beam that we investigated in Frame 14, if it is subjected to a sagging bending moment of 80 kN m. We will assume that the beam is transformed into an equivalent timber section. We will use subscripts s and t to denote the steel and timber respectively.

$$I_{t(equiv)} = 1432 \times 10^6 \text{ mm}^4 \quad \text{(see Frame 14)}$$

The maximum bending stress will occur at the outer fibres of the beam. Hence:

$$\sigma_{t(max)} = \frac{My_{(max)}}{I_{t(equiv)}}$$

$$= \frac{(80 \times 10^6) \times 125}{1432 \times 10^6}$$

$$= 6.98 \text{ N/mm}^2$$

Equation (11.4) is used to calculate the steel stress at the same outer level of the section:

$$\sigma_s = \frac{E_s}{E_t}\sigma_t$$

$$= \frac{200}{10}6.98$$

$$= 139.67 \text{ N/mm}^2$$

Now repeat this calculation but this time use the equivalent transformed steel section, the I value of which you calculated in Frame 15. The solution follows in the next frame.

$$\sigma_t = 6.98 \text{ N/mm}^2 \qquad \sigma_s = 139.67 \text{ N/mm}^2$$

If you didn't get these answers, as before look carefully again at your calculations and re-read this programme from Frame 9. You should have obtained the answer as below:

$$I_{s(equiv)} = 71.6 \times 10^6 \text{ mm}^4 \quad \text{(see Frame 15)}$$

The maximum bending stress will occur at the outer fibres of the beam. Hence:

$$\begin{aligned}
\sigma_{s(max)} &= \frac{My_{(max)}}{I_{s(equiv)}} \\
&= \frac{(80 \times 10^6) \times 125}{71.6 \times 10^6} \\
&= 139.67 \text{ N/mm}^2
\end{aligned}$$

Equation (11.4) is used to calculate the timber stress at the same outer level of the section:

$$\begin{aligned}
\sigma_t &= \frac{E_t}{E_s}\sigma_s \\
&= \frac{10}{200}\,139.67 \\
&= 6.98 \text{ N/mm}^2
\end{aligned}$$

From the previous two calculations you can see that it does not matter which material is chosen for the beam transformation, as the final answers for the stresses must be the same in both cases.

Th beam that we have been looking at had a steel plate on both sides of the timber section. If we now rotate the beam through 90° we will have a timber beam which is reinforced with two steel plates, one at the top and one at the bottom, effectively acting as two steel flanges. The approach to the analysis remains the same; we simply transform the beam into an equivalent one made of either timber or steel.

Can you sketch the shape of the section if it is transformed into an equivalent timber beam and show the appropriate dimensions on your sketch. (Remember that $E_s = 200 \text{ kN/mm}^2$ and $E_t = 10 \text{ kN/mm}^2$).

21

$$250 \times \frac{E_s}{E_t} = 250 \times \frac{200}{10} = 5000$$

The principle remains the same: *provided that the two materials are firmly bonded together* they must behave as a composite section and have a common neutral axis of bending, which in this case must, by symmetry, be at the mid height of the section. The principles established in Frame 13 are equally valid and the transformed section can be obtained by multiplying the *width* of the steel by the modular ratio (E_s/E_t).

Having identified the correct transformed section, the steps in the analysis to calculate the stresses in the materials remain the same.

If the transformed section shown above is subjected to a sagging moment of 80 kN m, what are the compressive stresses in the timber and steel at level (1) and the stress in the steel at the top of the beam at level (2)?

22

level (1) $\sigma_t = \quad 3.96$ N/mm^2	$\sigma_s = 79.20$ N/mm^2
level (2) $\sigma_s = 118.80$ N/mm^2	

Check your solution:

The second moment of area of the section about the neutral axis is given by:

For the web:

$$I_{xx} = bd^3/12 = 250 \times 100^3/12 \qquad\qquad = \quad 20.83 \times 10^6 \text{ mm}^4$$

For the flanges:

$$I_{xx} = 2(bd^3/12 + Ah^2) = 2(5000 \times 25^3/12 + 5000 \times 25 \times 62.5^2) = \quad 989.58 \times 10^6 \text{ mm}^4$$

Total I_{xx} for the whole section: $\qquad\qquad = 1010.41 \times 10^6 \text{ mm}^4$

At level (1):

$$\sigma_t = \frac{My}{I_{t(equiv)}}$$

$$= \frac{(80 \times 10^6) \times 50}{1010.41 \times 10^6}$$

$$= 3.96 \text{ N/mm}^2$$

The stress in the steel at the same level is given by:

$$\sigma_s = \frac{E_s}{E_t}\sigma_t = \frac{200}{10} \times 3.96 = 79.20 \ \text{N/mm}^2$$

Likewise at level (2):

$$\sigma_t = \frac{My}{I_{t(\text{equiv})}}$$
$$= \frac{(80 \times 10^6) \times 75}{1010.41 \times 10^6}$$
$$= 5.94 \ \text{N/mm}^2$$

The stress in the steel at the same level is given by:

$$\sigma_s = \frac{E_s}{E_t}\sigma_t = \frac{200}{10} \times 5.94 = 118.8 \ \text{N/mm}^2$$

You should realise that at level (2) the stress in the timber of 5.94 N/mm^2 is a fictitious stress which is only calculated as a step towards calculating the stress in the steel, which is the only material at the top of the section.

In Programme 6 you drew diagrams to show the distribution of strain and stress across the depth of a beam's section. These diagrams were linear, varying from zero stress and strain at the neutral axis to maximum stress and strain at the outer fibres.

Can you draw the strain and stress diagram for the beam section that we have just been considering, marking on your diagram the values of stresses and corresponding strains.

23

The strain diagram is identical to those you have seen before. The values of strain are obtained by dividing the calculated steel and timber stresses by the respective value of Elastic Modulus. The stress diagram is different from any that you have drawn before. It still shows a linear variation of stress but now there is a large discontinuity of stress at the interface of the timber and steel. The result could equally have been obtained by transforming the beam into an equivalent steel section; you should try this before you proceed.

24

As a further development of the same problem, consider the same beam but this time reinforced with just one 10 mm steel plate at the bottom face.

Sketch (a) the equivalent transformed timber beam and (b) the equivalent transformed steel beam. Show dimensions on your diagrams.

25

(a) Timber

(b) Steel

Not to scale

Again the procedure is the same. In diagram (a) the width of steel is multiplied by the modular ratio (E_s/E_t) and alternatively in diagram (b) the width of timber is multiplied by the modular ratio (E_t/E_s). The subsequent procedure for calculating beam stresses is the same as before with one additional step necessary in the calculations.

Can you say what this additional step is?

26

calculation of the position of the neutral axis

Determine the position of the neutral axis for figure (a), the equivalent timber section, and calculate the second moment of area of the section about the neutral axis.

27

$$\boxed{\text{23.33 mm from bottom of section } I_{xx} = 71.67 \times 10^6 \text{ mm}^4}$$

The calculations should be set out in the way in which you were shown in Programme 6 (Frame 24):

Part	Area (A) (mm^2)	y (mm)	Ay (mm^3)	I_{cc} $(bd^3/12)$ $(\text{mm}^4 \times 10^6)$	$h \, (= y - \bar{y})$ (mm)	Ah^2 $(\text{mm}^4 \times 10^6)$
1	50000	5	250000	0.42	18.33	16.80
2	25000	60	1500000	20.83	36.67	33.62
Totals	75000		1750000	21.25		50.42

$$\bar{y} = \frac{\Sigma Ay}{\Sigma A} = \frac{1\,750\,000}{75\,000}$$

$$= 23.33 \text{ mm}$$

$$\begin{array}{c} 50.42 \times 10^6 \\ 21.25 \times 10^6 \\ \hline \end{array}$$

$$I_{xx} = 71.67 \times 10^6 \text{ mm}^4$$

Did you remember to use the parallel axis theorem when working out the second moment of area? Once the position of the neutral axis is identified and the second moment of area of the transformed section is calculated, the subsequent calculations are identical to those that you have carried out in the previous frames. Move on to the next frame to try some problems.

28

PROBLEMS

1. A composite beam consists of a timber section 75 mm wide and 150 mm deep with a 6 mm steel plate securely fixed to the bottom face. Calculate the maximum stress in both timber and steel if the section is subjected to a sagging moment of 5 kN m; take $E_s = 200 \text{ kN/mm}^2$ and $E_t = 12 \text{ kN/mm}^2$.

Ans. (Timber: 10.95 N/mm² (compression), 4.51 N/mm² (tension). Steel: 85.58 N/mm² (tension))

2. If, in question 1, the steel stress must not exceed 120 N/mm² and the timber stress must not exceed 14 N/mm² (tension or compression), what is the maximum moment to which this beam can be subjected?

Ans. (6.39 kN m)

3. Repeat question 1 but transform the section to the alternative possible transformed section which you did not use in answering question 1.

29

REINFORCED CONCRETE SECTION

Reinforced concrete is a form of composite construction where steel reinforcing bars are bedded within a beam which is made out of concrete. When the wet concrete hardens it shrinks and bonds firmly to the steel, thus ensuring a composite behaviour under load. Concrete is a material which is weak in tension although strong in compression, and the reinforcing steel is normally located in the tension zone of the beam to contribute to the tensile strength of the section. Figure (a) shows the cross-section of a typical beam which is conventionally referred to as being *singly reinforced*, as it has reinforcement in the tension zone only.

Sometimes reinforcing steel is also provided in the compression zone, as shown in figure (b), and such a beam is referred to as *doubly reinforced*.

(a)   (b)

Reinforced concrete beams used to be commonly designed using similar assumptions to those that we have considered previously in this programme. The concept of elastic behaviour was fundamental to the design approach. This is no longer the case but nevertheless there are some aspects of concrete design (such as the calculation of deflections or crack widths in concrete beams or the design of water-retaining structures) which are based on concepts of elastic behaviour and the calculation of the transformed section of a reinforced concrete beam. For this reason the procedures for determining transformed sections in this unique form of composite construction are included in this programme.

Look at the beams in both diagrams below. Remembering that reinforcement is normally provided in the tension zone of a concrete beam, can you indicate the position where reinforcement should be provided in each case?

30

Consider the singly reinforced concrete section shown in figure (a) Frame 29. The symbols used on the diagram are those commonly used in practice for reinforced concrete design. The reinforcing bars shown have a total cross-sectional area A_s and are located at a depth d within the section. The section is assumed to behave elastically under load and the elastic moduli of both steel and concrete will be used in our calculations.

The commonly used assumption in the analysis of reinforced concrete is that concrete is so weak in tension that its contribution to the tensile strength of the beam can be ignored and it is assumed that all the tensile resistance of the beam will come from the steel reinforcement. *All the concrete below the neutral axis is therefore ignored in our calculations.*

Additionally, it is usual to transform the beam into an equivalent concrete section. The area of the steel reinforcement is replaced by an equivalent area of concrete which is the area of steel multiplied by the modular ratio ($E_{steel}/E_{concrete} \sim E_s/E_c$). This equivalent area is located at the same position as the steel reinforcement.

The transformed section is drawn by (a) neglecting all the concrete below the neutral axis and (b) replacing the reinforcement by an equivalent area of concrete using the modular ratio. Can you sketch the shape of the transformed section?

31

You may not have drawn this correctly but study the above diagram to see how it follows from the assumptions that we have made. To analyse such a transformed section the procedures are exactly the same as those that we have considered previously in this programme, with one small complication that we have not come across before.

If this were one of our previous problems the first step in your calculations would be to determine the location of the neutral axis, a procedure which by now should be quite familiar. However if you study the above diagram you will see that the area of concrete above the neutral axis depends on the position of the neutral axis and is hence a function of the distance x. You have not encountered this situation before and you will find that the equation you will write down to determine the position of the neutral axis will be more complicated than in previous problems.

To calculate the neutral axis depth x we will take moments of area in the usual way to locate the centroid of the transformed section but in this case the solution is obtained more easily if moments are taken about the neutral axis itself.

Can you write down the equation that would be used to calculate x; obtained by taking moments of area about the level of the neutral axis.

32

$$(bx)\frac{x}{2} - \frac{E_s A_s}{E_c}(d - x) = 0$$

As moments of area are taken about the neutral axis itself, it follows that the sum of the moments of the separate areas about this axis must be zero. (Remember, the neutral axis passes through the centroid of the area.)

That is:
$$(bx)\frac{x}{2} - \frac{E_s A_s}{E_c}(d - x) = 0$$

which can be rearranged to give the quadratic expression:

$$\tfrac{1}{2}bx^2 + \frac{E_s}{E_c}A_s x - \frac{E_s}{E_c}A_s d = 0 \qquad (11.5)$$

If we are analysing this beam all the terms in the equation would be known, other than x, and hence the quadratic can be solved to give the neutral axis depth, x.

Having located the neutral axis position, we now require the second moment of area of the transformed section about the neutral axis. The second moment of area of the concrete above the neutral axis is calculated using the parallel axis theorem and is given by:

$$I_{xx} = bx^3/12 + (bx)(0.5x)^2$$

$$= \frac{bx^3}{3}$$

The second moment of area of the transformed steel below the neutral axis, calculated using the parallel axis theorem but only including the Ah^2 term, as the other term is assumed to be small and negligible, is given by:

$$I_{xx} = \frac{E_s}{E_c}A_s(d - x)^2$$

Hence the total second moment of area of the transformed section is given by:

$$I_{xx} = \frac{bx^3}{3} + \frac{E_s}{E_c}A_s(d - x)^2 \qquad (11.6)$$

To analyse a singly reinforced concrete section we would therefore:

(a) Locate the position of the neutral axis using equation (11.5).
(b) Calculate the second moment of area of the section using equation (11.6).
(c) Calculate the stress in the concrete using the equation of the simple theory of bending ($\sigma = yM/I$).
(d) Calculate the stress in the steel using equation (11.4).

The final two steps, (c) and (d), are identical to our previous calculations. Now try a problem; the complete solution is in the next frame.

A rectangular singly reinforced concrete beam is 150 mm wide and has an overall depth of 300 mm. It is reinforced in the bottom face by two 20 mm diameter steel bars located at a depth of 270 mm. Calculate the maximum stress in both the concrete and steel if the beam is subjected to a moment of 35 kN m. Take the ratio of the elastic moduli of the two materials (E_s/E_c) as 15.

33

$$\boxed{\sigma_c = 15.67 \text{ N/mm}^2 \qquad \sigma_s = 246.45 \text{ N/mm}^2}$$

Refer to the figure showing the transformed section in Frame 31 when checking your solution:

The area of reinforcement, A_s is given by:

$$A_s = 2(\pi \times \text{diameter}^2/4) = 2 \times \pi \times 20^2/4 = 628 \text{ mm}^2$$

From equation (11.5) the neutral axis depth is given by:

$$\frac{bx^2}{2} + \frac{E_s}{E_c} A_s x - \frac{E_s}{E_c} A_s d = 0$$

$$\frac{150}{2} x^2 + 15 \times 628 \times x - 15 \times 628 \times 270 = 0$$

which solves as a quadratic to give positive root of $x = 131.8$ mm.

The second moment of area of the transformed section is given by the equation (11.6):

$$\begin{aligned}
I_{xx} &= \frac{bx^3}{3} + \frac{E_s}{E_c} A_s(d - x^2) \\
&= \frac{(150 \times 131.8^3)}{3} + (15 \times 628 \times [270 - 131.8]^2) \\
&= 294.4 \times 10^6 \text{ mm}^4
\end{aligned}$$

Hence to obtain the maximum compressive stress in the concrete at the top of the section we use the equation of the simple theory of bending as in all the previous flexural problems in this programme:

$$\begin{aligned}
\sigma_c &= \frac{My}{I_{c(\text{equiv})}} \\
&= \frac{(35 \times 10^6) \times 131.8}{294.4 \times 10^6} \\
&= 15.67 \text{ N/mm}^2
\end{aligned}$$

The stress in the concrete at the level of the reinforcement is obtained similarly:

$$\begin{aligned}
\sigma_c &= \frac{My}{I_{c(\text{equiv})}} \\
&= \frac{(35 \times 10^6) \times (270 - 131.8)}{294.4 \times 10^6} \\
&= 16.43 \text{ N/mm}^2
\end{aligned}$$

If you didn't get the correct answer before checking the solution, can you now say how to obtain the stress in the reinforcement?

> Multiply the stress in the concrete in the transformed section
> at the level of the reinforcement by the modular ratio.

This follows from equation (11.4) which, you should recall, is merely an expression of strain compatibility which says that the strains in the concrete and the steel at the same level must be the same. Hence:

$$\sigma_s = \frac{E_s}{E_c}\sigma_c$$
$$= 15 \times 16.43$$
$$= 246.45 \text{ N/mm}^2$$

We have calculated the maximum compressive stress in the concrete and the tensile stress in the steel using the transformed section. However you should recall from your studies in Programme 6 that for equilibrium of the cross-section, the total compressive force above the neutral axis (C) must equal the total tensile force below the neutral axis (T). (Look back at Programme 6, Frame 9 if you don't recall this.)

Can you calculate the total force in the concrete and steel in the previous problem and show that they are equal as a check on our previous calculations?

35

$$C = T = 154.8 \text{ kN}$$

$T = $ Area of steel $\times$ stress in steel $= 628 \times 246.45 \times 10^{-3} = 154.8$ kN
$C = $ Area of concrete $\times$ *average* stress in concrete
$= (150 \times 131.8) \times (0.5 \times 15.67) \times 10^{-3}$ $= 154.8$ kN

On to the next frame:

36

THERMAL STRESSES

So far in this book we have examined a wide range of problems involving the calculation of stresses in structural elements where the stresses are attributable to various forms of loading (axial, shear, torsion, bending) applied to the element. However there is a further situation where stresses can arise not from physical loading, but as a result of temperature changes in, or temperature variation across, a structure or an element of a structure.

A typical real example of such a problem is a bridge deck where, because of solar effects, the top deck may reach a very high temperature relative to the shaded underside. The differential expansion of the various layers of the deck can result in extremely high stresses being set up, which must be considered in practical design.

To complete this programme we are going to look at the elementary principle of such problems and learn how to analyse simple thermal stress problems.

What is the property of a material which would enable you to calculate its change of length when it is subjected to a temperature rise?

37

the coefficient of expansion α

The coefficient of expansion of a material is 'the change of length per unit length per degree rise in temperature', and is a term that should be familiar to you from your previous studies.

Now look at the bar shown in the figure which is sitting on a smooth surface and restrained from moving at one end. The bar is subjected to a temperature rise of $T°C$ and is allowed to expand freely. It has a coefficient of expansion α.

Write down an expression for the change in length of the bar, the strain in the bar and, if the Young's Modulus of Elasticity of the bar is E, state what the stress is in the bar.

38

> Change in length $\delta = \alpha LT$
> Strain $= \alpha T$
> Stress $=$ zero

The expression for the change in length of the bar follows from the definition of the coefficient of expansion given in Frame 37. The expression for strain follows from the definition of strain as change in length divided by original length ($\alpha LT/L = \alpha T$).

You should have obtained the first two answers without difficulty. Did you write down an expression for the stress in the bar based on the relationship that stress divided by strain is Young's Modulus, E? In this case that is wrong. If there are internal stresses there has to be an external force system to maintain equilibrium. In this case there is no such system and hence the internal stress must be zero.

It is important to recognise that if the bar is allowed to expand freely without any external restraint then it will strain without any corresponding stresses taking place.

Now let's look at the expanded bar as shown in figure (a) and let's apply a force P to the bar which is large enough to return the bar to its original length (figure (b)).

(a)   (b)

Write down an expression for P in terms of the cross-sectional area of the bar, A and Young's Modulus, E. Hence write down an expression for the stress (σ) in the bar.

39

> $P = AE\alpha T \qquad \sigma = E\alpha T$

From Hooke's Law: $\qquad \dfrac{\sigma}{\varepsilon} = E \quad$ where $\quad \sigma = \dfrac{P}{A}$

Hence: $\qquad\qquad P = A\sigma = AE\varepsilon = AE\alpha T$

and: $\qquad\qquad\quad \sigma = P/A = AE\alpha T/A = E\alpha T$

If a bar is subjected to a temperature change, what additional condition do you think must exist for stress to be set up in the bar?

40

What the equations in Frame 39 show us is that if the bar is restrained against expansion by some sort of restraining system capable of developing a restraining force (P), then stress ($E\alpha T$) will develop in the bar. If the bar is completely unrestrained, it will strain without being stressed. This is an important observation.

The logic of the steps that we have just followed can be applied in reverse order to the analysis of a restrained bar system subjected to a temperature rise. Consider the restrained bar shown in figure (a) where the supports are sufficiently rigid to prevent movement. Figure (b) shows the bar released at one end and allowed to expand freely. The amount of expansion can be calculated (αLT). Figure (c) shows the bar subjected to an applied force $P (= AE\alpha T)$ which is large enough to return the bar to its original length and induces a stress ($E\alpha T$) in the bar. The final figure (c) represents the completely analysed structure where the deformation are known (zero) and the internal stress and external forced have been calculated.

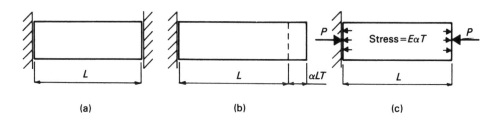

(a)   (b)   (c)

A steel beam, 4 metres long and forming part of a building structure, is firmly built into solid walls at either end. If the temperature rise in the beam on a summer day is 15°C, what stress is induced in the beam? Take $\alpha = 12 \times 10^{-6}$ per degree centigrade and $E = 200\ kN/mm^2$.

41

36 N/mm^2

The solution follows from the equation that we have just developed and ensuring, as in all problems, that the units are correct:

$$\sigma = E\alpha T = (200 \times 10^3) \times (12 \times 10^{-6}) \times 15 = 36\ \text{N/mm}^2$$

This is a compressive stress due to a temperature rise. If the stress was induced by a temperature fall then it would be a tensile stress. It is interesting to note that this is independent of the cross-sectional area of the beam and hence this stress will be induced in any beam in this situation, irrespective of its size.

This programme is concerned with composite sections, so let's extend our consideration of thermal stresses to composite problems. Consider the solid bar (material 1) contained within a hollow tube (material 2) shown below. Both materials have different cross-sectional areas and Young's Moduli, and the ends of the bar and tube are rigidly connected together.

If the bar and the tube are subjected to a temperature rise T, will there be stress induced in the bar if the coefficients of expansion (α_1 and α_2) are (a) the same, (b) different?

42

(a) no   (b) yes

If the materials have identical coefficients of expansion they will expand the same amount, and because there is no external restraint no stresses will be set up.

However if they have different coefficients of expansion they will attempt to expand by different amounts, although because they are fastened together they must expand together. Hence an *internal restraint* system will be set up as the material that is trying to expand the most pulls the other with a force (P_2) in the direction of expansion. Equally, the material trying to expand the least will pull back on the other with a force (P_1) against the direction of expansion. This is shown in the figure where it is assumed that the internal bar has the larger coefficient of expansion.

What is the relationship between P_1 and P_2?

43

$$P_1 = P_2 = P$$

As P_1 and P_2 are the only forces acting on the composite structure, they must represent a set of forces in equilibrium. The above equation therefore represents our *equilibrium expression*.

Now to analyse this structure to determine the internal stresses arising from the temperature rise, let's adopt the same approach that we used in Frame 40 to analyse the restrained bar made of one material. In other words, we will release one end of the system and allow both components to expand freely without restraint. This is shown in figure (a) where the free expansions δ_1 and δ_2 are given by:

$$\delta_1 = \alpha_1 LT \qquad \text{and} \qquad \delta_2 = \alpha_2 LT$$

The final position of the ends of both bar and tube must lie at the same point somewhere in-between the two extremities shown in figure (a). To bring them to the same position we will apply a compressive force P to the bar and an equal opposite tensile force to the tube, as shown in figure (b).

Using the notation shown in figure (b), can you write down an expression of displacement compatibility that relates the overall displacement of the bar to that of the tube?

44

$$\delta_2 + \delta_3 = \delta_1 - \delta_4$$

This equation follows from the geometry of figure (b). The displacement δ_3 and δ_4 are attributable to the application of the force P and can be expressed in terms of P by the equations:

$$\delta_3 = \frac{PL}{A_2 E_2} \quad \text{and} \quad \delta_4 = \frac{PL}{A_1 E_1}$$

Hence our equilibrium and compatibility equations for this problem are:

(a) *Equilibrium* $\qquad P_1 = P_2 = P$

and

(b) *Compatibility* $\qquad \delta_2 + \delta_3 = \delta_1 - \delta_4$

where:

$$\delta_1 = \alpha_1 LT \qquad \delta_2 = \alpha_2 LT \qquad \delta_3 = \frac{PL}{A_2 E_2} \qquad \delta_4 = \frac{PL}{A_1 E_1}$$

Re-arrange the above equations to eliminate the displacements δ_1 to δ_4 and produce an equation for the internal force P.

45

$$P = \frac{(\alpha_1 - \alpha_2) T A_1 E_1 A_2 E_2}{(A_1 E_1 + A_2 E_2)}$$

From the compatibility equation:

$$\delta_2 + \delta_3 = \delta_1 - \delta_4$$

Substituting:

$$\alpha_2 LT + \frac{PL}{A_2 E_2} = \alpha_1 LT - \frac{PL}{A_1 E_1}$$

which re-arranges to give:

$$P = \frac{(\alpha_1 - \alpha_2) T A_1 E_1 A_2 E_2}{(A_1 E_1 + A_2 E_2)} \qquad\qquad (11.7)$$

This equation gives the internal force between the bar and the tube. How would you calculate the stresses in the bar and the tube?

46

$$\sigma_1 = \frac{P}{A_1}, \ \sigma_2 = \frac{P}{A_2} \ \text{where } P \text{ comes from the equation (11.7)}$$

This should be fairly obvious (stress = force/area).

How would you calculate the final elongation of the structure? Write down the equation that would enable you to determine this.

343

47

$$\text{elongation} = \delta_2 + \delta_3 = \alpha_2 LT + \frac{PL}{A_2 E_2} \quad \text{where } P \text{ comes from equation (11.7)}$$

In Frame 44 and figure (b) in Frame 43 we identified the elongation as $\delta_2 + \delta_3$. Substituting the appropriate terms from Frame 44 gives the above equation. A corresponding equation in terms of material 1 would be obtained by taking the elongation as $\delta_1 - \delta_4$ and making the appropriate substitution.

The equations that we have identified above enables us to analyse any composite structure where *axial stress* is induced by temperature changes. In other types of structures, such as composite beams, temperature variations will induce flexural stresses and flexural deformations. The analysis of this situation is beyond the scope of this book although the principles involved are similar. Now try a problem, the solution to which is in the next frame:

A steel rod 20 mm in diameter is fitted inside an aluminum tube with internal and external diameters of 22 mm and 30 mm respectively. The ends of the rod are fastened to the tube by screwed connections. Calculate the stress induced in the rod if the temperature of both rod and tube is raised by 80°C. Take $E_{steel} = 200 \ kN/mm^2$, $E_{alum} = 70 \ kN/mm^2$, $\alpha_{steel} = 10 \times 10^{-6}/°C$ and $\alpha_{alum} = 23 \times 10^{-6}/°C$.

48

$$\sigma_s = 55.49 \text{N/mm}^2$$

Using subscripts 's' and 'a' to indicate the steel rod and aluminium tube respectively:

$$\text{Area } A_s = \pi \times 20^2/4 = 314.3 \text{ mm}^2 \colon \text{Area } A_a = \pi(30^2 - 22^2)/4 = 326.9 \text{ mm}^2$$

From equation (11.7):
$$P = \frac{(\alpha_a - \alpha_s)TA_aE_aA_sE_s}{(A_aE_a + A_sE_s)}$$

$$= \frac{(23 - 10)10^{-6} \times 80 \times 326.9 \times 70 \times 314.3 \times 200}{[(326.9 \times 70) + (314.3 \times 200)]}$$

$$= 17.44 \text{ kN}$$

Hence:
$$\sigma_s = \frac{P}{A_s} = \frac{17.44 \times 10^3}{314.3} = 55.49 \text{ N/mm}^2$$

A little bit of thought will tell you that this must be a tensile stress, as the steel has the lowest coefficient of expansion (look back at the diagram in Frame 43).

TO REMEMBER

The equations that we have derived in this programme are probably too complicated to commit to memory, but are quoted below for easy reference. What is important is that you can derive these equations from fundamental principles:

Axially Loaded Members:
$$\sigma_1 = \frac{PE_1}{A_1E_1 + A_2E_2} \qquad \sigma_2 = \frac{PE_2}{A_1E_1 + A_2E_2}$$

Flexural Member:
$$I_{1(\text{equiv})} = \left[t_1 + \frac{E_2}{E_1}t_2\right]\frac{d^3}{12} \qquad \sigma_2 = \frac{E_2}{E_1}\sigma_1$$

Reinforced Concrete:

neutral axis depth x comes from the equation:
$$\frac{bx^2}{2} + \frac{E_s}{E_c}A_sx - \frac{E_s}{E_c}A_sd = 0$$

$$I_{c(\text{equiv})} = I_{xx} = \frac{bx^3}{3} + \frac{E_s}{E_c}A_s(d-x)^2$$

Thermal Stresses:
$$P = \frac{(\alpha_1 - \alpha_2)TA_1E_1A_2E_2}{(A_1E_1 + A_2E_2)}$$

FURTHER PROBLEMS

1. A cable is made from 20 steel wires wound together with 20 aluminium alloy wires, each wire having a cross-sectional area of 12 mm². If the maximum stress in the steel and aluminium is limited to 130 and 85 N/mm² respectively, what is the maximum tensile load that can be carried by the cable? Take Young's Modulus for steel and aluminium as 200 and 70 kN/mm².

Ans. (42.12 kN)

2. A very rigid beam, 1 metre long, is supported in a horizontal position by two short columns made of different materials, one either end of the beam. A load of 100 kN is applied to the beam at a point 300 mm from the left-hand end and when supporting this load the beam remains horizontal. If the cross-sectional area of the left-hand column is 500 mm², calculate (a) the cross-sectional area of the right-hand column and (b) the stresses in both columns. Young's Moduli for the left- and right-hand columns are 200 and 100 kN/mm² respectively.

Ans. (428.57 mm², 140 N/mm² (left), 70 N/mm² (right))

3. A reinforced concrete column, 3.5 metres high, has a square section of side length 300 mm and supports an axial load of 1200 kN. If the shortening of the column after the application of the load has been estimated as 2 mm, what area of reinforcing steel is there in the column and what is the stress in both concrete and steel? Take the modular ratio (E_s/E_c) as 10 and $E_c = 20$ kN/mm^2.

Ans. (1666.67 mm^2, 11.43 N/mm^2 and 114.30 N/mm^2)

4. Figure Q4 shows a box section beam formed from two timbers and vertical steel plates. If the maximum stress in the steel is limited to 140 N/mm^2 and that in the timber to 10 N/mm^2, what is the maximum moment that can be resisted by this beam? Take $E_{steel} = 200$ kN/mm^2 and $E_{timber} = 10$ kN/mm^2.

Ans. (23.64 kN m)

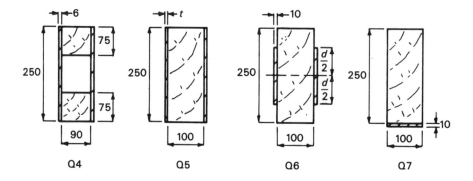

5. A timber beam 100 mm wide by 250 mm deep is supported over a span of 4 metres. It is to be strengthened to carry a load (inclusive of self-weight) of 10 kN/m by adding two vertical steel plates either side of the beam, as shown in figure Q5. What is the minimum thickness of plate required if $E_{steel} = 200$ kN/mm^2 and $E_{timber} = 10$ kN/mm^2, and if the allowable stresses are 125 N/mm^2 and 7.5 N/mm^2 for the steel and timber?

Ans. (5.18 mm)

6. During the strengthening of the beam in question 5 it was found that only 10 mm steel plate was available, and to economise it was decided to reinforce the sides as shown in figure Q6 where the beam is not reinforced to its full depth. Calculate the minimum height of steel d which needs to be used.

Ans. (182.53 mm)

7. As an alternative to the design in question 6, the beam could be reinforced by fixing the 10 mm plate to the underside of the beam (figure Q7). Is this alternative proposal adequate to meet the design specification? Calculate the load that the beam could carry if it is strengthened in this way.

Ans. (No, 6.53 kN/m)

8. A composite floor construction consists of a concrete slab supported on a steel beam as shown in figure Q8. The steel beam is propped during casting of the concrete so that when the concrete has hardened the props are removed and all loading is carried by composite beam action. If the beam spans 10 metres and carries a uniformly distributed load of 5 kN/m (inclusive of self-weight), determine the stress in the steel at level 1, the stress in the steel and concrete at level 2, and the stress in the concrete at the top of the slab at level 3. Take the modular ratio as 10 and for the steel beam the sectional area $= 5500$ mm^2 and $I_{xx} = 65.46 \times 10^6$ mm^4. (*Hint:* transform the section to an equivalent steel section)
Ans. (81.50, 4.59, 0.46, 3.77 N/mm^2)

Q8

Q12

9. A singly reinforced concrete beam is 250 mm wide and has three 32 mm diameter reinforcing bars located at a depth of 500 mm. It is simply supported over a span of 10 metres and carries a uniformly distributed load of 8 kN/m inclusive of self-weight. Calculate the maximum stress in both the steel and the concrete. Take the modular ratio as 15.
Ans. (100.42, 7.39 N/mm^2)

10. A steel rod 20 mm in diameter is placed inside a copper tube of external and internal diameters 30 and 25 mm respectively. The ends of the rod and tube are connected firmly together. When the temperature of the compound bar is raised by 250°C, the length of the bar is found to increase by 700×10^{-6} metres. Calculate the length of the original bar and the stress induced in both steel and copper. $E_{steel} = 200$ kN/mm^2, $E_{copper} = 120$ kN/mm^2, $\alpha_{steel} = 10 \times 10^{-6}$/°C and $\alpha_{copper} = 17 \times 10^{-6}$/°C.
Ans. (232.5 mm, 102.21, 148.67 N/mm^2; steel in tension, copper in compression)

11. If the bar in question 10 is maintained at the elevated temperature, determine what axial force would have to be applied to the compound bar to reduce the stress in the steel rod to zero.
Ans. (45.37 kN)

12. Determine the stresses in the compound bar shown in figure Q12 if the temperature of the bar is raised by 50°C. The left-hand part of the bar is made of solid steel circular section of diameter 20 mm and the right-hand part of solid copper of diameter 30 mm. Take the material properties from question 10.
Ans. (155.13 (steel), 68.95 (copper) N/mm^2, both in compression)

Programme 12

BEAM
DEFLECTIONS AND
ROTATIONS

On completion of this programme you will be able to:

- Explain why it is important to know the magnitude of beam deflections
- Define 'flexural stiffness'
- Calculate the magnitude of deflection and rotation at a point along a simply supported beam subject to different combinations of loading
- Calculate the magnitude of deflection and rotation at a point along a cantilevered beam subject to different combinations of loading
- State the formula for mid-span deflection of a simply supported beam subject to a centrally applied point load
- State the formula for mid-span deflection of a simply supported beam subject to a continuous uniformly distributed load

1

Most of the previous seven programmes have been concerned with the analysis of different forms of stress and the application of the theories which we have developed to the analysis and design of different forms of structure and structural components.

In the design of structures the primary requirement is to ensure that the structure or structural component can adequately resist the loading to which it is being subjected. Usually this means that under normal loading conditions the stresses set up within the structure must be less than some permissible value, that value being the failure stress of the material divided by some appropriate factor of safety. This aspect of design is concerned with designing for *strength*.

However, there are other aspects of design that are important, and in the case of the design of beams another consideration is the value of the vertical deflections that will occur when such a beam is loaded. This programme is intended as an introduction to the analytical techniques used for calculating deflections in beams and also for calculating the rotations at critical locations along the length of a beam.

Can you think why it is necessary to calculate the vertical deflections that take place when a beam is loaded?

2

> (a) To check that visual appearance is acceptable.
> (b) To ensure that no damage is likely to be caused to the supported structure.

A beam, in a building or forming part a bridge structure, is an important and often visible structural element. Excessive sagging of a beam can be visually distressing and any supported structure, such as a masonry wall, could be badly damaged if the beam on which it is supported deflects excessively.

To control deflections, the various British Standards and Codes of Practice for the design of beams in timber, steel or reinforced concrete usually specify a limit on the ratio of the maximum deflection to the overall span of the beam—typical limits are of the order of 1:360. Generally, this means that the designer must be able to assess deflections and compare them with allowable values. This may mean using standard formulae for standard loading cases or using analytical techniques if the loading case is non-standard.

A timber beam spans 4 metres. If the deflection is limited to 0.003 × the span length, what is the maximum allowable deflection?

3

$$\boxed{12 \text{ mm}}$$

That was fairly straightforward ($4000 \times 0.003 = 12$ mm). But where in practice does the maximum deflection that we are calculating take place? If you are to calculate the deflections to compare with allowable values, it is necessary to identify the critical location(s) where the maximum deflection may occur. The ability to sketch the deflected shape of the structure (see Programme 4) is therefore important as, unless you can envisage the deformations that are taking place in the structure, you may not correctly identify the critical position for calculating deflections.

Consider the beams shown below. Sketch the deflected shapes and mark the critical position where the maximum deflection will occur.

4

In the case of the simply supported beam, the maximum deflection must occur at mid span. For the simple cantilever, the maximum deflection will occur at the end of the cantilever. However, in the third diagram there are two locations where the deflection may be a maximum: at the end of the cantilever and at some position along the span AB. The maximum deflection within the span AB will not necessarily be at mid span and its position must be identified. Both deflections may have to be calculated. Now on to the next frame.

5

In Programme 6 we examined the structural efficiency of several shapes of beam cross-section and we came to some conclusions about the relative efficiency of the sections when considering their ability to resist the same value of bending moment. Let's consider the same cross-sections, which you may remember have identical cross-sectional areas.

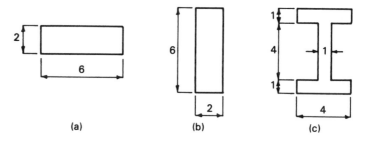

(a)   (b)   (c)

Three beams span identical distances, carry identical loads and are made of the same material. If they are made from three different sections as shown in the figures above, which beam would deflect the least?

6

the beam made from section (c)

You should have reasoned that section (b) is much *stiffer* then section (a) and will therefore deflect less. If in doubt about this, take your ruler and rotate both ends to cause it to bend. Now turn the ruler through 90° and repeat the exercise. Which direction offers greatest resistance to deformation?

Section (c) will deflect less than section (b). This may not be quite so obvious. If you guessed that this is the case, did you reason that section (c) deflects less because it has the greater second moment of area about the axis of bending (the neutral axis) and is therefore *stiffer*? The geometric property which we have identified as the second moment of area of a cross-section not only affects the strength of a beam but also the deformation under load.

If we now have two identical beams each having a cross-section as shown in figure (a) above and carrying the same load over identical spans but one being made of timber and the other of steel, which beam would deflect least?

7

> the steel beam

You probably guessed this, but do you know why? As the beams are identical in every respect except their material of construction, the stresses within the beams must be identical. However the *strains* will be different as the two materials will have different values of Young's Modulus of Elasticity. The amounts of straining in both tension and compression will be related to the beam deformations, and as steel has a modulus of elasticity approximately 10 to 20 times that of timber, the steel beam will deflect proportionately less.

What we have deduced is that the deflection of a beam is related to (a) the second moment of area of the beam section and (b) the modulus of elasticity of the material of which the beam is made. We will prove that these are important relevant properties later in this programme. The product of the modulus of elasticity (E) and the second moment of area (I) is known as the *flexural stiffness* (EI) of the beam. Remember this expression, as you will encounter it many times in your further studies of structural analysis. Note also that, in using a modulus of elasticity, we are again assuming *linear elastic behaviour* as we have done in all previous programmes.

Can you think of any factors other than the flexural stiffness which will affect the deflection of a beam?

8

> (a) the magnitude and distribution of the load
> (b) the length of the span

(a) should be obvious: double the load should give double the deflection, and so on (the principle of superposition again!).

The fact that the length of the span will affect the deflection should again be intuitively obvious: if there are two identical beams made from identical materials and sections and carrying the same total loading over different spans, the longer span must deflect the most. But in what way will the span length affect the deflections: does twice the span mean twice the deflections? That might seem a reasonable supposition but is in fact incorrect. We will see why and how later.

Let's now get down to the theory. There are in fact quite a number of methods for calculating beam deflections. Some are appropriate to manual calculation, some are more applicable to the use of modern computer methods. We will develop one manual method to introduce the basic concepts.

9

Consider the short length of beam shown in the figure which is subjected to loading, causing sagging bending moment M which results in downward deflection (v) of the beam. The bending moment along this short length is assumed to be constant even though the bending moments in the beam, of which this is part, may vary. This short length of beam is assumed to be bent into a circular arc of radius R and is defined in terms of an X/Y axis system where the y-axis is taken as positive downwards. Thus the direction of the loading that causes deflection and the vertical deflection (v) are both positive according to our chosen axis system.

From the work that you have done previously on the bending of beams, can you write down a relationship between the beam curvature $(1/R)$ and the moment M?

10

$$\frac{1}{R} = \frac{M}{EI}$$

In Programme 6 we developed the general theory of bending equalities:

$$\frac{\sigma}{y} = \frac{M}{I} = \frac{E}{R}$$

The equation at the top of this frame is merely a re-arrangement of the second two terms in this expression. Note that the term EI is, as previously stated, the flexural stiffness of the beam.

However there is a mathematical expression that defines the curvature $(1/R)$ at any point on a curve in terms of differential functions of v and x. The proof of this expression can be found in many mathematics textbooks and should be familiar to you from your studies of mathematics.

The expression is given as:

$$\frac{1}{R} = \frac{\pm \dfrac{d^2 v}{dx^2}}{[1 + (dv/dx)^2]^{1.5}}$$

In most methods of structural analysis we are concerned with *small* deflections on the assumption that the deflections of a structure are small when compared with the overall structural geometry. This implies that the slope at any point on the beam (dv/dx) is small, and is even smaller when squared and hence can be neglected in the above equation, which can then be written as:

$$\frac{1}{R} = \pm \frac{d^2 v}{dx^2}$$

Our moment–curvature relationship can then be written as:

$$\frac{1}{R} = \frac{M}{EI} = \pm \frac{d^2 v}{dx^2}$$

or
$$EI\frac{d^2 v}{dx^2} = \pm M \qquad (12.1)$$

Equation (12.1) is the fundamental expression that describes the vertical deflection (v) at any point distance x along the beam in terms of the applied bending moment (M) and the flexural stiffness (EI) of the beam. The intention of the analysis is to evaluate the vertical deflection, v, assuming that all the other parameters in the equation are known.

Can you write equation (12.1) in a way in which it can be used to evaluate the vertical deflection (v) at any distance (x) along the beam?

11

$$\boxed{v = \pm \iint \frac{M}{EI}\, dx\, dx} \qquad (12.2)$$

Mathematically, equation (12.1) can be solved in terms of v by re-arranging the terms in the expression and carrying out a double integration of the resulting expression. You are probably wondering whether the sign in front of the integral should be positive or negative. This depends on the choice of sign convention, which we will clarify a little later. However, before we consider sign conventions, move to the next frame to try a simple problem to illustrate the application of equation (12.2).

12

The beam in the figure is subjected to end couples of moment M causing sagging, as shown. For convenience, the origin of the axis system is taken through the left-hand end of the span. The problem is to find the maximum deflection which by symmetry must occur at mid span.

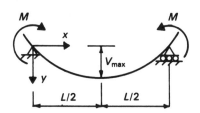

For the given loading configuration, the bending moment along the beam is constant and at any point distance x from the origin must equal M. If you are not sure about this, sketch the bending moment diagram and refer to Programme 4 if in doubt.

Hence from equation (12.1) the general equation to describe the deformation of the beam is:

$$EI\frac{d^2v}{dx^2} = M$$

where M is constant. For the time being, we will not worry about the sign of the moment M.

Can you now carry out a double integration of this expression?

13

$$\boxed{v = (0.5Mx^2 + Ax + B)/EI}$$

This standard integral equation gives the deflection v at any distance x along the span. However it contains two constants of integration (A and B) which must be evaluated before the equation can be used. To determine these two constants it is usual to apply the equation at *two known* locations (two values of x) where the deformations (v) are known, and hence A and B can be solved in terms of known deformations and distances. These are often referred to as the *boundary conditions*.

Can you identify appropriate boundary conditions for this problem?

14

$$\boxed{\begin{array}{l} x = 0: \ v = 0 \\ x = L: \ v = 0 \end{array}}$$

Both ends of the beam are supports where the deflections must be zero. The origin is located at the left-hand support, hence at $x = 0$, $v = 0$; and at the right-hand support, $x = L$ and $v = 0$. Substituting these boundary conditions into the equation in Frame 13, both A and B can be derived:

when $x = 0$: $v = 0 = (0.5M0^2 + A0 + B)/EI$:   Hence $B = 0$
when $x = L$: $v = 0 = (0.5ML^2 + AL + 0)/EI$: Hence $A = -0.5ML$

The complete solution to describe the deflection at any point along the span is therefore given by:

$$v = 0.5(Mx^2 - MLx)/EI$$

Now use this equation to calculate the deflection at mid span.

15

$$\boxed{v = -\frac{ML^2}{8EI}}$$

The equation gives the deflection at any point, distance x, along the span. To obtain the mid span deflection, simply substitute $x = L/2$ into the equation:

$$v = 0.5(Mx^2 - MLx)/EI$$
$$= 0.5(M \times [L/2]^2 - ML \times [L/2])/EI$$
$$= -\frac{ML^2}{8EI}$$

To obtain the deflection at any other location along the beam, simply substitute the appropriate value of x into the equation. The slope or *rotation* at any point can also be obtained from the first differential (dv/dx) of the equation.

The above equation describes the deflection of a simply supported beam for a standard loading case of two equal but opposite end moments. It is an equation that is useful to remember as it is used in many forms of structural analysis.

16

In solving the previous problem we did not consider whether the sign of the moment M should be positive or negative, and as a result the answer for the mid-span deflection came out as a negative quantity. However it is obvious that when the beam is subjected to end sagging moments it must deflect downwards in the direction of the positive y-axis. The mid-span deflections should therefore have been a positive quantity to be consistent with our chosen axis system, and to achieve this the moment should have been preceded by a negative sign in our original integral expression.

Let's reconsider the diagram that we originally drew in Frame 9 which defined the positive direction of beam deformation:

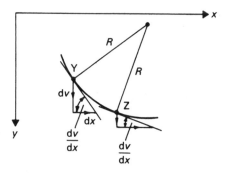

The curvature at any point distance x along the span can be written as:

$$\frac{1}{R} = \frac{\mathrm{d}^2 v}{\mathrm{d}x^2} = \frac{\mathrm{d}}{\mathrm{d}x}\left(\frac{\mathrm{d}v}{\mathrm{d}x}\right)$$

But as $(\mathrm{d}v/\mathrm{d}x)$ is the slope of the beam, then the curvature is the change of slope as x increases in the positive x direction. Hence as we move in the positive x direction from, say, point Y to point Z, the slope *decreases* and hence the *change of slope* must be *negative*. Hence to account for the negative curvature, the general expression that we originally developed should be written as:

$$\frac{1}{R} = \frac{M}{EI} = -\frac{\mathrm{d}^2 v}{\mathrm{d}x^2}$$

or

$$EI\frac{\mathrm{d}^2 v}{\mathrm{d}x^2} = -M$$

or

$$v = \int\int -\frac{M}{EI}\,\mathrm{d}x\,\mathrm{d}x$$

where *sagging* moments are taken as positive and *downward* deflections are taken as positive. Move on to the next frame.

Equations (12.1) and (12.2) were developed by considering a small length of beam curved to a circular shape as a result of the application of a constant sagging bending moment. The problem that we have just considered was of a beam subjected to constant moments and hence also deformed to a circular shape.

In most practical situations the bending moments along a beam will vary, and hence the curvature at different locations will also vary. The equations that we have developed for deflection calculations can however still be used provided that the variation of bending moment can be described mathematically at all positions along the beam span.

Let's look at another example and develop an expression for the maximum deflection which by symmetry must occur at mid span:

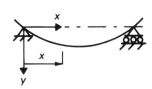

What is the correct expression for the sagging moment at the section distance x along the span?

$$M = \frac{wLx}{2} - \frac{wx^2}{2}$$

This follows from straightforward analysis of the beam. The end reaction R is given by:

$$R = wL/2$$

and the moment at the section being considered is obtained by taking moments of all forces to the left:

$$M = Rx - (wx)(x/2)$$
$$= \frac{wLx}{2} - \frac{wx^2}{2}$$

The general expression for the beam deformation is therefore:

$$v = \int\int -\frac{M}{EI}\,dx\,dx = \int\int -\frac{1}{EI}\left(\frac{wLx}{2} - \frac{wx^2}{2}\right)dx\,dx$$

Carry out the double integration of this expression.

19

$$v = \frac{1}{EI}\left(-\frac{wLx^3}{12} + \frac{wx^4}{24} + Ax + B\right)$$

Check your working:

$$\frac{d^2v}{dx^2} = \frac{1}{EI}\left(-\frac{wLx}{2} + \frac{wx^2}{2}\right) \tag{a}$$

Integrate once:

$$\frac{dv}{dx} = \frac{1}{EI}\left(-\frac{wLx^2}{4} + \frac{wx^3}{6} + A\right) \tag{b}$$

Integrate again:

$$v = \frac{1}{EI}\left(-\frac{wLx^3}{12} + \frac{wx^4}{24} + Ax + B\right) \tag{c}$$

Look at the diagram of the problem. Can you identify the boundary conditions that would enable you to solve for the constants A and B in the above expressions?

20

$$x = 0: v = 0 \text{ and } x = L: v = 0 \text{ also } x = L/2 \quad dv/dx = 0$$

To solve for the two constants of integration, we need only two boundary conditions. However in addition to the two end conditions of zero displacement, in this case there is a further boundary condition which could be more readily used. By symmetry the slope at mid span must be zero, and as the slope is the differential expression dv/dx we can say that when $x = L/2$, $dv/dx = 0$.

Substituting $x = 0$ and $v = 0$ in to equation (c) above will give the constant $B = 0$. To determine the constant A either substitute $x = L$ and $v = 0$ in to equation (c) or $x = L/2$ and $dv/dx = 0$ into equation (b). In either case the solution is:

$$A = wL^3/24$$

and the complete expression to describe the deformation of this beam is given by:

$$v = \frac{1}{EI}\left(-\frac{wLx^3}{12} + \frac{wx^4}{24} + \frac{wL^3x}{24}\right)$$

Use this expression to determine the mid-span deflection.

$$v = \frac{5wL^4}{384EI}$$

This follows from a straightforward substitution of $x = L/2$ into the final equation in Frame 20 and simplifying the answer. Make sure that you arrive at this solution before proceeding: it is only a matter of simple algebra.

This formula is a standard formula for a very common case of a uniformly distributed load on a simply supported span. It is worth remembering as it is widely used in the design of beams. Now try working through the next problem for yourself. The solution is in the next frame.

Determine the maximum end deflection and the rotation of the free end of the cantilevered beam shown. (Hint: take the origin at the left-hand end.)

$$\frac{WL^3}{3EI} \qquad -\frac{WL^2}{2EI}$$

Moment at distance x from origin $= -Wx$ (hogging moment; hence negative)

General equation given by:

$$\frac{d^2v}{dx^2} = -\frac{1}{EI}(-Wx)$$

Integrate once:

$$dv/dx = (Wx^2/2 + A)EI$$

Integrate again:

$$v = (Wx^3/6 + Ax + B)/EI$$

Boundary conditions: $x = L$: $v = 0$ and $x = L$: $dv/dx = 0$ (zero rotation at support)

Hence: $\qquad A = -WL^2/2$ and $B = WL^3/3$

The complete expression is given by:

$$v = (Wx^3/6 - WL^2x/2 + WL^3/3)/EI$$

Maximum deflection when $x = 0$: $\qquad v = WL^3/3EI$

Slope dv/dx given by $\qquad dv/dx = (Wx^2/2 + A)/EI = (Wx^2/2 - WL^2/2)/EI$

Slope when $x = 0$ given by $\qquad dv/dx = -WL^2/2EI$

23

PROBLEMS

1. For the cantilevered beam in figure Q1, calculate the deflection and rotation at the free end.
Ans. ($wL^4/8EI$, $-wL^3/6EI$)

2. The cantilevered beam in figure Q2 is subjected to a couple of magnitude M at the free end. Determine an expression for the deflection and rotation at the free end.
Ans. ($ML^2/2EI$, $-ML/EI$)

3. A cantilevered beam shown in figure Q3 carries a triangular load of maximum intensity w per unit length. Determine expressions for the deflection and rotation at the free end. (*Hint:* the load intensity at a distance x from the free end of the cantilever is wx/L. Remember, also, where the centroid of a triangular load acts when calculating the bending moment expression.)
Ans. ($wL^4/30EI$, $-wL^3/24EI$)

4. The simply supported beam in figure Q4 supports a triangular load of maximum intensity w per unit length. What is the mid-span deflection?
Ans. ($5wL^4/768EI$)

5. A timber beam of rectangular section 75 mm wide and 225 mm deep has been designed to carry a distributed load (including the self-weight of the beam) of 1.4 kN/m over a simply supported span of 4 metres. Calculate the mid-span deflection and if the maximum allowable deflection is limited to 0.003 times the span length state if the design is acceptable. Take $E = 10$ kN/mm^2.
Ans. (6.56 mm, allowable = 12 mm—acceptable)

So far we have looked at problems where the bending moment at any location along the beam span can be defined in terms of a single expression containing the variable x. This is not always the case, and in fact often will not be as beam loading is usually more complicated than indicated in the previous problems.

Consider the simply supported beam carrying a central point load:

Taking the origin at the left-hand support, write down the expression for the bending moment at sections X and Y.

$$
\begin{aligned}
&x \leq 0.5L: M_X = (W/2)x \\
&x \geq 0.5L: M_Y = (W/2)x - W(x - L/2) \\
&\qquad\qquad\quad = W(L/2 - x/2)
\end{aligned}
$$

Both of these equations are obtained by taking moments of forces to the left of the relevant points. What is important here is that we now have two equations describing the variation of bending moments in different regions of the beam, and hence it is necessary to apply equations (12.1) and (12.2) simultaneously to both regions of the beam.

Hence we can write:

$$\frac{d^2v}{dx^2} = -\frac{1}{EI}\frac{Wx}{2} \qquad \text{for } x \leq 0.5L$$

and:

$$\frac{d^2v}{dx^2} = -\frac{1}{EI}W\left(\frac{L}{2} - \frac{x}{2}\right) \qquad \text{for } x \geq 0.5L$$

Both of these equations can be integrated twice with respect to x:

Hence in the region $x \leq 0.5L$:

$$\frac{dv}{dx} = \frac{1}{EI}\left(-\frac{Wx^2}{4} + A\right) \qquad\qquad \text{(a)}$$

and

$$v = \frac{1}{EI}\left(-\frac{Wx^3}{12} + Ax + B\right) \qquad\qquad \text{(b)}$$

Complete the integration for the region $x \geq 0.5L$.

26

$$\frac{dv}{dx} = \frac{1}{EI}\left(\frac{Wx^2}{4} - \frac{WLx}{2} + C\right) \quad \text{(c)} \qquad v = \frac{1}{EI}\left(\frac{Wx^3}{12} - \frac{WLx^2}{4} + Cx + D\right) \quad \text{(d)}$$

The integration should present no problem but what is important is that now there are four constants of integration: two for each of the two separate equations.

How many boundary conditions must we find in order to eliminate the constants of integration?

27

Four

To eliminate four unknowns, we must effectively set up four equations making use of four known boundary conditions.

Looking back at the original diagram of the beam, can you identify four boundary conditions: two for each of the two sections of the beam?

28

For $x \leq 0.5L$: $x = 0$; $v = 0$ and $x = 0.5L$; $dv/dx = 0$
For $x \geq 0.5L$: $x = L$; $v = 0$ and $x = 0.5L$; $dv/dx = 0$

The two supports are obvious boundary conditions and because the beam is symmetrical the slope at mid span must be zero. The zero slope condition applies to both equations as the centre of the span is a point common to both equations.

Hence substituting the boundary conditions for $x \leq 0.5L$ into equations (b) and (a) of Frame 25:

$$v = \frac{1}{EI}\left(-\frac{Wx^3}{12} + Ax + B\right) \tag{b}$$

when $x = 0$; $v = 0$ gives $B = 0$.

$$\frac{dv}{dx} = \frac{1}{EI}\left(-\frac{Wx^2}{4} + A\right) \tag{a}$$

when $x = L/2$; $dv/dx = 0$ gives $A = WL^2/16$.

Hence substituting the constants of integration, A and B, back into equation (b):

$$v = \frac{1}{EI}\left(-\frac{Wx^3}{12} + \frac{WL^2x}{16}\right) \tag{e}$$

which defines the deformation of the beam in the region of the left-hand half of the span.

Can you complete this calculation by forming the equation which defines the deformations for the right-hand side of the span?

29

$$\boxed{v = \frac{1}{EI}\left(\frac{Wx^3}{12} - \frac{Wx^2L}{4} + \frac{3WL^2x}{16} - \frac{WL^3}{48}\right)}$$

Check your solution:

The equations for slope and deflection to the right of mid span were given in Frame 26 as:

$$\frac{dv}{dx} = \frac{1}{EI}\left(\frac{Wx^2}{4} - \frac{WLx}{2} + C\right) \tag{c}$$

and

$$v = \frac{1}{EI}\left(\frac{Wx^3}{12} - \frac{WLx^2}{4} + Cx + D\right) \tag{d}$$

Boundary condition: $x = L/2$; $dv/dx = 0$ substituted into equation (c) gives:

$$C = \frac{3WL^2}{16}$$

Boundary condition: $x = L$; $v = 0$ substituted into equation (d) together with the known expression for C gives:

$$D = -\frac{WL^3}{48EI}$$

Hence substituting for C and D into equation (d) gives:

$$v = \frac{1}{EI}\left(\frac{Wx^3}{12} - \frac{WLx^2}{4} + \frac{3WL^2x}{16} - \frac{WL^3}{48}\right) \tag{f}$$

We therefore have two equations defining the deflection of this beam. Equation (e) gives the deflections for $x \leq 0.5L$ and equation (f) for $x \geq 0.5L$.

Determine the expression for the mid-span deflection of this beam.

30

$$v = \frac{WL^3}{48EI}$$

This expression can be determined by substituting $x = L/2$ into either of the two equations that we have developed:

For example, from equation (e) in Frame 28:

$$v = \frac{1}{EI}\left(-\frac{Wx^3}{12} + \frac{WL^2x}{16}\right) \qquad \text{(e)}$$

Substitute $x = L/2$:

$$v = \frac{1}{EI}\left(-\frac{W(L/2)^3}{12} + \frac{WL^2(L/2)}{16}\right)$$

$$= \frac{WL^3}{EI}\left(-\frac{1}{96} + \frac{1}{32}\right)$$

$$= \frac{WL^3}{48EI}$$

Make the same substitution into equation (f) to convince yourself that the same result can be obtained.

Now use both equations to determine the rotations of the beam at either end.

31

Left-hand end $dv/dx = +\dfrac{WL^2}{16}$:   Right-hand end $dv/dx = -\dfrac{WL^2}{16}$

These results are obtained by substituting $x = 0$ into equation (a) (Frame 25) with the constant A being known, and by substituting $x = L$ into equation (c) (Frame 26) with the constant C being known. As you should expect, from the symmetry of the problem the rotations are of the same magnitude but opposite sign. This is a straightforward substitution and simplification of results, so no solution has been given; but do not proceed until you get these answers.

The formula for the mid-span deflection of the beam under a central point load is for another standard loading case and is worth remembering for use in your further studies of structural design.

32

Earlier in this programme we reasoned that the amount of deflection of a beam must depend upon (a) the magnitude of the load, (b) the flexural stiffness, EI and (c) the span of the beam. We also reasoned that as a consequence of the principle of superposition the deflection must be linearly related to the magnitude of the load, but posed the question whether such a linear relationship existed between the deflection and the span of the beam.

Examination of any of the formulae that we have derived shows that this is obviously not the case. For a beam supporting a central point load, the deflection formula was given by $WL^3/48EI$. In other words, the deflection is related to the third power of the span. A doubling of the span will result in eight times (2^3) the central deflection for the same load! This seemingly minor observation is of considerable importance when deflection control is considered in practical design.

Another simple observation is that the *form* of all the deflection equations that we have developed for all cases considered so far is practically the same. The only difference is the multiplying constant: 1/48 for a point load at mid span of a simply supported beam; 5/384 for a uniformly distributed load over the full span of a simply supported beam, and so on. Hence memorising a few of the formulae for the more common loading cases is not difficult, as only the constant needs to be committed to memory. Even if such constants cannot be remembered, they are readily available in commonly available design manuals.

Not all load cases are so simple that they can be considered to be standard cases, so let's develop the theory even further to see how to deal with more complex problems.

33

In the previous problem of a simply supported beam with a central point load, we derived two equations defining the beam's deflection and containing four constants of integration that had to be determined.

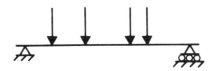

If similar methods were used to analyse the beam shown above, how many deflection equations would you have to develop and how many integration constants would you have to determine?

34

5 equations and 10 integration constants

If you sketch the bending moment diagram for this beam you will see that there are five separate straight lines which form the diagram, hence five separate moment equations must be written down and integrated within the five separate lengths of the beam. Each integration will result in two constants of integration, hence ten (2×5) constants.

It should be obvious that even for a fairly simple problem the amount of arithmetic will soon become unwieldy. The method can however be refined and simplified into a more powerful variation attributable to a mathematician called W. H. Macaulay. Hence you will often see the theory that we will now develop referred to as 'Macaulay's Method'.

To develop this variation, let's consider the case of a simply supported beam with a point load which on this occasion we will locate at some position other than the centre of the span. The support reactions, which are obtained in the usual way by taking moments, are shown on the diagram:

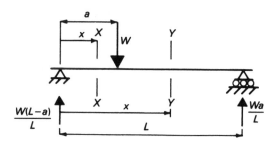

Write down the equation for the sagging moment at X and Y.

35

$$M_X = \frac{W(L-a)}{L}x \quad (x \le a): \qquad M_Y = \frac{W(L-a)}{L}x - W(x-a) \quad (x \ge a)$$

These equations are obtained by taking moments of the forces to the left of the respective points.

The second equation gives the bending moment to the right of the load for all values of $x \ge a$. If this second equation is compared with the first, you will see that the first term in the second equation also appears in the first equation.

The second equation can be written as:

$$M = \frac{W(L-a)}{L}x - W\{x-a\}$$

where the term in the curled brackets {} is ignored (or put equal to zero) if its value is negative (that is, for all values of $x \le a$) but included if its value is positive (that is, for all values of $x \ge a$). Hence this single expression can be used to describe the moments at any location along the span. For example:

if $x = a/2(< a)$

$$M = \frac{W(L-a)}{L}x - W\{x-a\}$$

$$M = \frac{W(L-a)}{L}a/2 - W\{a/2-a\}^0$$

Since the second term within brackets is negative, we can ignore it and M is given by:

$$M = \frac{W(L-a)a}{2L}$$

but if $x = L(> a)$

$$M = \frac{W(L-a)}{L}x - W\{x-a\}$$

$$M = \frac{W(L-a)}{L}L - W\{L-a\}$$

$$= WL - Wa - WL + Wa = 0$$

This technique of neglecting or including a term in brackets depending on which section of beam is being considered is an application of a mathematical technique known as step functions or discontinuity functions.

The general expression to describe the deformation of the beam can then be written as a single expression:

$$\frac{d^2v}{dx^2} = -\frac{1}{EI}\left[\frac{W(L-a)}{L}x - W\{x-a\}\right]$$

This expression can now be integrated and boundary conditions substituted in the usual way to solve the constants of integration. However the integration *must* be performed *without expanding the terms contained within the curled brackets*. These terms must only be expanded when after integration appropriate values of x are substituted, and the terms are then expanded only *if positive* but *neglected if negative*.

Integrate this expression and substitute the boundary conditions to obtain the deflection v. The complete solution is given in the next frame.

$$v = \frac{1}{EI}\left[-\frac{W(L-a)}{6L}x^3 + \frac{W\{x-a\}^3}{6} + \frac{Wa}{6L}(2L^2 - 3aL + a^2)x\right]$$

If you didn't get this answer, you probably made an error in integrating the general expression without expanding the curled brackets. Check your solution against the main steps given below.

The general equation is given by:

$$\frac{d^2v}{dx^2} = -\frac{1}{EI}\left[\frac{W(L-a)}{L}x - W\{x-a\}\right]$$

Integrate once

$$\frac{dv}{dx} = -\frac{1}{EI}\left[\frac{W(L-a)}{2L}x^2 - \frac{W\{x-a\}^2}{2} + A\right] \qquad (a)$$

Integrate again

$$v = -\frac{1}{EI}\left[\frac{W(L-a)}{6L}x^3 - \frac{W\{x-a\}^3}{6} + Ax + B\right] \qquad (b)$$

Boundary condition: $x = 0$; $v = 0$ gives:

$$0 = -\frac{1}{EI}\left[\frac{W(L-a)}{6L}0^3 - \frac{W\{0-a\}^3}{6} + A0 + B\right]$$

Hence:

$B = 0$ (note the term in curled brackets is negative and hence neglected)

Boundary condition: $x = L$; $v = 0$ gives:

$$0 = -\frac{1}{EI}\left[\frac{W(L-a)}{6L}L^3 - \frac{W\{L-a\}^3}{6} + AL + 0\right]$$

Hence:

$$A = -\left[\frac{W(L-a)}{6}L - \frac{W\{L-a\}^3}{6L}\right]$$

which simplifies to

$$A = -\frac{Wa}{6L}(2L^2 - 3aL + a^2)$$

Substituting for A and B in equation (b) gives:

$$v = \frac{1}{EI}\left[-\frac{W(L-a)}{6L}x^3 + \frac{W\{x-a\}^3}{6} + \frac{Wa}{6L}(2L^2 - 3aL + a^2)x\right]$$

This equation describes the deflection at all locations along the span provided that whenever a substitution is made for x the term contained within the curled brackets is ignored (or put equal to zero) if it is negative.

A simply supported beam spans 6 metres and carries a point load of 10 kN at a point 2 metres from the left-hand support. Use the above equation to calculate the deflection (a) at one metre from the left-hand end and (b) at mid span. Take EI = 20 × 10³ kN m².

> 1.06 mm   1.92 mm

Check your working if you didn't get the correct answer:
The equation for the beam deflection is given by:

$$v = \frac{1}{EI}\left[-\frac{W(L-a)}{6L}x^3 + \frac{W\{x-a\}^3}{6} + \frac{Wa}{6L}(2L^2 - 3aL + a^2)x\right]$$

where $L = 6$ metres, $a = 2$ metres and $W = 10$ kN. Hence:

$$v = \frac{1}{EI}\left[-\frac{10 \times (6-2)}{6 \times 6}x^3 + \frac{10 \times \{x-2\}^3}{6} + \frac{10 \times 2}{6 \times 6}(2 \times 6^2 - 3 \times 2 \times 6 + 2^2)x\right]$$

$$= [-1.11x^3 + 1.67\{x-2\}^3 + 22.22x]/EI$$

At one metre from the left-hand support $x = 1$ metre:

Hence
$$v = [-1.11 \times 1^3 + 1.67\{1-2\}^0 + 22.22 \times 1]/EI$$
$$= 21.11/EI$$
$$= 21.11 \times 10^3/20\,000$$
$$= 1.06 \text{ mm}$$

At mid span $x = 3$ metres:

Hence
$$v = [-1.11 \times 3^3 + 1.67\{3-2\}^3 + 22.22 \times 3]/EI$$
$$= 38.36/EI$$
$$= 38.36 \times 10^3/20\,000$$
$$= 1.92 \text{ mm}$$

The example given above illustrates the technique of using Macaulay brackets to develop a single expression to describe the displacements at every point along the span of a beam.

What are the two important features of the use of Macaulay brackets that you must always remember?

> (a) Terms within the curled brackets must be integrated without expanding the brackets.
> (b) On substituting values of the variable x, the bracketed terms are ignored or set equal to zero if negative.

Now on to the next frame:

39

Once the bending moments along the span of a beam are defined in terms of a Macaulay bracket expression, the mathematical procedures to be followed are practically the same in all cases although the appropriate boundary conditions used to evaluate the integration constants may vary.

To give you more practice at setting up the moment equations, let's look at a more complicated problem where the end reactions have been calculated by the usual methods involving taking moments and resolving forces.

To set up the moment equation you should start at the left-hand side and write down the general expression for a moment at any distance x from the origin at the left. To begin, cover up the sketch of the beam with a piece of card (or your hand) and move the card to the right until you expose the first force, which in this case is the end reaction. Write down the moment at a distance x from the origin:

$$M = 23x$$

Now move the card to the right and expose the next load. Write down the moment at a distance x from the origin which must include the term written above and a further term within curled brackets.

$$M = 23x - 50\{x - 5\}$$

Now repeat the operation until you reach the end of the beam at the right-hand side:

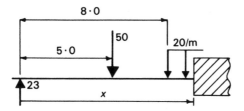

$$M = 23x - 50\{x - 5\} - \frac{20\{x - 8\}^2}{2}$$

Note the third term for the UDL and finally:

$$M = 23x - 50\{x - 5\} - \frac{20\{x - 8\}^2}{2} + 117\{x - 10\}$$

This final expression fully describes the bending moments at all positions along the span provided that the rules for the use of the terms contained within the curled brackets are remembered.

Carry out the integration of this expression and hence calculate the deflection under the 50 kN load. Take EI = 20 × 10³ kN m².

40

$$v = -\frac{1}{EI}\left[\frac{23x^3}{6} - \frac{50\{x - 5\}^3}{6} - \frac{20\{x - 8\}^4}{24} + \frac{117\{x - 10\}^3}{6} - 277.83x\right]$$

deflection under load = 45.50 mm

If you didn't get this answer check your integration, particularly the term relating to the uniformly distributed load. The solution is given in the next frame.

41

Check your solution:

The general equation is given by:

$$\frac{d^2v}{dx^2} = -\frac{1}{EI}\left[23x - 50\{x - 5\} - \frac{20\{x - 8\}^2}{2} + 117\{x - 10\}\right]$$

Integrate once:

$$\frac{dv}{dx} = -\frac{1}{EI}\left[\frac{23x^2}{2} - \frac{50\{x - 5\}^2}{2} - \frac{20\{x - 8\}^3}{6} + \frac{117\{x - 10\}^2}{2} + A\right]$$

Integrate again:

$$v = -\frac{1}{EI}\left[\frac{23x^3}{6} - \frac{50\{x - 5\}^3}{6} - \frac{20\{x - 8\}^4}{24} + \frac{117\{x - 10\}^3}{6} + Ax + B\right]$$

Boundary condition: $x = 0$; $v = 0$ gives $B = 0$ (all bracketed terms are negative and are ignored

Boundary condition: $x = 10$ metres; $v = 0$ gives

$$0 = -\frac{1}{EI}\left[\frac{23 \times 10^3}{6} - \frac{50\{10 - 5\}^3}{6} - \frac{20\{10 - 8\}^4}{24} + \frac{117\{10 - 10\}^3}{6} + A \times 10 + 0\right]$$

which solves to give $A = -277.83$.

Substituting for A and B gives:

$$v = -\frac{1}{EI}\left[\frac{23x^3}{6} - \frac{50\{x - 5\}^3}{6} - \frac{20\{x - 8\}^4}{24} + \frac{117\{x - 10\}^3}{6} - 277.83x\right]$$

and under the 50 kN load, $x = 5$ metres giving:

$$v = -\frac{1}{EI}\left[\frac{23 \times 5^3}{6} - \frac{50\{5 - 5\}^3}{6} - \frac{20\{5 - 8\}^4{}^0}{24} + \frac{117\{5 - 10\}^3{}^0}{6} - 5 \times 277.83\right]$$

$$= 909.98/EI = 909.98 \times 10^3/20\,000$$

$$= \underline{45.50 \text{ mm}}$$

42

$$\boxed{\text{PROBLEMS}}$$

1. A cantilevered beam carries a point load and a uniformly distributed load as shown in figure Q1. Calculate the deflection at the free end in terms of the flexural stiffness, *EI*.

Ans. (1278.67/*EI*)

2. Calculate the deflection and rotation of the free end of the cantilever shown in figure Q2. Express your answer in terms of the flexural stiffness, *EI*.
Ans. (1566.67/EI, −208/EI)

Q1

Q2

3. What is the deflection at the mid span of the beam shown in figure Q3? (*Hint:* this problem is symmetrical—can you make use of the symmetry?)
Ans. (23.05/EI)

4. Calculate the deflection under each of the point loads which act on the beam shown in figure Q4. Take $EI = 15 \times 10^3$ kN m^2.
Ans. (7.45 mm, 3.82 mm)

Q3

Q4

5. The cantilevered beam in figure Q5 carries a triangular load and a point load as shown. Determine an expression for the deflection at the free end of the cantilever.
Ans. (169WL^3/60EI)

6. A simply supported beam supports two identical point loads as shown in figure Q6. Determine a formula for the mid-span deflection of this beam. Use your formula to determine the standard formula for a point load acting at the centre of a simply supported span.

Ans. $\left(\dfrac{Wa(3L^2 - 4a^2)}{48EI}, \dfrac{WL^3}{48EI} \right)$

Q5

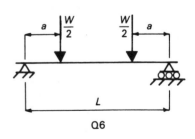

Q6

43

In all the problems that we have considered so far that involve uniformly distributed loads, these loads have all extended either across the whole span or, if partially covering the span, up to the extreme right-hand end of the span when our origin has been taken at the left.

If however this is not the case, as illustrated in figure (a) below, the expression that you would write down for the bending moment at, say, section Z–Z would not be correct.

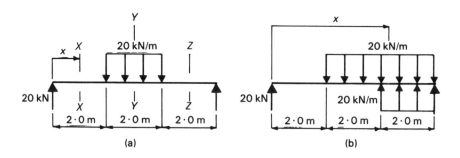

(a)   (b)

To show that this statement is valid we can write down the bending moment expressions at sections X–X, Y–Y and Z–Z:

$$M_{X-X} = 20x$$
$$M_{Y-Y} = 20x - 20\{x - 2\}^2/2$$
and
$$M_{Z-Z} = 20x - 40\{x - 3\}$$

from which it can be seen that the final equation cannot define the bending moments in the region of the UDL (say, section Y–Y) if the rules for the use of the Macaulay brackets are observed.

For example, substituting $x = 2.5$ metres (which lies within the UDL) into the equation for M_{Z-Z} gives:

$$M_{Z-Z} = 20x - 40\{2.5 - 3\}^0 = 20x$$

whereas from the equation for M_{Y-Y} when $x = 2.5$ metres:

$$M_{Y-Y} = 20x - 20\{2.5 - 2\}^2/2 = 20x - 20 \times 0.5^2/2$$

which is obviously not the same answer.

Hence we cannot find a single expression to define the moments at all points along the span. To avoid this problem we can introduce a dummy load of 20 kN/m over the right-hand span and an equal and upward-acting load over the underside of the span in the same region, as shown in figure (b). Because the two dummy loads are equal and opposite, they effectively cancel each other out and don't alter the overall behaviour of the structure.

However, both loads now extend to the right-hand support and hence the use of Macaulay brackets to this 'new' loading system will be correct. Once the bending moment expression is written down correctly, the integration procedures and general solution methods are exactly the same as for any other problem.

Write down the bending moment expression for diagram (b) using Macaulay brackets.

44

$$M = 20x - 20\{x - 2\}^2/2 + 20\{x - 4\}^2/2$$

The third term in this expression accounts for the upward-acting load on the right-hand 2 metres of the span, and the introduction of this term means that this single expression describes the complete variation of bending moment over the whole span provided that the rules of the Macaulay brackets are observed.

Any problem where the load does not extend to the extremities of the beam would be treated in the same way.

Now move to the next frame where we will discuss the final section of theory relating to beam deflections:

45

In Frame 36 you looked at a problem of a 6 metre span beam supporting a load of 10 kN at 2 metres from the left-hand support and calculated that the mid-span deflection was 1.92 mm.

*Is this mid-span deflection the **maximum** deflection that will take place in this beam?*

46

No

If the load had been in the middle of the span, by symmetry the mid-span deflection would have been the maximum. Generally, the maximum deflection will not occur at mid-span although usually in practice the mid-span deflection will be close to the maximum.

How can you determine the exact position where the maximum deflection will occur?

47

> It will occur where the slope of the beam is zero.

A little thought should tell you that this must be the case—if in doubt sketch a deflected beam and draw the tangents to the curve at different locations along the span. Mathematically we would say that the slope is zero when $dv/dx = 0$. In other words, once we have constructed the equation for v we can equate dv/dx to zero to find the location where the deflection is maximum. Once this position is known, we can then substitute the value of x at that position into the deflection equation to calculate the maximum deflection.

Refer back to the calculations in Frame 37 and find the position and value of the maximum deflection.

48

> 1.94 mm at 2.73 metres from the left support

The deflection equation for this beam was found to be:

$$v = [-1.11x^3 + 1.67\{x - 2\}^3 + 22.22x]/EI$$

Differentiating to obtain the slope (remembering *not* to expand the term within the curled brackets):

$$dv/dx = [-3.33x^2 + 5.01\{x - 2\}^2 + 22.22]/EI = 0 \text{ for maximum deflection}$$

By inspection, the point of maximum deflection must be somewhere in the middle region of the span, to the right of the point load. The solution is likely to be for some value of x greater than 2 metres. On this assumption, the term within the Macaulay brackets will be positive and hence *at this stage in the calculation* we can expand the brackets to solve for x. That is

$$-3.33x^2 + 5.01x^2 + 20.04 - 20.04x + 22.22 = 0$$

which solves to give $x = 2.73$ metres; hence the maximum deflection is given by:

$$v = [-1.11x^3 + 1.67\{x - 2\}^3 + 22.22x]/EI$$
$$= [-1.11 \times 2.73^3 + 1.67\{2.73 - 2\}^3 + 22.22 \times 2.73]/EI$$
$$= 38.73/EI$$
$$= 38.73 \times 10^3/20\,000 = \underline{1.94 \text{ mm}}$$

In this case the maximum deflection is very nearly the same as the mid-span deflection but occurs 0.27 metres from the centre of the span.

Although this is a simple example, it illustrates the technique to be adopted in all problems where the position of maximum deflection cannot be readily identified by inspection. Note that the equation for the slope $(\mathrm{d}v/\mathrm{d}x)$ will usually be obtained as one of the stages in the integration procedure used to develop the deflection equation. When this is the case, it is not necessary to differentiate the deflection equation to arrive at the equation for the slope. Some of the further problems at the end of this programme are based on this approach.

49

A final point to remember is that, in this programme, we have been considering the deflection of simply supported beams and have only taken into account deflections due to bending moments. The shearing forces in the beams also cause deformations and contribute to the total deflection. The study of shear deflections is beyond the scope of this book but in any case such deflections are usually only significant in deep beams which are not usually encountered in everyday design.

50

$$\boxed{\text{TO REMEMBER}}$$

$$EI\frac{\mathrm{d}^2v}{\mathrm{d}x^2} = -M$$

where sagging moments are taken as positive and deflections are positive downwards.

Maximum deflection occurs when $\mathrm{d}v/\mathrm{d}x = 0$.

Terms within Macaulay brackets *must* be integrated without expansion, and on substitution if a bracketed term is negative it is neglected in subsequent calculations.

For a simply supported beam supporting a UDL over the whole span:

$$v_{\max} = \frac{5wL^4}{384EI} \quad \text{where } w = \text{load/unit length.}$$

For a simply supported beam supporting a point load at mid span:

$$v_{\max} = \frac{WL^3}{48EI} \quad \text{where } W = \text{total load.}$$

51

1. A timber beam is to be designed as simply supported over a span of 4 metres and to carry a uniformly distributed load (inclusive of self-weight) of 5 kN/m. The maximum permissible bending stress is 12.5 N/mm^2 and the maximum deflection is limited to 0.003 of the span length. Calculate the minimum required depth of a rectangular beam section which is 150 mm wide. Take $E = 10$ kN/mm^2.

Ans. (223 mm)

2. Figure Q2 shows a timber beam which is to be designed to carry a maximum loading of 5 kN/m and a minimum loading of 1 kN/m. To calculate the maximum deflection of the cantilever section of the beam, the loading is located as shown. Calculate the end deflection of the cantilever in terms of the flexural stiffness (*EI*) of the beam section. (*Hint:* make use of the symmetry of the problem.)

Ans. (16.05/EI)

3. To determine the maximum mid-span deflection of the beam considered in question 2, the loading is repositioned as shown in figure Q3. Calculate the maximum mid-span deflection in terms of the flexural stiffness (*EI*) of the beam section. (*Hint:* make use of the symmetry of the problem.)

Ans. (14.42/EI)

Q2

Q3

4. The beam in figure Q4 supports three point loads as shown. Calculate the deflection under each load.

Ans. (30.14/EI, 5.15/EI, 46.30/EI)

5. Calculate the location of the point of maximum deflection of the beam shown in figure Q5 and determine the value of the maximum deflection. Take $EI = 170 \times 10^3$ kN m^2.

Ans. (26.52 mm at 10.05 metres from left support)

Q4   Q5

6. Calculate the maximum deflection of the steel beam shown in figure Q6. Take $E = 200$ kN/mm^2 and $I = 7500$ cm^4.

Ans. (6.77 mm)

Q6   Q7

7. Show that the central deflection of the beam in figure Q7 is given by the expression:

$$\delta = \frac{W}{384EI}(8L^3 - 4Lb^2 + b^3)$$

8. Determine an expression for the deflection at A and the rotation at B for the beam shown in figure Q8.

Ans. ($121wL^4/43\,740EI$, $-101wL^3/9720EI$)

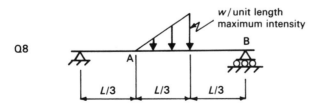

Programme 13

STRAIN ENERGY

On completion of this programme you will be able to:

- Explain the concept of Strain Energy
- State the units of energy
- Define 'Modulus of Resilience'
- Apply energy principles to calculate the deflection of a pin-jointed frame at the point of application of a single point load
- Explain the concept of Potential Energy
- Explain the concept of Kinetic Energy
- Calculate the stress and deformation conditions within simple structures subject to dynamic load conditions

1

In all forms of structure the application of an external loading system will give rise to an internal stress system and internal deformations. The work done by the loading system in deforming the structure will be conserved as energy within the structure. This energy, known as *strain energy*, is normally recoverable when the structure is unloaded provided that the elements of the structure have been stressed only within the elastic range.

You can illustrate this for yourself by stretching an elastic band and then letting go of the ends. In stretching the elastic the release of energy will be quite apparent particularly if you only let go of one end! A different form of strain energy due to bending can be illustrated by taking a plastic ruler and rotating both ends. If you then release the ends suddenly, the ruler will straighten as the stored energy is released.

There are a number of methods of structural and stress analysis which can be used to analyse structures subjected to static or dynamic loading and which are based on a study of the strain energy stored within the loaded structure. These methods are used to calculate structural deflections or to determine the forces within statically indeterminate structures. Although the study of these methods is too advanced to be included in this text, the fundamental concepts of strain energy are introduced in this programme to form the foundations of your further studies.

A spring is an elastic structure which when compressed absorbs energy which is stored as recoverable strain energy within the compressed spring. Can you think of any common situations where this property is made use of?

2

There are many possible answers to this question: did you think of the springs used in railway buffers to absorb the kinetic energy of moving rail carriages or those used in cars to deaden vertical rolling movement due to the vertical movement of the wheels? A more everyday situation is the release of strain energy in the spring of a mechanical watch which, as it slowly uncoils, provides the energy to operate the watch mechanism.

To begin our study of strain energy let's consider a bar of uniform cross-section, as shown below, which is made from a linearly elastic material.

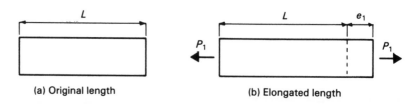

(a) Original length   (b) Elongated length

If the bar is subjected to an axial force P_1 it will elongate by an amount e_1 as shown in figure (b).

Remembering that the bar is linearly elastic and will obey Hooke's Law, can you plot a graph to show the variation of axial load with increasing values of axial extension?

3

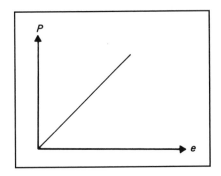

The form of this diagram should be quite familiar to you. For any *linearly elastic* material there will be a linear relationship between load and extension. Provided that the limit of proportionality is not exceeded, then this relationship will always apply.

Now let's suppose that the bar is extended by an amount e_1 by the application of a force P_1 as shown in figure (a), and that the bar is then stretched by a *small* additional displacement δe_1. This is shown on the load extension graph in figure (b).

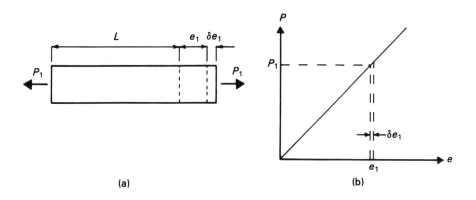

(a) (b)

What amount of work is done by this force P_1 in extending the bar by the small displacement δe_1?

4

$$\boxed{\text{work done} = P_1 \times \delta e_1}$$

This follows from the normal definition of work as force multiplied by the distance through which the force moves, and is represented graphically by the shaded area under the load–extension graph shown below.

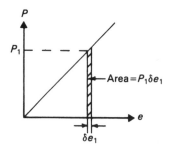

If the bar is extended from zero extension to a total extension e, the work done by the load P_1 as it increases from zero to a maximum value P is therefore given by

$$\text{work} = \int_0^e P_1 \, de_1$$

which must be represented by the total area under the load–extension graph between zero extension and the extension e. This work done by the externally applied load must be stored as *strain energy* within the bar, which will act like an elastic spring and will release the energy when the external load is removed and the bar is allowed to return to its original length. The strain energy stored (which is commonly given the symbol U) is therefore also given by the expression:

$$U = \int_0^e P_1 \, de_1 \tag{13.1}$$

By inspection of the load–extension graph above, can you evaluate equation (13.1) and express the strain energy in terms of the maximum load P and the overall extension e?

5

$$\boxed{U = \frac{P \times e}{2}}$$

We have already stated that the work done and hence strain energy stored is represented by the area under the load–extension graph between zero extension and the extension e. This area is triangular and the area of the triangle is one-half the base times the height: hence the above formula.

This formula can be derived alternatively by integration as follows:

$$U = \int_0^e P_1 \, de_1$$

But we know (from Hooke's Law) that for an elastic bar:

$$E = \frac{\text{stress}}{\text{strain}} = \frac{P_1/A}{e_1/L}$$

Hence:

$$P_1 = \frac{EAe_1}{L}$$

where: E = the elastic modulus and
A = the cross-sectional area of the bar.

Hence U can be expressed as:

$$U = \int_0^e \frac{EA}{L} e_1 \, de_1$$
$$= \frac{EA}{L} \left(\frac{e_1^2}{2}\right)_0^e = \frac{EA}{L}\frac{e^2}{2}$$

But as $P = EAe/L$ then

$$U = \frac{Pe}{2} \quad \text{(as before)} \tag{13.2}$$

If we alternatively substitute for e in equation (13.2) where $e = PL/AE$, then the equation for strain energy can be written as:

$$U = \frac{P^2 L}{2AE} \tag{13.3}$$

which is often a more usable expression for the strain energy stored in an elastic bar.

As the strain energy stored in the bar is equal to the work done to extend the bar, what do you think are the units of strain energy?

6

Joules (Newtons.metres)

You should recall this from the work we did in Programme 9.

A 50 mm diameter steel bar is 2 metres long and carries a tensile load of 100 kN. Determine the strain energy stored in the bar. Take $E = 200 \, kN/mm^2$.

$$\boxed{25.46 \text{ Joules}}$$

You should have used equation (13.3) to get this answer:

$$U = \frac{P^2 L}{2AE}$$

$$= \frac{(100 \times 10^3)^2 \times 2}{2 \times (\pi \times 0.05^2/4) \times (200 \times 10^9)}$$

$$= 25.46 \text{ Joules}$$

Note that all dimensions have been converted to Newtons and metres to express the final answer in Joules.

Equation (13.3) can be further re-arranged to give another formula for the strain energy stored as follows:

From equation (13.3): $\qquad\qquad\qquad\qquad\qquad U = \dfrac{P^2 L}{2AE}$

But P/A is the axial stress σ in the bar:

Hence: $\qquad\qquad\qquad U = \dfrac{P^2 L}{2AE} = \dfrac{P^2}{A^2}\dfrac{(A \times L)}{2E} = \dfrac{\sigma^2 (A \times L)}{2E}$

But as $A \times L$ is the volume of the bar, then the *strain energy per unit volume of material* is given by:

$$\text{strain energy/unit volume} = \frac{U}{(A \times L)} = \frac{\sigma^2}{2E} \qquad (13.4)$$

As σ/E is the strain ε in the bar, equation (13.4) can also be written as:

$$\text{strain energy/unit volume} = \frac{U}{(A \times L)} = \frac{\sigma^2}{2E} = \frac{\sigma}{2} \times \frac{\sigma}{E} = \frac{\sigma \times \varepsilon}{2} \qquad (13.5)$$

which shows that the strain energy per unit volume is represented by the area under the material stress–strain curve as shown in figure (a) on the next page. If the material is stressed to the proportional limit as shown in figure (b), then the area under the stress–strain curve up to the proportional limit ($\bar{\sigma}$) is given, from equation (13.4), as:

$$\text{strain energy/unit volume} = \frac{\bar{\sigma}^2}{2E} = \text{the } \textit{modulus of resilience} \qquad (13.6)$$

The *modulus of resilience* of a material is a measure of its ability to absorb energy while still remaining elastic and permitting full elastic recovery. The higher the modulus of resilience, the greater is the ability of the material to absorb energy whilst still remaining elastic.

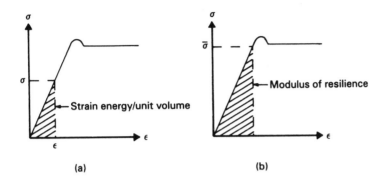

(a)   (b)

Which is the most resilient material—mild steel or high yield steel? The elastic modulus of steel is 200 kN/mm². The proportional limit of mild steel is 250 N/mm² and for high yield steel is 460 N/mm².

8

> high yield steel

Equation (13.6) shows that the modulus of resilience is proportional to the square of the proportional limit of the material. The elastic modulus is the same for both steels, and as high yield steel has a higher proportional limit it therefore has a greater ability to absorb energy while still remaining elastic.

In answering the question, you might have calculated the value of the modulus of resilience for both steels. The values that you should have obtained are:

Mild steel:

$$\text{modulus of resilience} = \frac{\bar{\sigma}^2}{2E} = \frac{(250 \times 10^6)^2}{2 \times 200 \times 10^3 \times 10^6} = 156.25 \times 10^3 \, \text{N/m}^2$$

High yield steel:

$$\text{modulus of resilience} = \frac{\bar{\sigma}^2}{2E} = \frac{(460 \times 10^6)^2}{2 \times 200 \times 10^3 \times 10^6} = 529.00 \times 10^3 \, \text{N/m}^2$$

Note that the modulus of resilience has been defined as the strain energy per unit volume. Its units must therefore be Joules per cubic metre ($\text{J/m}^3 = \text{Nm/m}^3 = \text{N/m}^2$). Hence the units for the modulus of resilience are the same as units of stress which we can take as N/m^2.

Now try some problems:

9

$$\boxed{\text{PROBLEMS}}$$

1. A steel bar is 500 mm long and is compressed by an axial force of 50 kN. The cross-sectional area of the bar is 1500 mm^2 and the elastic modulus of steel is 200 kN/mm^2. Calculate the strain energy stored in the bar.

Ans. (2.08 Joules)

2. A steel tie of circular cross-section is 2 metres long. For half its length the diameter of the bar is 50 mm and the remainder has a diameter of 40 mm. If the elastic modulus of steel is 200 kN/mm^2, calculate the strain energy stored in the bar when an axial force of 80 kN is applied.

Ans. (20.87 Joules)

3. To what value would the axial force in question 2 have to be increased to double the strain energy in the bar?

Ans. (113.14 kN)

4. Calculate the modulus of resilience of aluminium alloy which has an elastic modulus of 70 kN/mm^2 and a stress at the proportional limit of 270 N/mm^2.

Ans. (520.71 × 10^3 N/m^2)

5. Calculate the total strain energy stored in the two members of the pin jointed frame shown in figure Q5. Each member has a cross-sectional area of 1500 mm^2 and an elastic modulus of 200 kN/mm^2.

Ans. (23.57 Joules)

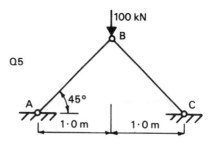

10

SINGLE LOAD DEFLECTIONS OF PIN JOINTED FRAMES

The principles of strain energy can be used to calculate the deflection under the point of application of a single load acting on a pin jointed plane frame. For example, consider problem 5 in Frame 9. Did you calculate the total strain energy stored in the frame as follows?

The force in each member of the frame is determined by resolving at joint **B** to give a compressive force in each member of $50\sqrt{2}$ kN. Equation (13.3) should then have been used to calculate the strain energy in one member, and as the members are identical the total strain energy is twice that in one member.

Hence the strain energy is given by:

$$U = 2 \times \frac{P^2 L}{2AE} = \frac{2 \times (50\sqrt{2} \times 1000)^2 \times \sqrt{2}}{2 \times (1500 \times 10^{-6}) \times (200 \times 10^9)}$$

$$= 23.57 \text{ Joules}$$

The vertical displacement of joint **B** is also the displacement of the 100 kN vertical load in the direction of the load. If this displacement is δ, then this external load will do work as the frame deforms under load.

Can you write down an expression for the work done by the external load?

11

$$\boxed{\text{work done} = \tfrac{1}{2} W \times \delta = \tfrac{1}{2} \, 100 \times 10^3 \times \delta = 50 \times 10^3 \times \delta \text{ Joules}}$$

Did you remember to include the multiplying factor of $\tfrac{1}{2}$? As we are dealing with the application of a gradually applied force to a linear elastic system, the external loads and internal forces will vary from zero to the maximum values. The load–displacement relationship will be as shown in the graph at the top of Frame 3, and the work expression will therefore be given by equation (13.2) in Frame 5.

We have therefore established that the internal energy stored in the frame is 23.57 Joules and the work done by the external load is $50 \times 10^3 \delta$.

Can you calculate the vertical deflection of joint B?

12

$$\boxed{0.47 \text{ mm}}$$

From the principle of conservation of energy you can equate the external work done to the internal strain energy:

$$50 \times 10^3 \times \delta = 23.57$$

Hence:
$$\delta = 0.47 \times 10^{-3} \text{ matres} = 0.47 \text{ mm}$$

On to the next frame:

13

The method that we have employed in the previous problem can be extended to any statically determinate pin jointed frame *subjected to a single load*. The steps in the method can be summarised as:

(1) calculate the forces in each member using the method of resolution at joints
(2) calculate the strain energy in each member $(P^2 L/(2AE))$
(3) calculate the total strain energy stored in the structure by summing the strain energy components for all members:

$$U = \sum \frac{P^2 L}{2AE}$$

(4) calculate the external work done by the *single load*, W:

$$U = \frac{W \times \delta}{2}$$

(5) equate internal and external work done:

$$U = \frac{W \times \delta}{2} = \sum \frac{P^2 L}{2AE}$$

or

$$\delta = \frac{1}{W} \sum \frac{P^2 L}{AE} \qquad (13.7)$$

In practice, equation (13.7) can be used directly to calculate the deflection of the load. When using this equation it is best to set out your work in tabular form, as this is another situation where a systematic approach to presentation is more workmanlike and less prone to error. Try the following problem, the answer to which is given in the next frame:

Calculate the vertical deflection of joint E in the frame shown below. Take the elastic modulus as 200 kN/mm² and the cross-sectional areas as 1500 mm² for all members.

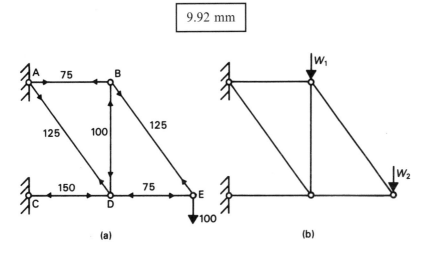

9.92 mm

(a)   (b)

Member	P (kN)	Length (metres)	$P^2 \times L$
AB	75	3	16 875
BE	125	5	78 125
CD	−150	3	67 500
DE	− 75	3	16 875
AD	125	5	78 125
BD	−100	4	40 000

$$\Sigma = \underline{297\,500}$$

From equation (13.7):

$$\delta = \frac{1}{W}\sum \frac{P^2 L}{AE}$$

$$= \frac{1}{W \times AE}\sum P^2 L = \frac{297\,500 \times 10^6 \times 10^3}{(100 \times 10^3) \times (1500 \times 10^{-6}) \times (200 \times 10^9)} = 9.92 \text{ mm}$$

Note that in tabulating the axial forces in each member, tensile forces have been taken as positive. However, as in this instance the force term in equation (13.7) is squared, the signs of the forces do not really matter. This is not always the case and it is good practice to adopt and show the sign convention being used. Note also that in this problem the AE terms are identical for all members. If this is not the case, then include additional columns in the table when evaluating equation (13.7).

This method permits the calculation of the deflection at the point of application of the load when a frame is subjected to a single load.

In figure (b) above, the deflection under load W_1 is δ_1 and under load W_2 is δ_2. Can δ_1 and δ_2 be evaluated using the strain energy methods that we have been using?

393

15

$$\boxed{\text{NO}}$$

You should have reasoned that the external work done is the sum of the work done by the two individual forces, that is:

$$\text{external work} = \tfrac{1}{2}(W_1 \times \delta_1) + \tfrac{1}{2}(W_2 \times \delta_2)$$

and this must equal the internal strain energy, U, stored within the members of the frame, that is:

$$\tfrac{1}{2}(W_1 \times \delta_1) + \tfrac{1}{2}(W_2 \times \delta_2) = U$$

The right-hand side of this expression can be easily evaluated from the member forces. The left-hand side of the expression however contains two unknown deflections (δ_1 and δ_2) and the complete expression cannot be solved for either of these two unknowns. To solve for these we would need to write down another equation relating δ_1 and δ_2, and no such equation exists.

This demonstrates that, as already stated, the method that we have used is limited to a single external load system and only the deflection of the external load itself can be evaluated. The principle of strain energy does however form the basis of several structural energy theorems that enable us to evaluate the deflection at any point on a loaded structure subjected to any number of loads. You will probably come across these energy theorems in your more advanced studies of structural analysis. In Programme 14 we will introduce another technique of analysis known as the method of Virtual Work, which is another more powerful technique of solving deflection problems.

On to the next frame:

16

DYNAMIC LOADING

So far we have looked at strain energy problems where the structure is subjected to gradually applied loads. There are many situations where the loading applied to a structure is dynamic in nature; that is, the load varies with time or is applied suddenly. For example, a structure supporting a heavy piece of machinery with moving parts may be subjected to vibratory loading. A pile driver used to drive foundation piles will subject the pile to a suddenly applied impact loading. To analyse these and many other problems, the principles of strain energy that we have developed must be extended further.

To illustrate the principles involved, let's consider a short column made from a linear elastic material. The column is rigidly fixed at one end and a heavy load W is dropped on to it from a height h. Assuming that the load was initially at rest and on impact remains in contact with the column and does not rebound, the energy of the falling load will be converted into strain energy within the column as the column shortens by an amount *e*.

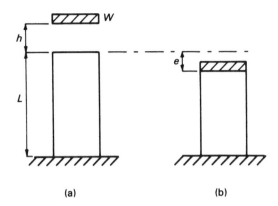

(a)   (b)

What form of energy does the load possess just before it comes in contact with the column?

17

Potential and Kinetic Energy

You should be familiar with these two concepts of energy. Potential energy is the energy possessed by a body by virtue of its height above some datum point. Kinetic energy is the energy possessed by a body by virtue of its motion.

The load will initially be at rest and its kinetic energy will be zero, but it will possess potential energy. However, as it falls, some of its potential energy will be converted to kinetic energy as its velocity increases and after impact this kinetic energy will be converted to strain energy within the column. This assumes that none of the energy of the falling load is converted to any other form of energy such as heat.

As the load causes shortening of the column it will eventually come to rest at the point of maximum shortening. After this, the column will recover elastically and will attempt to return to its original length, resulting in longitudinal oscillations until a state of equilibrium is eventually reached.

Consult figure (b) above and assume that the shortening e represents the maximum shortening of the column. Can you write down expressions for (i) the loss of potential energy of the load, (ii) the kinetic energy of the load and (iii) the strain energy in the column in terms of the axial force P within the column at the instant of maximum shortening?

18

$$\boxed{\text{(i) } W(h+e) \qquad \text{(ii) zero} \qquad \text{(iii) } \frac{P^2 L}{2AE}}$$

The potential energy expression follows from the definition of potential energy. The kinetic energy is zero as the load is just at rest at the instant of maximum shortening, and the equation for the strain energy follows from equation (13.3) in Frame 5.

The principle of conservation of energy allows us to equate the energy lost by the falling load to the strain energy gained by the column, that is:

$$W(h+e) = \frac{P^2 L}{2AE}$$

but we know from Hooke's Law that the elastic shortening e is given by:

$$e = \frac{PL}{AE}$$

and hence combining the above equations:

$$W\left(h + \frac{PL}{AE}\right) = \frac{P^2 L}{2AE} \tag{13.8}$$

Equation (13.8) is a quadratic equation in P which can be solved to give the *instantaneous* axial force in a member which is subjected to the impact of a falling load. This axial force occurs at the instant that the falling load first comes to rest. The corresponding *instantaneous axial stress* can be calculated once the axial force is known (stress = force/area). Try the following example, the answer to which is in the next frame.

A load of 1 kN falls from a height of 0.25 metres on to a short column which has a length of 0.4 metres and a circular section of 100 mm diameter. Calculate the maximum instantaneous stress set up in the column if the elastic modulus is 25 kN/mm².

19

$$\boxed{63.21 \text{ N/mm}^2}$$

The cross-sectional area: $A = \pi \dfrac{d^2}{4} = \pi \dfrac{100^2}{4} = 7853.99 \text{ mm}^2$

From equation (13.8)

$$W\left(h + \frac{PL}{AE}\right) = \frac{P^2 L}{2AE}$$

Substituting known values and working in units of kN and mm:

$$1 \times \left(250 + \frac{(P \times 400)}{(7853.99 \times 25)}\right) = \frac{P^2 \times 400}{2 \times (7853.99 \times 25)}$$

which simplifies to give:

$$P^2 - 2P - 245.44 \times 10^3 = 0$$

The positive root of this equation is $P = 496.42$ kN

From which the axial stress in the column $= \dfrac{P}{A} = \dfrac{496.42 \times 10^3}{7853.99} = 63.21 \text{N/mm}^2$

Assuming that instead of dropping the load from a height it is applied gradually to the top of the column, what would be the stress in the column and what do you deduce from these calculations about the effect of dynamic loading?

20

$$\boxed{0.13 \text{ N/mm}^2}$$

This answer follows from the fact that for a static loading the stress is simply given by:

$$\sigma = \frac{W}{A} = \frac{1000}{7853.99} = 0.13 \text{ N/mm}^2$$

By comparing the answers for the stress due to static (0.13 N/mm^2) and dynamic (63.21 N/mm^2) loading, it is obvious that dynamic loading acting on any structure will produce instantaneous stresses many times higher than the stresses arising from a static load of comparable magnitude. The instantaneous deformation of the structure under the application of dynamic loading will also be correspondingly higher.

The theory that we have been developing assumes that the material remains elastic and that full elastic recovery is possible. It should be obvious that if dynamic loading is applied to a structure, the very high instantaneous stresses arising in the structure make it much more likely that the proportional limit will be exceeded and material failure is more likely to occur.

Move to the next frame.

21

The use of equation (13.8) is complicated by the fact that it results in a quadratic expression. However it can be simplified for the case where the height through which the load is dropped is large compared with the elastic shortening of the column.

Equation (13.8) was originally written in the form:

$$W(h + e) = \frac{P^2 L}{2AE} \qquad \text{(see Frame 18)}$$

However if the height of the drop h is large compared with the shortening e, this can be written as:

$$Wh = \frac{P^2 L}{2AE}$$

The static stress in the column if the load were applied gradually without dropping is given by:

$$\sigma_{\text{stat}} = W/A$$

and the stress due to the dynamic loading if the load is dropped through a height h is given by:

$$\sigma_{\text{dyn}} = P/A$$

hence if the energy equation

$$Wh = \frac{P^2 L}{2AE}$$

is written as:

$$\frac{W}{A} = \frac{[P]^2}{[A]^2} \frac{L}{2Eh}$$

then

$$\sigma_{\text{stat}} = \sigma_{\text{dyn}}^2 \frac{L}{2Eh}$$

which re-arranges to give:

$$\sigma_{\text{dyn}} = \left[\sigma_{\text{stat}} \times \frac{2Eh}{L} \right]^{\frac{1}{2}} \qquad (13.9)$$

This equation is much easier to apply to problems of this nature and gives only slightly inaccurate answers.

Use equation (13.9) to solve the problem given in Frame 18.

22

$$\boxed{63.0 \text{ N/mm}^2}$$

The static stress is given by: $\sigma_{\text{stat}} = \dfrac{W}{A} = \dfrac{1000}{7853.99} = 0.127 \text{ N/mm}^2$

Hence the dynamic stress is given by:

$$\sigma_{dyn} = \left[\sigma_{stat} \times \frac{2Eh}{L}\right]^{\frac{1}{2}}$$

$$= \left[0.127 \times \frac{2 \times 25\,000 \times 250}{400}\right]^{\frac{1}{2}}$$

$$= 63.0 \text{ N/mm}^2$$

Further examination of equation (13.9) will lead to some interesting conclusions about the nature of stress induced by dynamic loading. The equation can be expressed in the following form:

$$\sigma_{dyn} = \left[\sigma_{stat} \times \frac{2Eh}{L}\right]^{\frac{1}{2}}$$

$$= \left[\frac{W}{A} \times \frac{2Eh}{L}\right]^{\frac{1}{2}}$$

$$= \left[Wh \times \frac{2E}{AL}\right]^{\frac{1}{2}}$$

$$= \left[\text{the available potential energy of the load} \times \frac{2 \times \text{elastic modulus}}{\text{volume of the column}}\right]^{\frac{1}{2}}$$

Study the above equation and write down your conclusions about the factors that affect dynamic stresses and how these differ from the factors that affect static stresses.

23

Dynamic stresses are increased by (1) increased energy in the impacting load
(2) higher values of elastic modulus.
Dynamic stresses are reduced by increasing the volume of material in a structure.

Did you come to these conclusions? They show that for an axially loaded member of constant cross-section, dynamic stresses are influenced by different factors from those that affect static stresses. Static stresses in such members are independent of both the elastic modulus and the volume of material.

Two short steel piles are fabricated from identical steel universal column sections. Pile A is twice as long as pile B and both piles are driven into the ground until they reach solid rock at their bases. If they are driven by dropping a heavy drop hammer from a constant height above the top of each pile, which one do you think is most likely to be damaged if it is subjected to continuous driving after it has hit solid rock?

24

pile B

Pile A is longer and has greater volume. It therefore has a greater capacity to absorb the energy of the drop hammer, with resulting smaller dynamic stresses. The stresses in pile B will be higher and, if excessive, could exceed the proportional limit of steel and result in plastic deformation or even fracture of the steel section.

Let's assume that the stresses induced in pile B are so large that they do exceed the proportional limit. The figure shows two possible stress-strain graphs for two differing grades of steel (X and Y) out of which the pile section could be made.

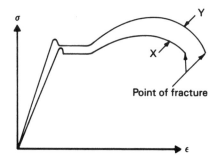

Which of the two steels would be the better steel for fabrication of the pile. Give your reasons.

25

steel Y

The figure shows that both steels have the same fracture stress but the area contained under the curve for steel Y is much greater than that for steel X. Remembering that the area under the stress–strain curve represents the energy absorbed by the material, this means that steel Y has a greater capacity for absorbing energy than steel X, and can therefore sustain higher impact loading before material failure takes place.

The stress–strain diagrams above are typical of ductile materials which undergo large strains before failure. Generally, materials which are ductile are more able to resist impact loading than brittle materials because the area contained under the stress–strain graph for a ductile material is usually large, indicating a high energy absorption capacity.

26

SUDDENLY APPLIED LOADS

To complete our study of impact loadings, let's consider the case of a load W which is placed on top of our column at zero height ($h = 0$) and then released suddenly. This is a different case from static loading where it is assumed that the load is applied gradually until static equilibrium is reached. In the case we are now considering, the load will introduce a sudden shock loading to the column. When the column and the load reach a state of static equilibrium, the stress in the column will be the static stress ($= W/A$). However, as we have seen already, the maximum stress in the column is likely to be greater than this figure. Let's look again at equation (13.8):

$$W\left(h + \frac{PL}{AE}\right) = \frac{P^2 L}{2AE}$$

If we put h equal to zero the equation becomes:

$$W\frac{PL}{AE} = \frac{P^2 L}{2AE}$$

or

$$W = P/2$$

What does this tell you about the magnitude of the instantaneous dynamic stress in the column?

27

$\boxed{\text{instantaneous dynamic stress} = 2 \times \text{static stress}}$

You should have deduced from the above equation that the internal force induced in the column when the load is suddenly applied is twice the magnitude of the applied load ($P = 2 \times W$). Hence it follows that the dynamic stress (P/A) must be twice the static stress (W/A) which develops when the load is applied gradually. Once the column ceases to oscillate and a state of static equilibrium is reached, the stress will stabilise at the static value (W/A).

This conclusion has practical implication in many design situations. For example, in the fabrication of a precast concrete beam it may be necessary to design lifting hooks so that the beam can be hoisted by a crane into its correct location on site. The hooks must be strong enough to support the beam as it is lifted and if there are, say, two hooks, each will carry half the total weight of the beam. However if the beam is lifted roughly, the sudden jerking of the beam (snatch loading) will induce dynamic loading in the lifting hooks which should be allowed for by multiplying the static force in the hook by a factor of two.

28

SPRINGS

At the beginning of this programme we made reference to a spring as an elastic structure which has a capacity for absorbing and releasing energy. Structurally a spring behaves in a similar manner to an elastic bar, and with minor modifications the equations that we have developed in this programme can be applied to similar problems involving springs.

From Hooke's Law we know that for an elastic bar:

$$P = \frac{EAe}{L}$$

where the symbols have the usual meaning. This can be written as:

$$P = Ke$$

where:

$$K = \frac{EA}{L} = \text{the } \textit{elastic stiffness} \text{ of the bar} = \frac{P}{e}$$

The *elastic stiffness* is defined as the force required for a unit extension (or compression) of the bar ($K = P/e = P/1 = P$). This concept of *stiffness* is one that you will come across often in your further studies of structural analysis. In the case of a spring, the elastic properties of the spring are also described in terms of a *spring stiffness* which is the force required to produce a unit extension or compression of the spring.

The spring stiffness will have units of kN/m or N/mm, or other appropriate force/length units.

Two springs are subjected to equal compressive forces. Spring A has a stiffness of 10 N/mm and spring B has a stiffness of 20 N/mm. Which spring will compress the least?

29

> spring B

As $P = Ke$, then $e = P/K$. Hence the spring with the largest stiffness will compress the least. It is important to remember that the *stiffer* the spring then the smaller will be the deformation under load.

To analyse problems involving springs it is therefore only necessary to replace the stiffness term EA/L in the equations that we have developed by the appropriate value of spring stiffness. For example, equation (13.8) was developed as:

$$W\left(h + \frac{PL}{AE}\right) = \frac{P^2 L}{2AE}$$

and in terms of spring stiffness this can be written as:

$$W\left(h + \frac{P}{K}\right) = \frac{P^2}{2K}$$

The solution of this equation will give the force in the spring from which the compression (or extension) of the spring can be calculated.

Most problems involving energy stored within springs can be analysed without remembering equations such as this, but by applying fundamental energy principles and simply recalling that the spring stiffness is given by the expression $P = K \times e$. Try the following problem: the complete solution is in the next frame.

A railway truck with a mass of 1000 kg is travelling at a constant speed of 10 km/ hour when it collides with a buffer which brings it to rest. The buffer consists of a spring with a stiffness of 200 kN/m. Calculate the maximum shortening and the maximum force developed in the spring. (Hint: this is not a problem involving a falling body: you will have to decide what form of energy is possessed by the truck.)

30

196 mm   39.20 kN

Speed of truck $= 10$km/hour $= \dfrac{10 \times 1000}{60 \times 60} = 2.78$ metres/second

Kinetic energy of truck $= \frac{1}{2}mv^2 = \frac{1}{2} \times 1000 \times 2.78^2 = 3864.20$ Joules

Strain energy stored in spring $= \frac{1}{2}P \times e = \frac{1}{2}(K \times e) \times e = \frac{1}{2}Ke^2$

$$= \frac{1}{2} \times 200\,000 \times e^2$$

$$= 100\,000e^2 \text{ Joules}$$

Hence equating energy: energy gained by spring $=$ energy lost by truck

$$\therefore \quad 100\,000e^2 = 3864.20$$

Hence: $\qquad\qquad\qquad e = 0.196$ metres

$$= 196 \text{ mm}$$

The maximum force developed $=$ spring stiffness $(K) \times$ compression (e)

$$= 200 \times 0.196$$

$$= 39.20 \text{ kN}$$

31

| TO REMEMBER |

The strain energy stored in an elastic bar $U = \dfrac{P \times e}{2} = \dfrac{P^2 L}{2AE}$

Strain energy/unit volume $= \dfrac{\sigma^2}{2E} = \dfrac{\sigma \times \varepsilon}{2}$

Modulus of resilience $= \dfrac{\bar{\sigma}^2}{2E}$

Deflection at the point of application of a single load applied to a pin jointed frame (in the direction of the load)

$$= \frac{1}{W} \sum \frac{P^2 L}{AE}$$

To calculate forces and stresses due to a falling load use:

$$W\left(h + \frac{PL}{AE}\right) = \frac{P^2 L}{2AE}$$

or

$$\sigma_{\text{dyn}} = \left(\sigma_{\text{stat}} \times \frac{2Eh}{L}\right)^{\frac{1}{2}}$$

Dynamic stress due to suddenly applied load (not falling) $= 2 \times$ static stress

For a spring: force in spring $=$ spring stiffness $\times$ compression (or extension)
$$P = K \times e$$

32

| FURTHER PROBLEMS |

1. A concrete beam has a rectangular section of width 250 mm and depth 500 mm and is 5 metres long. It is stressed by the application of a horizontal compressive force of 2000 kN which acts along the longitudinal central axis of the section. If the elastic modulus (E) of the concrete is 25 kN/mm^2, calculate the strain energy stored in the beam.

Ans. (3200 Joules)

2. Calculate the vertical deflection of the joint marked 'A' in the pin-jointed frame shown in figure Q2. Take the cross-sectional area of all members as 1000 mm^2 and $E = 200$ kN/mm^2.

Ans. (19.35 mm)

3. If the load on the frame in figure Q2 is applied horizontally instead of vertically, what would be the horizontal deflection at joint A?

Ans. (33.05 mm)

Q2

Q4

4. Calculate the vertical deflection of the joint marked 'A' in figure Q4. The cross-sectional area of each external member of the frame is 2000 mm² and that of all the internal members is 1500 mm². Take $E = 200$ kN/mm². (*Hint:* because of the symmetry of the problem, the deflection at A will equal the deflection at B.)

Ans. (3.57 mm)

5. A vertical steel bar is 1 metre long and has a circular section of diameter 25 mm. It is fitted with a collar at the bottom end (figure Q5) and is firmly held at the top. A load of 0.4 kN falls through a height of h metres and is arrested by the collar. If the maximum instantaneous stress in the bar is not to exceed 75 N/mm², what is the maximum allowable value of h? Take $E = 200$ kN/mm². (*Hint:* use equation 13.9.)

Ans. (17.27 mm)

6. The load in figure Q5 is lowered until it just touches the collar and then suddenly released. Calculate the largest value of load that can be applied if the maximum stress of 75 N/mm² is not to be exceeded.

Ans. (18.41 kN)

Q5

7. A railway truck with a mass of 500 kg starts from rest at the top of an incline and rolls freely down as shown in figure Q7. At the bottom of the slope it is brought to rest when it collides with a buffer which consists of a spring with stiffness of 400 kN/m and a reaction frame. Calculate the maximum shortening of the spring and the maximum force transmitted to the reaction frame.

Ans. (145.92 mm, 58.37 kN)

8. A metal stamping machine consists of a load of 5 kN which falls through a vertical height of 1.5 metres. The machine sits on a rigid foundation which is rectangular in plan and supported in each corner on bearings which can be idealised as spring supports as shown in figure Q8. If the vertical movement of the foundation must not exceed 3 mm, what value of spring stiffness is required and what is the maximum reactive force transmitted to the ground below each bearing?

Ans. (417 500 kN/m, 1252.50 kN)

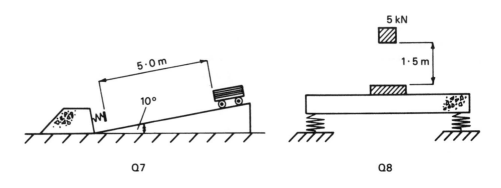

Q7   Q8

Programme 14

VIRTUAL WORK

On completion of this programme you will be able to:

- State the principle of Virtual Work
- Apply the principle of Virtual Work to the calculation of deflections in pin-jointed plane frames
- Apply the principle of Virtual Work to the calculation of deflections and rotations of statically determinate beams using integration techniques
- Apply the principle of Virtual Work to the calculation of deflections and rotations of statically determinate beams using approximate summation techniques

1

This programme is intended as an introduction to a method of structural analysis which is arguably the most powerful analytical technique for solving a very wide range of structural problems. We will examine the basis of the method and its application in the calculation of deflections of framed structures and beams. We have already, in Programme 12, looked at one method of determining the deflections of beams. The method that we will now investigate is an alternative and more versatile technique known as the *Principle of Virtual Work*.

2

The general principle of Virtual Work can be seen by considering a set of concurrent forces acting on a small particle at a point P as shown below. The forces will have a resultant R, also shown on the diagram.

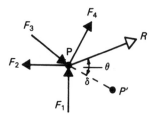

Now let's assume that the particle is displaced by a *small* distance, δ, in any arbitrary direction to the point P'. We will assume that because the displacement is small, the directions of the forces and their resultant will remain unchanged.

Because the particle has moved, each force will do work. The work done by the individual forces must also be equal to the work done by their resultant, R.

Write down an expression for the work done by the resultant force R in terms of δ and the angle θ shown in the figure.

3

$$\boxed{\text{work done} = R \times \delta \cos \theta}$$

Did you remember that work is equal to the product of the force and the distance moved *in the direction of the force*? Hence, to obtain the work done by the resultant R

the displacement δ must be resolved in the direction of the resultant force. Hence the expression given for work done.

If the forces acting on the particle are in equilibrium, what can you say about the resultant, R, and hence about the work done during this small displacement?

4

> R = zero: hence the total work done equals zero

You know that any set of forces in equilibrium cannot have a resultant force. It follows therefore that for the set of forces given, the work done during this small displacement of the particle must be zero.

The displacement δ was a small arbitrary displacement in any direction which can be referred to as a *virtual* or *imaginary* displacement: it is a displacement that does not necessarily occur in practice but one that the analyst imagines to take place for the purpose of structural analysis.

The general principle of what we have shown is known as *the Principle of Virtual Work* (sometimes referred to in this context as the Principle of Virtual Displacements) which can be stated as:

If a system of forces is in equilibrium and undergoes any arbitrary virtual displacement then the total work done by the system of forces is zero.

This may not seem a very impressive statement, but forms the foundation of a powerful method of structural analysis. Move on to the next frame and we will see how to apply the method to a common structural analysis problem.

5

For the time being we will confine our consideration of the method to the calculation of the deflections at the nodal points of frame structures such as those that we have already considered in Programme 3.

Indeed, in calculating the deflections of such structures it is necessary to determine the axial forces in the members of the structures; so you may need to revise Programme 3 before proceeding. In particular, make sure that you are confident that you can carry out the method of 'Resolution at Joints', as the ability to do this quickly and accurately is fundamental to what we are about to do in this programme.

Let's begin by looking at the simple pin jointed frame shown overleaf:

6

Use the method of 'Resolution at Joints' to calculate the forces in the members of the frame and hence calculate the change in length of each member. Take the cross-sectional areas of each member as 600 mm² and E = 200 kN/mm².

7

> AB: force = 200 kN (tension), extension = 8.33 mm
> BC: force = 120 kN (comp.), shortening = 3 mm
> AC: no force or change in length

The forces are shown in figure (a) and the change in lengths of the two members are shown on figure (b). These changes in lengths (δ) are calculated using the formula $\delta = PL/AE$.

For example, for member AB

$$\delta = \frac{PL}{AE} = \frac{200 \times 10^3 \times 5000}{200 \times 10^3 \times 600} = 8.33 \text{ mm}$$

and likewise for member BC.

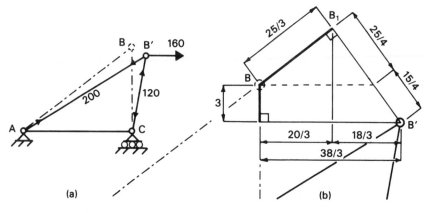

(a)   (b)

The geometrical construction in figure (b) shows the structural displacements and also that the external force of 160 kN moves through a horizontal distance of

38/3 mm. The construction assumes that member AB extends by 25/3 mm to B_1 and the member then rotates about the fixed end A, so that for *small displacements* B_1 moves along a line *at right angles* to AB to the displaced position B'. A similar consideration applies to member BC which shortens by 3 mm. The three displacements (38/3, 25/3 and 3 mm) represent a *compatible set of displacements* consisting of an external deflection in the direction of the load and internal axial deformations. These are a set of geometrically possible *small* displacements which will ensure that the structure deforms but retains its original configuration.

Figure (a) shows a *set of forces in equilibrium* and figure (b), as already stated, shows a compatible set of displacements. These happen to be the real or actual displacements of the structure but we could also think of them as a set of arbitrary virtual displacements as the principle of virtual work allows us to consider *any* *compatible set* of displacements, real or otherwise. Now, remembering that the general definition of work is force times displacement in the direction of the force, let's calculate the total work done by the internal forces shown in figure (a) when multiplied by the corresponding displacements shown in figure (b):

Internal Work done = work done by force in AB + work done by force in BC

$$= 200 \times (25/3) + (-120) \times (-3)$$

$$= 6080/3 \text{ kN mm}$$

Note that in the calculation we have taken tensile forces as positive and axial extensions also as positive. Conversely, axial compression and axial shortening are both negative. Note also that we are taking the axial loads as a set of *constant* forces multiplied by a set of virtual displacements so, unlike the method of Strain Energy in Programme 13, there is no multiplying factor of $\frac{1}{2}$.

Calculate the work done by the external force and state the relationship between internal and external work done.

8

6080/3 kN mm: internal and external work are the same

The external work is given by the product of the external force and its corresponding displacement $(160 \times 38/3)$. The fact that the two sets of work are equal may not be too surprising and is just another way of saying that the total work done by the system of internal and external forces is zero.

However, let's look again at the problem, but this time take the same set of forces in equilibrium and a set of *virtual displacements* which are *not the real displacements of the structure*.

9

The virtual displacements (δ and $4/5\delta$) in figure (b) are a set of *imaginary* displacements *which are not the real displacements of the structure under the given load system*. However they are geometrically compatible, ensuring that the original configuration of the structure is unchanged, and they are *small* in relation to the geometry of the structure such that the directions of the forces acting on and within the actual structure do not effectively change. The vertical member BC does not change in length but rotates about C, with B moving along a line at right angles to BC to the deformed position B′ which is at the same horizontal level as B. Although for clarity we have shown rather exaggerated displacements, they are in fact very small.

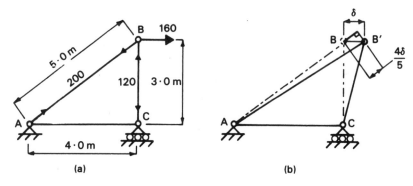

(a)   (b)

Taking the work done as the product of the forces in figure (a) and the corresponding displacements in figure (b), calculate the internal work and external work done.

10

$$\boxed{160\delta \text{ in both cases}}$$

Check your working:

Internal work done = work done by force in AB + work done by force in BC

$$= 200 \times (4/5\delta) + (-120) \times 0$$

$$= 160\delta$$

External work done = external force × displacement

$$= 160 \times \delta = 160\delta$$

What we have done is to take a set of real forces, multiply them by a set of virtual (imaginary) displacements and show that the internal work done is equal to the external work done *even though the set of displacements was virtual not real*. We refer to the work done as being *virtual work* as it is associated with virtual displacements. The fact that the internal work done is equal to the external work done is just another way of saying that the total work done is zero.

This is an application of the *Principle of Virtual Work* which for a pin jointed frame structure can be stated as:

If a pin jointed frame structure is in equilibrium under the action of a set of external loads and is subjected to a set of virtual joint displacements, then the virtual work done by the external loads must equal the virtual work done by the internal forces.

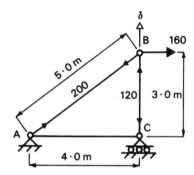

If the same frame with the same loading as before is imagined to have a small virtual displacement δ in the direction of the member BC, as shown above, calculate the internal and external virtual work and show that they both are zero. The solution is in the next frame.

11

Check your solution:

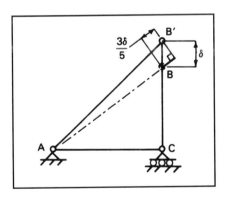

$$\text{Internal work} = 200 \times (3/5\delta) + (-120) \times \delta$$
$$= 0$$
$$\text{External work} = 160 \times 0$$
$$= 0$$

The principle of virtual work is thus seen to be valid for yet another set of virtual displacements.

Now on to the next frame.

12

So far we have considered only an example involving a simple frame, but the principles that we have established are valid for any pin jointed frame. In general, we may have a frame consisting of n members subjected to any number of forces (or components of force) applied at the joints of the frame. In such a case we can express the principle of virtual work in the form of a general expression which equates the total work done by the external loads to the work done by the internal forces:

$$\underbrace{\sum W\delta}_{\text{force set}} = \overbrace{\sum^{n} Pe}^{\text{displacement set}} \tag{14.1}$$

where: W and P are a set of external and internal forces respectively which must be in equilibrium.

δ and e are a set of virtual joint displacements and compatible axial extensions (or shortenings) of members which must be geometrically possible. The joint displacements must be in the direction of the corresponding joint forces.

Now state whether the following are true or false:

(a) In using equation (14.1) the set of joint displacements (δ and e) must be the real displacements of the structure under the action of the set of loads, W and P.

(b) If any set of compatible displacements can be identified they can be used in equation (14.1).

(c) The magnitude of the displacements used in equation (14.1) do not matter.

(d) If you can identify any set of real displacements of a structure, the forces causing those displacements are the only forces that you can associate with that set of displacements in the use of equation (14.1)

(e) You can use any set of forces in equation (14.1).

13

> (a) false   (b) true   (c) false   (d) false   (e) false

Did you get these all correct? If not, check your reasoning against the explanation given on the next page, and if in doubt re-read the previous frames before proceeding:

(a) You shouldn't have got this wrong. We have already extensively worked through an example showing that *any* set of virtual displacements which are geometrically compatible will satisfy the principle of Virtual Work. They do not have to be the real displacements under the set of loads being used.

(b) Let's restate this again—*any* set of virtual displacements which are geometrically compatible can be used.

(c) The displacements *must* be small such that the magnitude and direction of the force systems can be considered to be unchanged by the displacement system.

(d) Any set of real displacements must be a set of compatible displacements. Hence if the structure is being analysed under the action of any set of forces (set I), the set of real displacements arising from any other set of forces (set II) can be used as the virtual displacements together with the first set of forces, set I, in equation (14.1).

(e) Any set of forces *which are in equilibrium* can be used. If they are not in equilibrium, they do not satisfy the basic concepts of the principle of virtual work.

Make sure that you fully understand these answers before proceeding to the next frame.

14

So far we have established some principles but haven't yet applied them to any problems. Let's see how the Virtual Work method can lead to an elegant solution technique to calculate the deflections of pin jointed frame structures:

Consider the frame shown below where all members have identical properties (EA). Let's assume that we want to calculate the horizontal deflection of joint C(δ_c) under the action of the given load system.

The internal member forces are as shown and have been obtained by resolving at the joints. The member displacements are calculated from the formula ($\delta = PL/AE$) and are shown in figure (b). Note that the length of member AC is $\sqrt{(2)}L$. Let's call this set of member and compatible external joint displacements, set I.

(a)   Set I   (b)

Check the values shown in the figures before proceeding.

15

Now let's take the same structure and consider it under the action of a single point load of 1 unit acting at joint C in the direction of the displacement δ_c which we are trying to calculate. This load is often referred to as a dummy load. If this load were acting we could calculate the member forces by resolving at the joints to give the internal forces shown. Note that we are working in terms of symbols; if we were using numerical values then the unit load would have to be in consistent units (for example, if units were kN, then unit load $= 1$ kN).

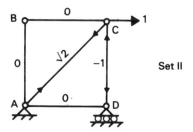

Let's say that this set of forces is set II. Now if we are to apply the principle of virtual work to this load system we would make use of equation (14.1):

displacement set

$$\sum W\delta = \sum^{n} Pe$$

force set

where the set of forces (W and P) can be taken as those of set II in the above diagram, and the set of virtual displacements (δ and e) can be taken as *any* compatible set of small displacements. Let's take the displacements from set I as the compatible set.

Hence, as there is only one external load (of unit value) acting at joint C in the horizontal direction, equation (14.1) can be written as:

set I

$$1 \times \delta_c = \sum^{n} Pe$$

set II

or

$$\delta_c = \sum^{n} P_{II}e$$

The solution of the right-hand side of this equation will therefore give δ_c, the horizontal deflection of joint C in the originally loaded frame that we were trying to analyse. Values of P_{II} will come from the figure above and corresponding values of e from figure (b) in Frame 14.

Can you calculate δ_c? (Tabulate your calculations.)

$$\delta_c = 4[1 + \sqrt{(2)}]\frac{WL}{AE}$$

Check your solution against the tabulated solution given below. Study the presentation of the table and learn to set out all your subsequent calculations in this way. Also take careful note of the sign convention that we are using, which we first introduced in Frame 7. (Signs are important!)

Member	Length	P (set I)	$e(PL/AE)$ (set I) ($\times WL/AE$)	P_{II} (set II)	$P_{II} \times e$ (set II)$\times$(set I) ($\times WL/AE$)
AB	L	$-2W$	-2	0	0
BC	L	$-W$	-1	0	0
CD	L	$-4W$	-4	-1	4
AD	L	0	0	0	0
AC	$\sqrt{(2)}L$	$2\sqrt{(2)}W$	4	$\sqrt{2}$	$4\sqrt{2}$

$$\delta_c = \sum^n P_{II}e = 4[1 + \sqrt{(2)}]\frac{WL}{AE}$$

This method is often referred to as the *Unit Load Method* for calculating deflections and can be applied to any similar type of problem in an identical way. If the members had different sectional areas and/or elastic moduli, this would be accounted for by having an additional column(s) in the above table, otherwise the procedure remains the same.

The steps in the method can be summarised as:

(a) Analyse the structure to give the member forces under the given load system.
(b) Use these forces to calculate the change in member lengths ($e = PL/AE$). (set I).
(c) Apply a unit load to the structure in the direction of the displacement that is to be calculated (no other loads on the structure) and calculate the member forces (P_{II}—set II).
(d) Use the principle of virtual work to give the displacement:

$$\delta = \sum^n P_{II}e$$

Note that if a deflection comes out as a negative answer, this simply means that the deflection is in the opposite direction to the direction of your chosen unit force. All calculations should be tabulated as this leads to a systematic solution with less chance of arithmetic error.

Calculate the vertical deflection of joint C in the structure in Frame 14.

17

$$4\frac{WL}{AE} \text{ in a downward direction}$$

The solution is given below and is again presented in tabular form. The first four columns of the table are identical to our previous calculation and normally need only be tabulated once. If more than one deflection is being calculated, all calculations can therefore be done in one large table using the first four columns as common to all calculations.

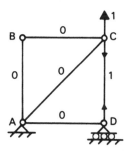

Member	Length	P (set I)	$e(PL/AE)$ (set I) ($\times WL/AE$)	P_{II} (set II)	$P_{II} \times e$ (set II)$\times$(set I) ($\times WL/AE$)
AB	L	$-2W$	-2	0	0
BC	L	$-W$	-1	0	0
CD	L	$-4W$	-4	1	-4
AD	L	0	0	0	0
AC	$\sqrt{(2)}L$	$2\sqrt{(2)}W$	4	0	0

$$\delta_c = \sum^n P_{II}e = -4\frac{WL}{AE}$$

Our previous two calculations have given us the horizontal and vertical components of the deflection at joint C. If the magnitude and direction of the actual deflection at C is required, how can you obtain this information?

18

by combining the deflection components as vectors

Displacements are vectors and hence the component displacements can be combined vectorially by the methods discussed in Programme 1.

Now move to the next frame and try some problems for yourself:

19

PROBLEMS

1. Calculate the horizontal and vertical deflections of joint C in the frame shown in figure Q1 if all members have identical properties (AE).

Ans. $(4.54WL/AE, 15.90WL/AE)$

2. Repeat question 1 but this time assume that the members on the outside of the frame have a cross-sectional area of $2A$ and the internal members a cross-sectional area of A.

Ans. $(2.27WL/AE, 10.71WL/AE)$

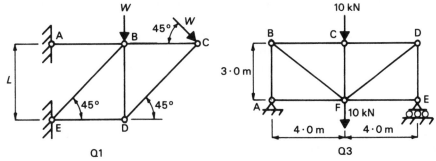

Q1   Q3

2. Calculate the vertical deflection of joint F in the frame in figure Q3 where for all members $AE = 200 \times 10^3$ kN. (*Hint:* make use of symmetry.)

Ans. (*1.2 mm*)

20

The application of the Principle of Virtual Work to determine the deflections of statically determinate pin jointed frame problems forms only an introduction to this powerful method of analysis. It can equally be used to analyse statically indeterminate frames, but this is outside the scope of this text and will not be considered here.

To complete our introductory study of this method of analysis we will, however, look at the application of the method to the determination of the deflections and rotations of statically determinate beam problems.

In the case of a frame structure, when applying the virtual work equation, we can use any set of axial forces in equilibrium with external linear loads and any set of compatible internal axial displacements and linear external displacements.

If the Principle of Virtual Work is to be used to calculate the deflection of, say, a simply supported beam supporting a uniformly distributed load, can you say what we would mean by the 'set of forces' and what we would mean by the 'set of virtual displacements'?

21

> Set of forces = external loads (and couples) and internal bending moments
> Set of displacements = external linear displacements and rotations
> and internal rotations

From our previous study of deflection in beams we have already established that the beam deflections are a function of the bending moments within the beam arising from the external loading system. Deflections can result from shearing forces but these are usually small and neglected. The response of a beam to loading is to deflect, causing a rotation at all points along the beam, and it is these rotations that we must take into account in our application of the Principle of Virtual Work.

Consider a simply supported beam carrying a system of loads:

(a) Bending moments   (b) Real displacements   (c) Virtual displacements

Figure (b) shows the real displacement of the beam and a *small* element of beam at a distance x from the left-hand end where the bending moment is M. Figure (c) shows the beam subjected to *small* additional *virtual* displacements such that the load W deflects through a small vertical displacement δ and the small element rotates through a small virtual rotation $\delta\theta$. These deflections and rotations must be a compatible set.

Write down the expressions for the virtual work done by the external load W and the internal virtual work done by the moment M.

22

External work = $W \times \delta$;   Internal work = $M \times \delta\theta$

The expression for the external work is quite familiar and follows from the general definition of work as being equal to force times displacement in the direction of the force. The expression for internal work may not be quite so familiar but follows from a similar concept of work done as being equal to the product of the moment and the angular distance through which the moment rotates. The total virtual work done by all the external forces is hence given by:

$$\text{External work} = \sum W\delta$$

And the total internal work done by the bending moments is given by:

$$\text{Internal work} = \int_0^L M \, d\theta$$

where, in this case, the summation of terms for all elements along the beam span has to be carried out as an integration over the span length.

Hence if we now apply the Principle of Virtual Work we can equate the above two expressions to give

$$\overbrace{\sum W\delta}^{\text{displacement set}} = \underbrace{\int_0^L M \, d\theta}_{\text{force set}} \qquad (14.2)$$

where: W and M are a set of external forces (and couples) and internal bending moments which must be in equilibrium.
  δ and $d\theta$ are a compatible set of virtual beam displacements and rotations which must be geometrically possible.

Is it necessary for the set of displacements used in equation (14.2) to have been caused by the set of forces also used in equation (14.2)?

23

NO

We have already established this principle quite firmly in our study of the method applied to pin jointed frames. The load set can be any set of forces (and/or moments) *in equilibrium* and the displacement set can be any *compatible* set of geometrically possible small displacements.

Make sure that you are sure about this before proceeding.

24

We are interested in using the Principle of Virtual Work to calculate the deflections of beams subject to any loading system. For example, let's say that we want to calculate the deflection of point A in the beam shown below. The loads shown are the actual loads which will give rise to a set of deformations (deflections and rotations). Let's call this set I.

Set I

Based on what you have learnt about how to calculate deflections in pin jointed frames, what do you think that you should do to calculate the vertical deflection of point A?

25

$$\boxed{\text{apply a vertical load at A}}$$

The principle is exactly the same as before. If we apply a unit load at A (which should be vertical if we are to calculate a vertical deflection) this unit load will giver rise to a set of internal moments and deformations which we can call set II.

Hence if we make use of equation (14.2):

$$\sum W\delta = \overbrace{\int_0^L M\,\mathrm{d}\theta}^{\text{displacement set}}$$

$$\underbrace{\phantom{\sum W\delta = \int_0^L M\,\mathrm{d}\theta}}_{\text{force set}} \tag{14.2}$$

we can take the set of forces (W and M) as those from set II and the set of virtual displacements as *any* compatible set, which we will take as those from set I. Hence, as in set II there is only one external unit load acting at A, equation (14.2) can be written as:

$$1 \times \delta_A = \overbrace{\int_0^L \underbrace{M_{\mathrm{II}}\,\mathrm{d}\theta_{\mathrm{I}}}_{\text{set II}}}^{\text{set I}}$$

or

$$\delta_A = \int_0^L M_{II} \, d\theta_I \qquad (14.3)$$

The solution of the right-hand side of this equation will give δ_A which is the vertical deflection of the point on the original structure that we are trying to determine.

If we wanted to calculate the rotation of the support in the beam in Frame 24, what would you do?

26

apply a unit *moment* at the support

The principle is the same—if you wish to calculate a deflection you apply a single unit force at the point where the deflection is to be calculated and in the direction of required deflection. If you wish to calculate a rotation you apply a single unit moment at the point where the rotation is required.

Before we can make use of equation (14.3) it is necessary to re-arrange the equation into a form that makes it more readily usable:

$$\delta_A = \int_0^L M_{II} \, d\theta_I = \int_0^L M_{II} \frac{d\theta_I}{dx} \, dx$$

But from our previous work we know that the term $(d\theta/dx)$ is the change of slope at the point we are considering at a distance x along the beam, and is therefore equal to the beam curvature $(1/R)$ at that point. Look back at Frame 16 of Programme 12 if you are in doubt about this.

We also know that the beam curvature is related to the bending moment at that point by the flexural stiffness term (EI).

Hence
$$\frac{d\theta_I}{dx} = \frac{1}{R} = \frac{M_I}{EI}$$

and substituting into the above equation:

$$\delta_A = \int_0^L M_{II} \frac{d\theta_I}{dx} dx = \int_0^L M_{II} \frac{M_I}{EI} \, dx$$

which for a prismatic beam of constant section properties can be written as:

$$\delta_A = \frac{1}{EI} \int_0^L M_{II} M_I \, dx$$

423

27

Let's summarise this last equation but omit the subscript to the δ term, as now the equation can be totally general to give the deflection at any point on the beam.

$$\delta = \frac{1}{EI} \int_0^L M_{\text{II}} M_{\text{I}} \, dx \qquad (14.4)$$

where: M_{II} = the bending moment at a point along the beam due to a unit load applied at the location where the deflection δ is to be calculated (sagging moments are taken as positive)

$\quad\quad M_{\text{I}}$ = the bending moment at the same point due to the actual loads applied to the beam (sagging moments taken as positive)

$\quad\quad EI$ = the flexural stiffness of the beam.

The form of this expression is very similar to that used for the calculation of deflections in pin jointed frames, but whereas previously we were able to perform our calculations as a tabulated summation of terms now we have to use integration techniques to evaluate the expression.

See if you can use equation (14.4) to calculate the end deflection of a cantilever beam of span L which carries a uniformly distributed load of w per unit length. (Hint: write down M_{II} and M_{I} at a distance x from the free end of the cantilever, multiply the terms together and integrate over the beam length.) The complete solution is in the next frame.

28

$$\boxed{\delta = wL^4/8EI}$$

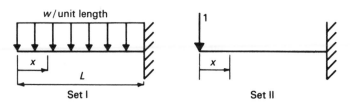

Set I : moment at distance x due to actual loads $= -wx^2/2$
Set II : moment at distance x due to unit load $\quad = -x$

Hence

$$\delta = \frac{1}{EI} \int_0^L M_{\text{II}} M_{\text{I}} \, dx = \frac{1}{EI} \int_0^L x \frac{wx^2}{2} \, dx = \frac{w}{EI} \int_0^L \frac{x^3}{2} \, dx = \frac{w}{EI} \left(\frac{x^4}{8} \right)_0^L = \frac{wL^4}{8EI}$$

If the deflection is required at any other point on this beam, then the principle is the same; apply a unit load at that point and carry out the integration given by equation (14.4). For example, if we require the deflection at the centre of the cantilever that we have just considered, we would carry out the following calculation:

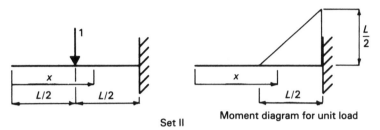

Set II   Moment diagram for unit load

Set I : moment at distance x along the span (M_I) $= -wx^2/2$
        (as before)
Set II: moment at distance x due to unit load (M_{II}) $= 0$
        (for the length of beam between $x = 0$ and $x = L/2$
        and $= -1(x - L/2)$
        (for the length of beam where $X > L/2$

As there are no moments due to the unit load over the left-hand half of the span, then no work can be done in this region when a set of virtual displacements are imposed on the structure. Hence, when using equation (14.4) it is only necessary to carry out the integration over the right-hand half of the span for integration limits from $L/2$ to L. Hence:

$$\delta = \frac{1}{EI}\int_{L/2}^{L} -\left(x - \frac{L}{2}\right)\left(-\frac{wx^2}{2}\right)\,dx = \frac{w}{2EI}\int_{L/2}^{L}\left(x^3 - \frac{Lx^2}{2}\right)\,dx$$

$$= \frac{w}{2EI}\left(\frac{x^4}{4} - \frac{Lx^3}{6}\right)_{L/2}^{L}$$

$$= \frac{17wL^4}{384EI}$$

If we are interested in the rotation at, say, the end of the cantilever, we can apply a unit positive (sagging) moment at the free end. The moment M_{II} at any distance x along the span will be given by $M_{II} = +1$, and the integration of equation (14.4) will give the end rotation.

To use equation (14.4) you had to evaluate the product integral $\int M_{II}M_I\,dx$ where both M_I and M_{II} were functions of the distance x. Generally, these functions will be linear (in the case of point loads), parabolic (in the case of UDLs) or constant (in the case of moments). To aid the integration of equation (14.4), tables of *product integrals* for the most common functions are available which will obviate the necessity to carry out direct integration of equation (14.4). You will find these tables of product integrals in other, more advanced, textbooks which give more comprehensive coverage of this subject.

29

PROBLEMS

1–4. Calculate the rotation of the free end of the cantilever shown in figure Q1, and the deflection and rotation of the free end of the cantilever in figures Q2 to Q4. Also calculate the deflection at the mid span of the cantilevers in figures Q2 and Q3.

Ans. ((1) $-wL^3/6EI$ (2) $WL^3/3EI$, $-WL^2/2EI$, $5WL^3/48EI$ (3) $ML^2/2EI$, $-ML/EI$, $ML^2/8EI$ (4) $wL^4/30EI$, $-wL^3/24EI$)

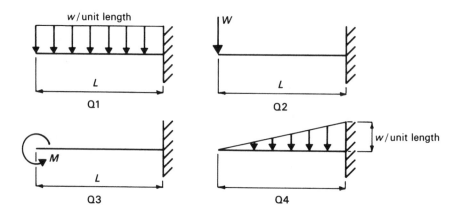

30

Having completed the last few simple examples, you should now be familiar with the concept of calculating beam deflection using virtual work. The steps in the method, which can be applied to any problem are:

(a) At a general point on the beam, determine an expression for the bending moments M_I due to the actual loads (forces and couples) acting on the structure (set I). It will often be helpful to sketch the bending moment diagram first.

(b) Divide this expression by the beam stiffness term (EI) to give the virtual curvature of the point. If the beam has uniform properties, EI will just be a constant value which is the case considered in this book.

(c) Determine the expression for the bending moment M_{II} at the general point due to a single unit load (force or moment) applied at the point on the beam where the deflection (or rotation) is to be determined (set II).

(d) Use the principle of virtual work to give the displacement (deflection or rotation) by evaluating the expression:

$$\delta = \frac{1}{EI}\int_0^L M_{II} M_I \, \mathrm{d}x$$

Although the above steps are generally applicable, not all problems are as straightforward as those considered so far. Move on to the next frame and we will look at how to tackle a slightly more complicated example.

31

Consider the beam shown below. The units shown are kN and metres, but we will not show units in the calculations, so as to concentrate on the principles involved. Figure (a) shows the actual bending moments in the beam (set I).

The requirement is to calculate the mid-span deflection of the beam and hence figure (b) shows the bending moment diagram due to unit load applied at the mid-span position (set II).

From the above diagram you will note that there are three separate lengths over which the integration of equation (14.4) must be performed (A to B, B to C, and C to D). This is because there is a change in one or the other of the two bending moment diagrams in each of these regions, and it is not possible to find a general equation for the bending moments over the full length of the beam. Hence the problem will be split into three parts and three separate integrations performed. The results of the three integrations will be summed to give the final answer. Because equation (14.4) is essentially an evaluation of internal work and is independent of the chosen origin of integration, we can choose separate and the most convenient origins to evaluate the three integrals. Hence in region A–B, A will be taken as the origin; in B–C, B will be taken; and in C–D, D will be taken with the integration performed from D to C.

Write down the expressions for M_I and M_{II} in each of the three regions.

427

32

	M_{I}	M_{II}	
A–B	$8x$	$0.5x$	(moments taken to left of point)
B–C	$8(5+x)$	$\frac{1}{2}(5-x)$	(M_{I}: moments taken to left of point)
			(M_{II}: moments taken to right of point)
D–C	$12x$	$0.5x$	(moments taken to right of point)

Use these expressions and equation (14.4) to evaluate the mid-span deflection of the beam in terms of its stiffness properties (EI). The complete solution is in the next frame.

33

$$\delta = \frac{393.33}{EI}$$

Check your solution:

Section	Origin	M_{I}	M_{II}	$M_{\mathrm{II}}M_{\mathrm{I}}$	Limits	$\int M_{\mathrm{II}}M_{\mathrm{I}}\,dx$	$\delta\ (\times 1/EI)$
A–B	A	$8x$	$0.5x$	$4x^2$	0–5	$4x^3/3$	166.67
B–C	B	$8(5+x)$	$\frac{1}{2}(5-x)$	$(100-4x^2)$	0–1	$100x - 4x^3/3$	98/66
C–D	D	$12x$	$0.5x$	$6x^2$	0–4	$6x^3/3$	128.00

$$\text{deflection} = \frac{393.33}{EI}$$

This approach to calculating deflections can be applied to any problem where the variation of bending moment from both loading systems (set I and set II) can be established over discrete lengths of the beam. As in all similar problems, it will always help if you sketch the shape of the bending moment diagrams before proceeding to do any calculations.

When the loading on a beam consists of a series of loads giving rise to a complicated bending moment diagram, the evaluation of the integral expression in equation (14.4) by normal methods of integration may be difficult and in some cases almost impossible. To finish this programme, let's see how we can carry out an approximate analysis of a beam under a complicated system of loads using the same principles of Virtual Work.

The problem is to evaluate equation (14.4) without direct integration. From your study of mathematics you should know that we can carry out numerical integration using *Simpson's Rule* as equation (14.4) is of the general form $\int y\,dx$. The general approach to such a calculation is to split the range over which integration is to be performed into an even number of equal increments and evaluate the 'ordinates' of the equation given by the function y at each increment. The application of Simpson's Rule will give the value of the integral, the accuracy of which will depend on the number of increments used in the calculation.

Can you recall and write down Simpson's Rule?

34

$$\int y\,dx = \frac{s}{3}[(\text{sum of first and last ordinate}) + 4(\text{sum of even ordinates})$$
$$+ 2(\text{sum of remaining odd ordinates})]$$
where s = length of each increment

If you are not sure about this, consult reference 1 at the end of this programme. Let's apply Simpson's Rule to the problem shown below where the requirement is to evaluate the deflection at point A. The calculations will be performed by dividing the beam into 10 equal increments of 2 metres, calculating M_{II} and M_I at each increment, and evaluating the deflection by applying Simpson's Rule. The stiffness of the beam (EI) is 170×10^3 kN m².

At each section of the beam, spaced 2 metres apart, calculate (by taking moments) the value of the bending moments M_I and M_{II} and apply Simpson's Rule to obtain the deflection at point A. Present your work in a table and check your answer against the complete solution in the next frame.

35

$$\boxed{\delta = 21.5 \text{ mm}}$$

Check your working against that given in the table. The units shown for bending moments are kN m.

Section	1	2	3	4	5	6	7	8	9	10	11
M_I	0	28.0	56.0	84.0	112.0	115.0	108.0	91.0	64.0	32.0	0
M_II	0	1.4	2.8	4.2	3.6	3.0	2.4	1.8	1.2	0.6	0
$M_\mathrm{II} \times M_\mathrm{I}$	0	39.2	156.8	352.8	403.2	345.0	259.2	163.8	76.8	19.2	0

Hence applying Simpson's Rule:

$$\delta = \frac{1}{EI}\int M_\mathrm{II} M_\mathrm{I}\, \mathrm{d}x = \frac{1}{EI}\frac{2}{3}[(0+0) + 4(39.2 + 352.8 + 345.0 + 163.8 + 19.2)$$
$$+ 2(156.8 + 403.2 + 259.2 + 76.8)]$$
$$= \frac{3648}{EI}$$
$$= \frac{3648}{170 \times 10^3} \text{ metres}$$
$$= 21.5 \text{ mm}$$

This approach to numerical integration is a little tedious to carry out manually, but does provide a means of calculating deflections when the shape of the bending moment diagram makes it difficult to carry out straightforward integration techniques.

This numerical approach which is essentially a repetitive calculation does however lend itself to computer applications, and in reference 2 at the end of this programme you will find a programme listing for beam deflection calculations which is based on the Virtual Work methods that we have studied.

Can you now calculate the mid-span deflection of the beam that we have just been considering using the same numerical integration techniques? You do not need to calculate the moments M_I again, as they are already given in the table above.

36

$$\boxed{\delta = 27.0 \text{ mm}}$$

Check your solution against the working given on the next page:

Your solution should have been as follows:

Section	1	2	3	4	5	6	7	8	9	10	11
M_I	0	28.0	56.0	84.0	112.0	115.0	108.0	91.0	64.0	32.0	0
M_{II}	0	1.0	2.0	3.0	4.0	5.0	4.0	3.0	2.0	1.0	0
$M_{II} \times M_I$	0	28.0	112.0	252.0	448.0	575.0	432.0	273.0	128.0	32.0	0

Hence applying Simpson's Rule:

$$\delta = \frac{1}{EI}\int M_{II}M_I\,dx = \frac{1}{EI}\frac{2}{3}[(0+0) + 4(28.0 + 252.0 + 575.0 + 273.0 + 32.0)$$
$$+ 2(112.0 + 448.0 + 432.0 + 128.0)]$$
$$= \frac{4586.7}{EI}$$
$$= \frac{4586.7}{170 \times 10^3} \text{ metres}$$
$$= 27.0 \text{ mm}$$

37

TO REMEMBER

The statement of the Principle of Virtual Work:
If a system of forces is in equilibrium and undergoes any arbitrary virtual displacement, then the total work done by the system of forces is zero.

For a pin jointed frame:
$$\delta = \sum^n P_{II}e$$
where: P_{II} = member forces due to unit load
e = change in member lengths due to the actual load system (PL/AE)
where the unit load is applied at the joint at which the deflection δ is required and in the direction of the required deflection.

For a beam:
$$\delta \text{ (or } \theta) = \frac{1}{EI}\int M_{II}M_I\,dx$$
where: M_{II} = bending moments due to unit load (or moment)
M_I = bending moments due to actual load system
where the unit load (or moment) is applied at the point along the beam where the deflection δ (or rotation θ) is required and in the direction of the required deflection (or rotation).

38

1. Determine the vertical and horizontal deflection of joint A in the pin jointed frame shown in figure Q1. The cross-sectional area of all members is 1000 mm² and $E = 200$ kN/mm².
Ans. ($\delta_V = 31.15$ mm, $\delta_H = 39.73$ mm)

2. A truss is loaded as shown in figure Q2. Calculate the horizontal and vertical deflection of joint A if the cross-sectional areas of the compression members is 1000 mm² and that of the tension members is 750 mm². $E = 200$ kN/mm².
Ans. ($\delta_H = -2.50$ mm, $\delta_V = 14.82$ mm)

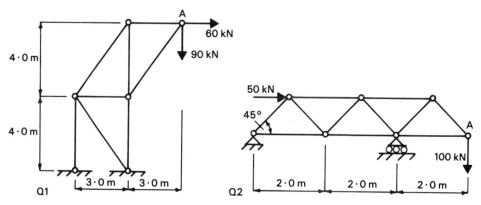

3. All the members of the warren girder shown in figure Q3 have identical sectional areas. Determine the minimum required cross-sectional area if the mid-span deflection is limited to 10 mm. $E = 200$ kN/mm².
Ans. (3354.0 mm²)

4. Use the principle of virtual work to calculate the mid-span deflection and support rotations of a beam of span L which carries a uniformly distributed load w per unit length.
Ans. ($5wL^4/384EI$, $\pm wL^3/24EI$)

5. Calculate the mid-span deflection of the beam shown in figure Q5.
Ans. ($5wL^4/768EI$)

6. The cantilevered beam in figure Q6 carries a point load W at the end of the cantilever. Determine an expression for the deflection at the end of the cantilever. (*Hint:* sketch the shape of the bending moment diagrams for the actual and the unit load systems before doing any calculations—you will have to integrate over two lengths of the beam.)

Ans. ($WL^3/8EI$)

7. Calculate the deflection at the end of the cantilever and at the point marked 'A' for the beam shown in figure Q7. (*Hint:* sketch the bending moment diagrams and take account of the signs of the moments.)

Ans. ($wL^4/128EI$, $wL^4/192EI$)

8. Make use of Simpson's Rule to calculate the deflection of the point marked 'A' for the beam shown in figure Q8. Take $EI = 30 \times 10^3$ kN m^2.

Ans. (1.41 mm)

REFERENCES

1. K. A. Stroud, *Engineering Mathematics*, Macmillan, London, 3rd edn, 1987, p. 621.
2. R. Hulse and W. H. Mosley, *Reinforced Concrete Design by Computer*, Macmillan, London, 1986, pp. 17–26.

Programme 15

MOMENT DISTRIBUTION
AND INDETERMINATE
STRUCTURES

On completion of this programme you will be able to:

- Recall the conditions for statical determinacy
- Explain the nature of indeterminacy
- Recognise and identify simple forms of statically indeterminate structures
- Calculate the support reactions for statically indeterminate structures
- Calculate the bending moments within such structures
- Draw the bending moment and shear force diagrams for continuous beams subject to various load systems
- Draw the bending moment diagrams for propped portal frame structures
- Draw the bending moment diagrams for unpropped portal frame structures

1

So far we have looked at how to solve problems of a structural nature where the structures have been *statically determinate*: that is, they can be solved using the fundamental principles of statics and the equations of statics. Many structures are not as simple, in form or construction, as those that we have considered and it is not possible to apply the techniques that we have considered to their solution. However, to design these structures it is still necessary to establish the magnitude and direction of the supporting reactive forces and the magnitude and nature of the internal shear forces and bending moments which develop to resist the external loading system.

This programme will therefore introduce one of the most common techniques for the solution of structures which are not statically determinate. The particular method that we will develop is known as the method of *moment distribution* and is a manual solution method. In a later programme we will develop another solution technique which is the basis of modern computer methods of analysis.

Can you recall and write down the necessary condition for a structure to be considered to be externally statically determinate?

2

> The number of unknown support reactions cannot exceed the number of independent equations of statics

If you are not sure about this then you should revise Programme 2, Frame 29 before proceeding.

Which of the two structures shown below is externally statically determinate?

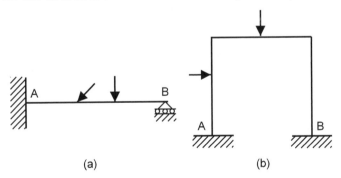

(a)   (b)

3

> Neither of them

Both of these structures are externally statically *indeterminate*.

Figure (a) is a propped cantilever which has three components of reaction at A (V_A, H_A, M_A) and a single reaction at B (V_B): a total of four unknowns. We can only write down three independent equations of statics: this structure is therefore externally statically indeterminate. Similarly, figure (b), a rigidly jointed portal frame structure, has three components of reaction at each support (V_A, H_A, M_A, V_B, H_B, M_B); a total of six unknowns and is also externally statically indeterminate.

Before moving on to the next frame ensure that you are confident that you can identify whether or not a structure is determinate or indeterminate. Remember that if it is externally statically determinate you can use the equations of equilibrium to calculate the support reactions. If not, then other methods, such as those that we will develop in this programme must be used. Now on to the next frame.

4

Before we can carry out any form of analysis using the *Moment Distribution Method*, it is necessary to establish a number of fundamental relationships and principles that will form the basis of the analysis. To do this we will consider the case of a propped cantilever as shown in figure (a) which is built-in at A, supported on a roller at B and subject to an externally applied moment M as shown. You will recall from the previous frame that this is a statically indeterminate structure and we cannot analyse it by simple statics. Let's see how we can, however, determine the vertical reaction at B, V_B, using new concepts and principles.

In Programme 3, Frame 41 we introduced the concept of the *Principle of Superposition*. If we make use of this principle we can consider the propped cantilever as equivalent to the two cantilevered structures shown in figures (b) and (c) . Both of these are statically determinate structures and *superimposed* (i.e. added together) they give us the original structure that we are attempting to solve. The unknown reaction V_B is shown as an upward point load in figure (c).

The end of the cantilever in figure (b) will deflect downwards, whilst the corresponding deflection in figure (c) will be upwards.

What can you say must be the relationship between the cantilever deflections in both figure (b) and figure (c)?

5

They must be equal in magnitude

As the *superposition* of these two figures results in the original propped cantilever that we are attempting to solve the net deflection at B must be zero as B is a support and does not deflect. Hence the downward deflection in (b) must equal the upward deflection in (c).

If you look back at the examples that you did in Programme 14, Frame 29 you will see that figures Q3 and Q2 are almost identical to the two problems that we are considering except that the point load on the end of cantilever (c) is acting upwards and not downwards. Taking the answers from these two problems you can see that:

The downward deflection in figure (b) $= ML^2/2EI$

and the upward deflection in figure (c) $= V_B L^3/3EI$

where EI, as before, is the flexural stiffness of the beam.

Remembering what we said about the relationship between the two deflections, can you now calculate the vertical reaction at B in terms of the applied moment M?

6

$$V_B = 3M/2L$$

This is quite simple as, if we equate the two equations we get:

$$V_B L^3/3EI = ML^2/2EI$$

$$\therefore \quad V_B = 3M/2L$$

Hence we have calculated the unknown end reaction in this statically indeterminate structure using the concept of *compatibility of displacements* and *superposition*. Once this end reaction is known then the structure is now *statically determinate* as we now have only three unknowns (V_A, H_A, M_A) and there are still three independent equations of statics that we can write down and solve.

Draw a free-body diagram of the propped cantilever and use normal principles of statics to calculate V_A, H_A and M_A.

7

$$\boxed{V_A = -V_B = -3M/2L: \quad H_A = 0: \quad M_A = M/2}$$

If you did not get the correct answers then check your calculations against those given below. Take particular note of the equation for moments about A.

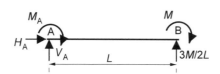

$$\Sigma H = 0: \qquad\qquad\qquad \therefore \ H_A = 0$$
$$\Sigma V = 0: \qquad\qquad V_A + V_B = 0$$
$$\qquad\qquad\qquad \therefore \ V_A = -V_B = -3M/2L$$
$$\Sigma M_A = 0: \qquad M_A + M - (V_B \times L) = 0$$
$$\therefore \ M_A = (V_B \times L) - M = (3M/2L \times L) - M$$
$$= M/2$$

Having performed these calculations the important result to note, which will form the basis of future work in this programme, is that:

Principle 1: the moment at the built-in end of a propped cantilever is exactly one-half of that applied at the free end and acts in the same direction.

A propped cantilever has a clockwise moment of 10 kN m applied at the propped end. What is the moment induced at the built-in support?

8

$$\boxed{\text{5 kN m clockwise}}$$

Let's consider the same propped cantilever and consider the _rotation_ at the propped end at B.

How would you calculate the rotation at B?

439

9

> Add together, algebraically, the rotation in figure (b) and that in figure (c)

If you look back again at the examples that you did in Programme 14, Frame 29 the solutions to figures Q2 and Q3 give the end rotations as:

End rotation in figure (b) $= +ML/EI$ (clockwise)

and end rotation in figure (c) $= -V_B L^2/2EI$ (anti-clockwise)

Hence in the original propped cantilever (figure (a)) the net end rotation is given by:

$$\theta_B = ML/EI - V_B L^2/2EI = ML/EI - (3M/2L) \times L^2/2EI$$
$$= ML/4EI$$

And re-arranging the above equation gives: $\quad M = \dfrac{4EI}{L}\theta_B = k \times \theta_B$

where k is the *rotational stiffness* of the beam which is defined as the moment that causes an end rotation of unit value. For simplicity we can call k the beam *stiffness*.

Now consider the simple structure shown below which has a moment M applied at joint B. The two parts of the structure, AB and BC, are identical and have the same stiffness values ($4EI/L$).

Sketch the deflected shape of the structure.

10

(i) State the relationship between the rotation at the right-hand end of AB and the left-hand end of BC and (ii) identify, conceptually, what proportion of the moment M is carried by AB and by BC.

> (i) They are the same as they are the same point on the structure
> (ii) Both parts of the beam will carry an equal proportion of moment $= M/2$

As both AB and BC are identical, and therefore have identical stiffness properties $(4EI/L)$, and, as at the central propped support they must rotate by the same amount, then we can say that the moment induced in each part of the structure at B is given by:

$$M_{BA} = M_{BC} = \frac{4EI}{L} \times \theta_B$$

where M_{BA} denotes the moment at B in the part of the structure between B and A. Likewise M_{BC} denotes the moment at B in the part of the structure between B and C. Consideration of equilibrium will tell us that the external moment M applied at B must be *distributed* equally between the two parts of the structure. That is:

$$M = M_{BA} + M_{BC} = \frac{4EI}{L}\theta_B + \frac{4EI}{L}\theta_B = \frac{8EI}{L}\theta_B$$

or put another way:

$$M = \sum \text{the stiffness of each member at the joint} \times \theta_B$$

and $\quad M_{BA} = \dfrac{\text{stiffness of member BA}}{\sum \text{stiffness of each member at the joint}} \times M = \dfrac{4EI/L}{8EI/L} \times M = \dfrac{M}{2}$

Likewise for member BC. This equation is the basis for the second fundamental principle that we will use:

Principle 2: At a joint in a structure moments are distributed into each member in proportion to the stiffnesses of the members meeting at that joint.

Can you use the above principle to determine the moment distributed into each of the members of the three-member frame? All members have the same stiffness value k.

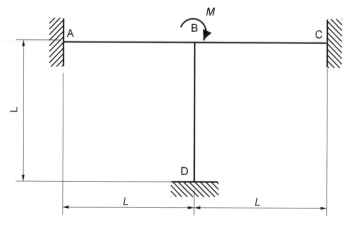

12

$$M_{BA} = M_{BC} = M_{BD} = M/3$$

As each member has an equal stiffness value it follows from *Principle 2* that the moment in each member must be equal.

The members of the two-span beam shown have identical EI values. Calculate the moments in each member at joint B and the moments induced in the built-in ends at A and C. (Hint: look back at both Principle 1 and Principle 2.)

13

$$M_{BA} = M/3: \quad M_{BC} = 2M/3: \quad M_{AB} = M/6: \quad M_{CB} = M/3$$

Did you calculate these as follows?

$$\text{Stiffness of AB} = k_{AB} = 4EI/L$$
$$\text{Stiffness of BC} = k_{BC} = 4EI/(L/2) = 8EI/L$$

From *Principle 2*:

$$M_{BA} = \frac{\text{stiffness of member AB}}{\sum \text{stiffness of each member at the joint}} \times M$$

$$= \frac{4EI/L}{(4EI/L + 8EI/L)} \times M = \frac{4}{12} \times M = \frac{M}{3}$$

and

$$M_{BC} = \frac{\text{stiffness of member BC}}{\sum \text{stiffness of each member at the joint}} \times M$$

$$= \frac{8EI/L}{(4EI/L + 8EI/L)} \times M = \frac{8}{12} \times M = \frac{2M}{3}$$

From *Principle 1*:

$$M_{AB} = \frac{1}{2} \times M_{BA} = \frac{1}{2} \times \frac{M}{3} = \frac{M}{6}$$

and

$$M_{CB} = \frac{1}{2} \times M_{BC} = \frac{1}{2} \times \frac{2M}{3} = \frac{M}{3}$$

Did you remember that a moment applied at the free end of a propped cantilever induces a moment at the built-in end of one-half that at the free end? This continuous beam is not a propped cantilever but each part (AB and BC) can be independently considered as such with the end moment being that proportion of the applied joint moment that is carried by each part of the structure according to its stiffness.

The factor of one-half that relates the moment at the propped end to that at the fixed end is referred to as the *carry-over factor*. In other words, one-half of the moment at the propped end is *carried over* to the fixed end. The ratio of the stiffness of a member to the sum of the stiffnesses of all members at a joint is referred to as the *distribution factor*. Hence in this problem the distribution factors for **AB** and **BC** are 1/3 and 2/3, respectively.

The calculation conveniently lends itself to a tabular presentation as follows:

Joint	A		B		C
Distribution factor		1/3		2/3	
Applied moment at joint			M		
Distributed moment		M/3		2M/3	
	0.5			0.5	
Carry-over moment	M/6				2M/6
Final moments	M/6	M/3		2M/3	M/3

Study this presentation and then attempt the problems in frame 14, setting out your work in tabular form.

14

PROBLEMS

1. Calculate the moments at A, B and C in the continuous beam shown in figure Q1. Member AB has an *EI* value twice that of member BC.

Ans. ($M_{BA} = 5.0$, $M_{BC} = 5.0$, $M_{AB} = 2.5$, $M_{CB} = 2.5$ kN m)

2. Calculate the end moments in both members of the steel frame shown in figure Q2. The two members are constructed from the same steel section.

Ans. ($M_{BA} = 9.0$, $M_{BC} = 6.0$, $M_{AB} = 4.5$, $M_{CB} = 3.0$ kN m)

15

So far we have considered the case of a moment applied to a joint of a rigidly jointed structure and shown that the moment is distributed between the members at the joint proportional to the member stiffnesses ($4EI/L$). We have also shown that a moment applied at the end of a member which is free to rotate results in a 'carry-over' moment at the fixed end of one-half of the applied moment. Now let's look at the more realistic case of a structure which is subject to loading along the length of its members such as the continuous beam shown in (a) below.

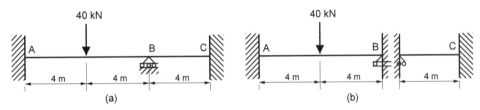

To analyse this problem we will use an analytical technique whereby we 'clamp' joint B which is, in reality, free to rotate such that initially all the joints of the frame are imagined to be prevented from rotating. This is shown in figure (b) where, for illustrative purposes, joint B has been separated to show both parts of the structure. Both AB and BC are now *built-in* beams with fixing moments at either end.

Can you sketch the deflected shape of each part of the structure in figure (b), indicating which side of the beam is in tension and hence sketch the shape of the bending moment diagram for both parts?

16

As beam AB is fixed at both ends and it must enter the built-in supports at right angles, its deflected shape indicates a change of curvature along its length such that at mid-span it is subject to sagging bending moments and at the ends to hogging bending moments. Hence the shape of its bending moment diagram as shown. Beam BC is not subject to any loads, does not deflect and is not subject to any bending moments.

The bending moment diagram for AB is in fact that of a standard load case of a centrally placed point load acting on a fixed-ended beam. The fixed-end hogging moments are given by a standard well-known formula ($WL/8$) where W is the magnitude of the load and L is the span of the beam. We are not going to prove this formula at this stage but we are going to use it.

Calculate the fixed-end moments of the beam AB and indicate their directions (clockwise or anti-clockwise) on a sketch of the beam.

17

$$FEM_A = -40 \times 8/8 = -40 \text{ kN m} \quad FEM_B = +40 \times 8/8 = +40 \text{ kN m}$$

− 40 kN m   +40 kN m

A   B

The fixed-end moment at B will be clockwise to prevent the tendency of the beam to rotate in an anti-clockwise direction at that support. Likewise the fixed-end moment at A will be anti-clockwise. We will adopt the sign convention that *clockwise moments are taken as positive.*

If we return to the original diagram of the continuous beam we can now see that to prevent rotation at joint B we have in fact introduced an imaginary moment of 40 kN m at B in a clockwise direction as shown in (b) below. This is an artificial situation as in reality we know that the structure is free to rotate at B under load. To allow this rotation to take place we can therefore apply an anti-clockwise moment of 40kNm at B as shown in figure (c). The *superposition* of figure (c) onto figure (b) will cancel out the two equal and opposite moments of 40kNm at B and will return us to the original continuous beam problem that we are trying to solve.

(a) Original structure   (b) restrained system – no joint rotation   (c) released system – joint B free to rotate

Can you see how figure (c) now looks very much like the problems that we solved in the earlier parts of this programme? As the structure rotates at B the anti-clockwise moment of 40 kN m will be distributed into the two parts of the structure (BA and BC) according to their relative stiffnesses and there will be a *carry-over moment* of one-half of the distributed moment to the respective fixed ends.

Using the principles established in the earlier parts of this programme calculate the distributed and carry-over moments in figure (c). Then superimpose these moments on to those shown in figure (b) to obtain the actual set of bending moments in the beam at each joint.

$$M_{AB} = -46.67, \; M_{BA} = +26.67, \; M_{BC} = -26.67, \; M_{CB} = -13.33 \text{ kN m}$$

Check your working. For figure (c):

Joint	A		B		C
Distribution factor		0.33		0.67	
Applied moment at joint			−40		
Distributed moment		−13.33		−26.67	
Carry-over moment	−6.67				−13.33
End moments	−6.67	−13.33		−26.67	−13.33

But these results need to be superimposed onto the bending moment values for figure (b) to give the final set of end moments:

Joint	A		B		C
End moments for: Figure (c) Figure (b)	−6.67 −40.00	−13.33 +40.00		−26.67 0	−13.33 0
Final moments	−46.67	+26.67		−26.67	−13.33

Note that the final moments at joint B in the two members are equal in magnitude and opposite in direction as there is no net external moment at this joint in the actual structure that we have solved. If there had been an actual external joint moment then the two internal moments would have summed to the value of the external moment. The two sets of calculations performed above can conveniently be combined into a single set of calculations in a *moment distribution table* as follows:

Joint	A		B		C	
Distribution factor		0.33		0.67		
Fixed-end moments	−40.00	+40.00				(i)
Balancing moment Distributed moment		−13.33	(−40.00)	−26.67		(ii) (iii)
Carry-over moment	−6.67				−13.33	(iv)
Final moments	−46.67	+26.67		−26.67	−13.33	(v)

Note that:
  (i) Fixed-end moments are calculated that prevent the joints from rotating
  (ii) At joint B a 'balancing moment' is applied that releases the joint and allows joint rotation

(iii) This moment is distributed into the members and

(iv) one-half of the distributed moment is *carried over* to the fixed ends.

(v) The moments in each column are summed to give the final set of moments at each support.

19

DRAWING BENDING MOMENT DIAGRAMS

The technique for drawing the bending moment diagram from the calculated end-moments is somewhat different to that which we have previously used. The usual technique is to again make use of the *Principle of Superposition*. The bending moment diagram for the calculated end-moments can be *superimposed* onto the bending moment diagram for the loadings based on each span being treated separately as a simply supported beam. The simply supported beam case will consist of a series of statically determinate beams and normal principles of statics can be used to draw, what is often referred to as, the *free bending moment diagram*. This is illustrated in the figure below:

(a) Support moments   (b) Simply supported moments   (c) Final bending moment diagram

You can see from the addition of figures (a) and (b) that the:

$$\text{moment at A} = -46.67 + 0 = -46.67 \text{ kN m (hogging)}$$
$$\text{moment at B} = -26.67 + 0 = -26.67 \text{ kN m (hogging)}$$
$$\text{moment at C} = +13.33 + 0 = +13.33 \text{ kN m (sagging)}$$
$$\text{moment at D} = +80.00 - (46.67 + 26.67)/2 = +43.33 \text{ kN m (sagging)}$$

You can use this technique of superposition to draw the bending moment diagrams for all the problems in this programme. Ensure that you translate the calculated moments which are based on a sign convention of *clockwise positive* into a correct support moment diagram which is based on the conventional sign convention of *sagging moments positive*. Sketching the deflected shape of the structure under the action of the end moments, as shown below, should help you to get this correct.

Now on to the next frame

20

CALCULATION OF FIXED-END MOMENTS

The key to the technique of moment distribution that we have considered so far is to be able to calculate the correct values of *fixed-end moments* (FEM) for each loading case under consideration. The table below shows the standard formulae for the most common load cases. Where there is more than one load acting on a span the fixed-end moments for each load can be calculated separately and the total fixed-end moments can be calculated by adding together (superimposing) the individual moments. The sign convention of the moments shown in the table conforms to our convention that *clockwise moments are positive.*

Case	Loading condition	FEM at end A	FEM at end B
1	Central point load	$-\dfrac{WL}{8}$	$+\dfrac{WL}{8}$
2	Non-central point load	$-\dfrac{Wab^2}{L^2}$	$+\dfrac{Wa^2b}{L^2}$
3	Uniformly distributed load	$-\dfrac{wL^2}{12}$	$+\dfrac{wL^2}{12}$
4	Right-hand support settlement	$-\dfrac{6EI\delta}{L^2}$	$-\dfrac{6EI\delta}{L^2}$
5	Left-hand support settlement	$+\dfrac{6EI\delta}{L^2}$	$+\dfrac{6EI\delta}{L^2}$

Calculate the FEMs for a beam spanning 4 m and supporting a uniformly distributed load of 24 kN/m over the whole span.

$$\pm \frac{wL^2}{12} = \pm \frac{24 \times 4^2}{12} = \pm 32 \text{ kN m}$$

We are now going to work through a complete illustrative example to demonstrate the technique of moment distribution. Study each step in the distribution table carefully.

Joint	A	B		C		D	
Distribution factor		0.33	0.67	0.67	0.33		
Fixed-end moments	−40.00	+40.00	−32.00	+32.00	−20.00	+20.00	
Balancing moment		(−8.00)		(−12.00)			(ii)
Distributed moment		−2.67	−5.33	−8.00	−4.00		(iii)
Carry-over moment	−1.33		−4.00	−2.67		−2.00	(iv)
Balancing moment		(+4.00)		(+2.67)			(v)
Distributed moment		+1.33	+2.67	+1.78	+0.89		
Carry-over moment	+0.67		+0.89	+1.34		+0.45	
Balancing moment		(−0.89)		(−1.34)			
Distributed moment		−0.29	−0.60	−0.89	−0.45		
Carry-over moment	−0.15		−0.45	−0.30		−0.23	
Balancing moment		(+0.45)		(+0.30)			
Distributed moment		+0.15	+0.30	+0.20	+0.10		(vi)
Final moments							(vii)

Note that:
(i)   Fixed-end moments are calculated that prevent the joints from rotating. The appropriate formula in the FEM table in Frame 20 is used in each case.
(ii) At joint B the 'net' FEM is $40 - 32 = 8$ kN m. A balancing moment of -8 kN m is applied that releases the joint and allows joint rotation. Likewise a balancing moment of -12 kN m is applied at C.
(iii) These moments are distributed into the members and
(iv) one-half of each distributed moment is carried over to the fixed ends.
(v)   This introduces a new out-of-balance moment of -4.00 kN m and -2.67 kN m at B and C respectively. The process of distribution is repeated until
(vi) the final distribution represents a relatively small set of values compared with the original set of fixed-end moments and the solution has progressively converged.
   Complete the table by summing the moments in each column to give the final set of end moments to be inserted in (vii).

22

	A		B		C		D	
Final moments	−40.81	38.52	−38.52	23.46	−23.46	18.22	(vii)

Did you ensure that you only added together the terms on either side of each line? The bracketed moment terms are not included and are only there to clarify the balancing moments that need to be applied at each stage of the distribution. Note also that the final moments on either side of column B and column C are equal in magnitude and opposite in direction indicating the state of internal equilibrium at each joint.

Using the calculated end moments draw the bending moment diagram for the structure indicating key values on the diagram. Use the principle of superimposing the end moment diagram and the free bending moment diagram as illustrated in Frame 19.

23

(a) Support moments

(b) Simply supported moments

$WL^2/8 = 48$ kN m

$WL/4 = 80$ kN m

$WL/4 = 40$ kN m

(c) Bending moment diagram

$(40.81 + 38.52)/2 = 39.67$

$(38.52 + 23.46)/2 = 30.99$

$(23.46 + 18.22)/2 = 20.84$

$80 - 39.67 = 40.33$

$48 - 30.99 = 17.01$

$40 - 20.84 = 19.16$

PROBLEMS

In each of the following problems you are required to calculate the end moments then to draw the bending moment diagram. As in Programme 4, you should sketch the deflected shape of each beam and use the deflected shape to determine which side of the beam is in tension as a guide to the shape of the bending moment diagram before attempting any calculations. Remember that where the beam supports point loads only the diagram will consist of straight lines; where it supports a uniformly distributed load the diagram will be in the form of a parabolic curve.

Assume that for each problem each span of the beam has an identical EI value.

Check your answers against the solutions given below. The bending moment diagrams are in sketch format and are not to an exact scale. Maximum sagging moments are given directly under the point loads and at the centre of the UDLs.

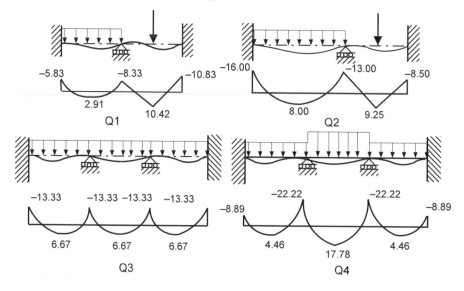

25

CALCULATION OF SUPPORT REACTIONS

To calculate the support reactions of any of the problems that we have considered so far the normal principles of statics can be used and applied separately to each part of the structure. Let's consider again the problem in Frame 21 which has been reproduced below:

The *free-body diagram* of each part of the structure can be drawn with all known loads, moments and unknown reactive forces indicated as shown. The unknown reactions are given the normal symbol V and the subscripts refer to the joint with L referring to the left of the joint and R to the right.

Consider the free-body diagram of AB only and use normal principles of statics to calculate both V_A and V_{BL}.

26

$$\boxed{V_A = 20.26 \text{ kN} \quad V_{BL} = 19.74 \text{ kN}}$$

Check your calculations:

$\Sigma M_A = 0$ $-40.62 + 38.52 + (40 \times 4) - (V_{BL} \times 8) = 0$
$$\therefore \quad V_{BL} = 19.74 \text{ kN}$$

$\Sigma M_B = 0$ $-40.62 + 38.52 - (40 \times 4) + (V_A \times 8) = 0$
$$\therefore \quad V_A = 20.26 \text{ kN}$$

and as a check:

$\Sigma V = 0$: $V_A + V_{BL} - 40 = 20.26 + 19.74 - 40 = 0$

When taking moments did you include both end-moments in your calculations?
Now consider BC only and calculate the reactive forces for span BC.

27

$$\boxed{V_{BR} = 51.77 \text{ kN} \quad V_{CL} = 44.23 \text{ kN}}$$

Check your calculations:

$\Sigma M_B = 0$ $\qquad -38.52 + 23.46 + (24 \times 4) \times 4/2 - (V_{CL} \times 4) = 0$

$$\therefore \quad V_{CL} = 44.23 \text{ kN}$$

$\Sigma M_C = 0$ $\qquad -38.52 + 23.46 - (24 \times 4) \times 4/2 + (V_{BR} \times 4) = 0$

$$\therefore \quad V_{BR} = 51.77 \text{ kN}$$

and as a check:

$\Sigma V = 0$: $\qquad V_{BR} + V_{CL} - (24 \times 4) = 51.77 + 44.23 - 96 = 0 \quad \text{(OK)}$

Now consider CD only and calculate the reactive forces for span CD.

28

$$\boxed{V_{CR} = 10.65 \text{ kN} \quad V_D = 9.35 \text{ kN}}$$

Check your calculations:

$\Sigma M_C = 0$ $\qquad -23.46 + 18.22 + (20 \times 4) - (V_D \times 8) = 0$

$$\therefore \quad V_D = 9.35 \text{ kN}$$

$\Sigma M_D = 0$ $\qquad -23.46 + 18.22 - (20 \times 4) + (V_{CR} \times 8) = 0$

$$\therefore \quad V_{CR} = 10.65 \text{ kN}$$

and as a check:

$\Sigma V = 0$: $\qquad V_{CR} + V_D - 20 = 10.65 + 9.35 - 20 = 0$

Consider the vertical equilibrium of joint B. What is the external support reaction at B?

29

$$\boxed{R_B = V_{BL} + V_{BR} = 19.74 + 51.77 = 71.51 \text{ kN}}$$

It should be apparent that the *external reaction* at B must equal the sum of the two reactive forces V_{BL} and V_{BR} which represent the *internal shear forces* at the right- and left-hand ends of spans AB and BC, respectively.

Calculate the external support reactions at A, C and D.

30

$$\boxed{V_A = 20.26 \text{ kN} \quad V_C = 54.88 \text{ kN} \quad V_D = 9.35 \text{ kN}}$$

Did you note that at A there is only one shear force which must equal the external reaction at A? Likewise at D. At C the external reaction must equal the sum of the two internal shear forces either side of the joint. This is shown in the figure below:

How can you check the solution to ensure that we have the correct external reactions?

31

> Check that the total of the reaction forces equates to the total downward load on the structure
> i.e. $\Sigma V = 0$

Total reaction forces $= 20.26 + 71.51 + 54.88 + 9.35 = 156$ kN

Total downward loads $= 40 + (24 \times 4) + 20 \qquad = 156$ kN

which is correct!

Now can you draw the shear force diagram for the whole beam? Remember that when drawing the shear force diagram you should start from the left-hand side of the beam and construct the diagram working from left to right.

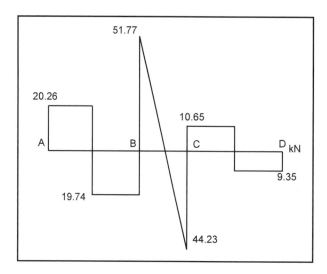

Did you remember that where there are only point loads acting on a span the shear force diagram consists of horizontal lines and where there is a uniformly distributed load acting it consists of sloping lines as in span BC ?

Note also that in span BC the point of zero shear force *which is also the point of maximum bending moment* is not at the mid-span point of BC. Hence the mid-span moment, calculated and shown on the bending moment diagram in Frame 23, is not the true maximum value. To obtain the true maximum value the point of zero shear should be obtained from the shear force diagram and moments taken to the left or right of that point to determine the maximum bending moment which will occur at that point. Now try some more problems.

33

PROBLEMS

Calculate the support reactions of the beams shown and draw the shear force diagrams. These are two of the problems from Frame 24. Refer back to your solution to those problems for the end-moment calculations.

Check your answers against the solutions given at the end of the programme.

34

BEAMS WITH SIMPLY SUPPORTED ENDS

The end supports of the problems that we have considered so far have all consisted of *built-in* supports and in the process of moment-distribution, moments have been distributed to these supports. When, however, an end support is *simply supported* or *pinned*, in other words the moment at that support is zero, then it is necessary to ensure that inherently within our calculations the moment at that support is maintained as zero.

In Frame 8 we showed that a propped cantilever has a stiffness k given by $4EI/L$ as shown in figure (a) below. If the built-in end at A is replaced by a pinned support, as shown in figure (b), Macaulay's method (Programme 12) or the Virtual Work method (Programme 14) can be used to show that the stiffness of this beam is reduced to $3EI/L$.

(a) $M = 4EI/L\theta_B$ (b) $M = 3EI/L\theta_B$

This fact enables us to modify our moment-distribution calculations by using a stiffness value of $3EI/L$ in calculating the *distribution factors* in our distribution table rather than $4EI/L$ whenever a member has a pinned or simply supported end.

Let's demonstrate this by considering the problem in Frame 15 with the support at A replaced by a simple support:

Can you calculate the distribution factors at joint B?

35

$$DF_{BA} = 0.27 \quad DF_{BC} = 0.73$$

Check your working:

Stiffness of AB $= k_{AB} = 3EI/8$ Stiffness of BC $= k_{BC} = 4EI/4$

From *Principle 2*:

$$DF_{BA} = \frac{\text{stiffness of member AB}}{\sum \text{stiffness of each member at the joint}} = \frac{3EI/8}{(3EI/8 + 4EI/4)} = 0.27$$

and

$$DF_{BC} = \frac{\text{stiffness of member BC}}{\sum \text{stiffness of each member at the joint}} = \frac{4EI/4}{(3EI/8 + 4EI/4)} = 0.73$$

Let's complete the distribution table for this problem:

Joint	A		B	C	
Distribution factor		0.27	0.73		(i)
Fixed-end moments	−40.00	+40.00			
Release A	+40.00 ⟶ +20.00				(ii)
Balancing moment		(−60.00)			(iii)
Distributed moment		−16.20	−43.80		(iv)
Carry-over moment				−21.90	(v)
Final moments	0	+43.80	−43.80	−21.90	(vi)

Note that:
(i)   Distribution factors are calculated, as above, allowing for the modified stiffness of AB;
(ii) Joint A is released by the application of a balancing moment of 40 kN m to reduce the moment at A to zero. One-half of this value is carried over to B;
(iii) Joint B is now out of balance by +60 kN m (40 + 20) and a balancing moment of −60 kN m is applied that releases the joint and allows joint rotation;
(iv) This balancing moment is distributed into the members, as normal, in the ratio of the distribution factors;
(v)   One-half of the distributed moment is carried over to the fixed end at C *but not to A* which maintains zero moment at A;
(vi) The moments in each column are summed to give the final set of moments at each support.

If this were a more extensive problem the process of redistribution and carry over would be repeated as necessary until the solution is found. The distribution would be carried out as normal but with no carry over to support A at any stage. If the right-hand end support were simply supported then the moment at that joint would be also released before distribution commences and no moment would be redistributed back to that joint.

Once the final set of end moments are calculated the process of calculating support reactions and drawing shear force and bending moment diagrams would be as previously studied in this programme.

36

BEAMS WITH CANTILEVERED ENDS

Let's look at the situation of a beam with a cantilevered end such as the example shown below which is a further development of the problem in frame 34.

What is the moment at A due to the load applied to the cantilever?

37

$$\boxed{-5 \text{ kN m}}$$

The 5 kN load at 1 m from A exerts an anti-clockwise moment about A of 5 kN m. This is a fixed value and the solution routine must ensure that the moment at A is maintained at -5 kN m. The solution method is practically identical to that of the previous example except that the moment at A must be reduced to a value of -5 kN m and not zero as it is no longer a support of zero-moment.

Joint	A		B		C	
Distribution factor		0.27	0.73			(i)
Fixed-end moments	-40.00	$+40.00$				
Release A	$+35.00 \longrightarrow$	$+17.50$				(ii)
Balancing moment			(-57.50)			(iii)
Distributed moment		-15.52	-41.98			(iv)
Carry-over moment				$\searrow$	-20.98	(v)
Final moments	-5.00	$+41.98$	-41.98	-20.98		(vi)

Note that:
(i) Distribution factors are calculated again allowing for the modified stiffness of AB;
(ii) Joint A is released by the application of a balancing moment of 35 kN m to reduce the moment at A to -5 kN m. One-half of this value is carried over to B;
(iii) Joint B is now out of balance by $+57.50$ kN m $(40 + 17.50)$ and a balancing moment of -57.50 kN m is applied that releases the joint and allows joint rotation;
(iv) This balancing moment is distributed into the members, as normal, in the ratio of the distribution factors;
(v) One-half of the distributed moment is carried over to the fixed end at C *but not to* A which maintains a fixed moment at A of -5 kN m.
(vi) The moments in each column are summed to give the final set of moments at each support.

Now try some more problems.

38

$$\boxed{\text{PROBLEMS}}$$

In each of the following problems you are required to calculate the end moments then to draw the bending moment diagram. You should sketch the deflected shape of each beam and use the deflected shape to determine which side of the beam is in tension as a guide to the shape of the bending moment diagram before attempting any calculations. You should then calculate support reactions and draw the shear force diagram.

Assume that for each problem each span of the beam has an identical EI value.

Check your answers against the solutions given at the end of the programme.

39

SETTLEMENT OF SUPPORTS

In the continuous beam problems considered so far the supports have been assumed to be rigid and to remain at the same level. However, if such a continuous beam is subject to differential settlement whereby the supports settle by differing amounts this will cause relative movement in the vertical direction between the ends of any given span. This settlement will result in the development of bending moments and shear forces which will be additional to those generated by the load system. This effect can be seen in the fourth and final diagram of fixed-end moments in the table in Frame 20.

Hence, if the relative settlement movement δ is known together with the span (L) and the stiffness value of the member (EI) the fixed-end moments due to settlement can be calculated from the formula in the table. These fixed-end moments are added to those from the loading and the total fixed-end moments are then distributed in the usual way.

40

PORTAL FRAME STRUCTURES

One common form of structure is the rigidly jointed portal frame constructed in reinforced concrete or in structural steelwork with jointed connections which are designed to transmit both forces and moments. Such frames are also classified as *sway* where the resistance to lateral load comes from the sway action of the structure or *non-sway* where lateral load resistance is as a result of the frame being braced by shear walls, steel bracing or reacting against another rigid structure, such as a lift shaft, to prevent sway movement. Both types are illustrated in the figures below.

(a) Sway frame   (b) Non-sway frame

Let's start by looking at an example of a non-sway frame as shown below. The stiffness of each column is taken to be twice that of the beam for illustrative purposes.

What are the distribution factors at joint B and joint C?

$$DF_{BA} = 0.5 \quad DF_{BC} = 0.5: \qquad DF_{CB} = 0.5 \quad DF_{CD} = 0.5$$

The principle of calculating distribution factors at a joint is exactly the same as that considered for the continuous beam problems. The fact that one of the members is a vertical column does not affect the way that the moment is distributed between the members. Check your working:

Stiffness of AB $= k_{AB} = 3(2EI)/6 = EI$ \qquad Stiffness of BC $= k_{BC} = 4EI/4 = EI$

(Note that joint A is pinned so we have used the modified stiffness of AB)

From *Principle 2*:

$$DF_{BA} = \frac{\text{stiffness of member AB}}{\sum \text{stiffness of each member at the joint}} = \frac{EI}{(EI + EI)} = 0.5$$

and

$$DF_{BC} = \frac{\text{stiffness of member BC}}{\sum \text{stiffness of each member at the joint}} = \frac{EI}{(EI + EI)} = 0.5$$

Similarly for joint C.

Calculate the fixed-end moments for AB, BC and CD.

$$FEM_{AB} = -15 \text{ kN m} \qquad FEM_{BA} = +15 \text{ kN m} \qquad FEM_{BC} = -7.5 \text{ kN m}$$
$$FEM_{CB} = +22.5 \text{ kN m} \qquad FEM_{CD} = 0 \text{ kN m} \qquad FEM_{DC} = 0 \text{ kN m}$$

Note that column CD is not subject to any loads and therefore has no fixed-end moments. Did you get the values for the beam BC correct? You should have used the formulae in the second row of the table in Frame 20 as follows:

$$FEM_{BC} = -\frac{Wab^2}{L^2} = -\frac{40 \times 3 \times 1^2}{4^2} = -7.5 \text{ kN m}$$

and

$$FEM_{CB} = +\frac{Wa^2 b}{L^2} = +\frac{40 \times 3^2 \times 1}{4^2} = +22.5 \text{ kN m}$$

The fixed-end moments for the column AB are calculated using the standard formula for the uniformly distributed load case ($\pm wL^2/12$).

Can you now set up a distribution table and calculate the joint moments?

43

$M_A = 0$ kN m	$M_B = \pm 17.52$ kN m	
$M_C = \pm 9.96$ kN m	$M_D = 0$ kN m	

Check your calculations:

Joint	A	B		C		D	
Distribution factor		0.5	0.5	0.5	0.5		
Fixed-end moments	−15.00	+15.00	−7.50	+22.50	0	0	
Release A	+15.00 ⟶ +7.50						(i)
Balancing moment		(−15.00)		(−22.50)			
Distributed moment		−7.50 | −7.50		−11.25 ⊢11.25			
Carry-over moment		| −5.63		−3.75 |			
Balancing moment		(+5.63)		(+3.75)			
Distributed moment		+2.81 | +2.81		+1.88 | +1.88			
Carry-over moment		| +0.94		+1.40 |			
Balancing moment		(−0.94)		(−1.40)			
Distributed moment		−0.47 | −0.47		−0.70 | −0.70			
Carry-over moment		| −0.35		−0.23 |			
Balancing moment		(+0.35)		(+0.23)			
Distributed moment		+0.17 | +0.17		+0.12 | +0.12			
Final moments	0	+17.51	−17.53	+9.97	−9.95	0	(ii)

Note that:
(i) Joint A is released by the application of a balancing moment of +15 kN m to reduce the moment at A to zero as it is a pinned support. One-half of this moment is carried over to B. There are no fixed-end moments for CD so it is not necessary to do this for the pinned joint at D.

(ii) The small differences in the moments either side of the two joints, B and C, are due to rounding-off errors in the calculations.

Having calculated the end moments we can now calculate the values of the support reactions and the magnitude of the horizontal propping force at C which is preventing the structure from swaying.

The figure on the next page shows the whole structure and the free-body diagram of each column with calculated end moments and unknown support reactions as shown. The support reactions are assumed to act in the opposite direction to the horizontal loading on AB: if this assumption were incorrect then, on calculation, they would be found to have negative values. It can be seen from the first figure that the propping force can be calculated by considering the horizontal equilibrium of the whole structure providing that the horizontal reactions at the base of each column are known. These horizontal reactions can be calculated by considering the equilibrium of the free-body diagrams of the columns AB and CD respectively by taking moments about the top of each column in turn.

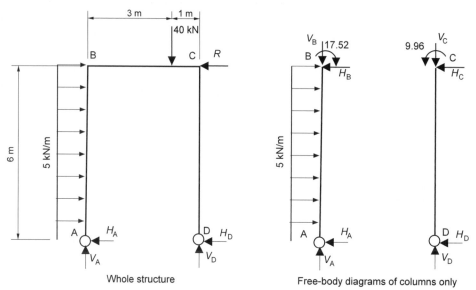

Whole structure   Free-body diagrams of columns only

Can you calculate the horizontal reaction at the base of each column, H_A and H_D, and hence the propping force R?

44

$$H_A = 12.08 \text{ kN} \quad H_D = 1.66 \text{ kN} \quad R = 16.26 \text{ kN}$$

Check your calculations:

For the left-hand column:

ΣM_B: $\qquad +17.52 + (H_A \times 6) - (5 \times 6) \times 6/2 = 0$

$\qquad\qquad\qquad\qquad\qquad \therefore \quad H_A = 12.08 \text{ kN}$

For the right-hand column:

ΣM_C: $\qquad\qquad\qquad -9.96 + (H_D \times 6) = 0$

$\qquad\qquad\qquad\qquad\qquad \therefore \quad H_D = 1.66 \text{ kN}$

For the whole structure:

ΣH: $\qquad\qquad (5 \times 6) - H_A - H_D - R = 0$

$\qquad\qquad (5 \times 6) - 12.08 - 1.66 - R = 0$

$\qquad\qquad\qquad\qquad\qquad \therefore \quad R = 16.26 \text{ kN}$

Using the calculated end moments and the principle of superimposing the end moment diagram and the free bending moment diagram sketch the bending moment diagram for the structure.

45

(a) Joint moments   (b) Simply supported moments   (c) Bending moment diagram

In drawing the bending moment diagram the usual convention of drawing sagging moments as positive on the under-side of a beam applies to member BC but for the columns AB and CD the concept of sagging moments has no directly equivalent meaning. In this case we draw the bending moment diagram on the *inside* of the frame if the moments cause tension on the *inside* and on the *outside* of the frame if they cause tension on the *outside*. The above diagrams are consistent with this convention.

Note that at joint B the moment at the top of column AB is equal in magnitude to the moment at the left-hand end of BC (17.52 kN m) and acts in the opposite direction. This reflects the fact that, for equilibrium, the two internal moments at this joint must sum to zero as there is no external moment acting at the joint. If there was an external moment the sum of the two internal moments would be a moment equal in value to the external moment and acting in the opposite direction to that moment. Both moments are drawn on the *outside* of the frame indicating that joint B is in tension on the outside and compression on the inside. Likewise for joint C.

Note also that in column AB the moment of 13.74 kN m has been calculated at the mid-height position of the column. This is not necessarily the maximum bending moment causing tension on the right-hand side of this column.

What is the principle that we could use to calculate the position and true value of maximum bending moment between A and B?

46

The maximum moment occurs at the point of zero shear

Let's see how to calculate the maximum moment. The position of the point of maximum moment is unknown. Let's assume that it acts at X which is at an unknown distance x from A as shown in the figure below. The figure shows the free-body diagram of the part of the structure between A and the point of maximum bending moment. We know that at X the shear force must be zero for maximum moment.

By equating horizontal forces can you calculate the position of point X?

47

$$x = 2.42 \text{ m}$$

Check your working. If at X the shear force is zero that implies that forces acting to the right on AX must equal those acting to the left. In other words:

ΣH: $\qquad\qquad (5 \times x) - 12.08 = 0$

$\qquad\qquad\qquad \therefore \quad x = 12.08/5 = 2.42 \text{ m}$

Now calculate the bending moment at X.

48

$$14.59 \text{ kN m}$$

Check your calculations:

ΣM_X: $\qquad 12.08 \times 2.42 - (5 \times 2.42 \times 2.42/2) - M_X = 0$

$\qquad\qquad\qquad \therefore \quad M_X = 29.33 - 14.64 = 14.59 \text{ kN m}$

This is greater than the moment of 13.74 kN m, previously calculated at mid-height of the column and represents the true maximum moment in the column.

We have yet to calculate the vertical reactions at A and D. Move to the next frame to see how to complete these calculations.

49

The figure below shows the free-body diagrams of both the columns and the beam BC. It can be seen that for vertical equilibrium of the left-hand column V_A must be equal in magnitude and opposite in direction to V_B. Likewise V_D must be equal in magnitude and opposite in direction to V_C. Hence the support reactions at A and D can be calculated by considering the equilibrium of the beam BC and calculating V_B $(= V_A)$ and V_C $(= V_D)$.

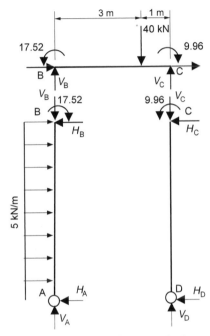

Free-body diagrams of beam and column

Consider the beam BC only and calculate the magnitude of V_B and V_C.

50

$$\boxed{\begin{array}{l} V_B \ (= V_A) = 11.89 \text{ kN} \\ V_C \ (= V_D) = 28.11 \text{ kN} \end{array}}$$

You should have calculated as follows:

ΣM_B:
$$-17.52 + 9.96 + (40 \times 3) - (V_C \times 4) = 0$$
$$\therefore \quad V_C = 28.11 \text{ kN}$$

ΣV:
$$V_B + V_C - 40 = 0$$
$$\therefore \quad V_B = 40 - V_C = 40 - 28.11 = 11.89 \text{ kN}$$

Now try some more problems yourself.

PROBLEMS

1. Calculate the values of the reactions at A and D and the value of the propping force for the portal frame shown in figure Q1. Sketch the shape of the bending moment diagram marking all important values on the diagram. Determine the position and value of the maximum moment in column AB.

Ans. ($H_A = -16.31$ kN, $V_A = 3.38$ kN, $H_D = 0.56$ kN, $V_D = -3.38$ kN, $R_C = -20.25$ kN)

2. Calculate the values of the reactions at A and D and the value of the propping force for the portal frame shown in figure Q2. Sketch the shape of the bending moment diagram marking all important values on the diagram. *(Hint: (i) when calculating the FEMs for BC calculate the values for each of the 40 kN loads in turn and then superimpose the two sets of results (ii) note that the supports at A and D are fixed. When distributing the FEMs at B and C the distributed moments will have to be carried over to A and D respectively.)*

Ans. ($H_A = 4.29$ kN, $V_A = 40.0$ kN, $H_D = -4.29$ kN, $V_D = 40.0$ kN, $R_C = 0$ kN)

Check your bending moment diagrams against the solutions given at the end of the programme.

Now let's look at how to analyse *sway* frames.

52

PORTAL FRAMES WITH SWAY

Portal frames which are free to sway are slightly more complex to analyse. Figure (a) shows a typical sway frame which is considered equivalent to an identical frame propped by a reactive force R, shown in figure (b), together with the identical sway frame in figure (c), which is subject to a single equal and opposite force R, as shown. The *superposition* of figure (b) and figure(c) will result in the two equal and opposite forces, R, cancelling out and will return us to the original sway problem that we are trying to solve.

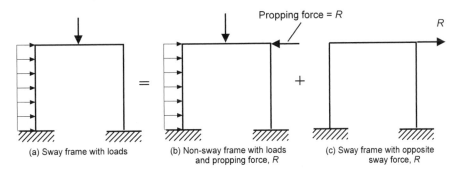

(a) Sway frame with loads

(b) Non-sway frame with loads and propping force, R

(c) Sway frame with opposite sway force, R

We have already seen how to analyse a non-sway frame, as in figure (b), and how to calculate the value of the propping force. What remains is to see how to analyse the sway frame of figure (c) which, whilst a sway frame, is only subject to a single load, R. Once we have analysed figure (c) the reactions and bending moments calculated for this figure can be superimposed (simply algebraically added) onto those for figure (b) to give us the solution that we are looking for. Let's see how to do this by reanalysing the previous example which we will now take to be a sway frame as shown. All other parameters, loads etc., remain the same.

(d)

(e)

(f)

The propping force of 16.26 kN was determined in our previous calculations as was the distribution of bending moments in figure (e). The application of this force to the sway frame in figure (f) will cause the frame to sway, but by an indeterminate amount.

Let's now assume that it sways as shown below and let's clamp each joint to prevent joint rotation as the first stage in a moment distribution analysis. The amount of sway is unknown and we can assume that it has an arbitrary value, δ.

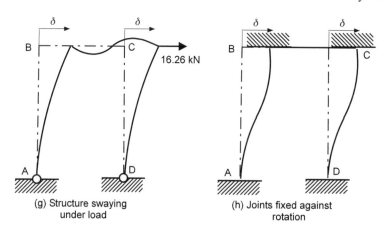

(g) Structure swaying under load

(h) Joints fixed against rotation

What are the values of the fixed-end moments for figure (h)? (Hint: look back at the table of standard FEM formulae in Frame 20.)

53

$$\boxed{-0.33EI\delta}$$

The fourth standard case in the table in Frame 20 gives the fixed-end formula for the equivalent situation to this sway case as $-6EI\delta/L^2$. As L for each column has a value of 6 m and EI is given as a value of $2EI$ substitution into this formula will give the equation for the fixed-end moments for the situation in figure (h) above. As column AB is identical to column CD and the sway value δ is the same the FEMs will be the same at the ends of both columns. Unfortunately this does not help very much as the FEMs depends on the value of δ which is unknown.

What we can therefore do is to assume that the FEMs have values *of the same order as the moments calculated for the non-sway* case (see the bending moment diagram in figure (c) in Frame 45) and then release and distribute these moments. Once we have done this we can analyse what we have done to see if our assumption is valid. Let's assume FEMs of, say, -12 kN m. Move on to the next frame to carry out this analysis.

54

Let's set up the usual distribution table with the, previously calculated, distribution factors.

Joint	A		B		C		D
Distribution factor		0.5	0.5	0.5	0.5		
Fixed-end moments	−12.00	−12.00				−12.00	−12.00
Release A and D	+12.00	+6.00				+6.00	+12.00
Balancing moment		(+6.00)			(+6.00)		
Distributed moment		+3.00	+3.00	+3.00	+3.00		
Carry-over moment			+1.50	+1.50			
Balancing moment		(−1.50)		(−1.50)			
Distributed moment							
Carry-over moment							
Balancing moment							
Distributed moment							
Final moments							

Complete the distribution table and calculate the final moments.

55

Joint	A		B		C		D
Final moments	0.00	−3.61	+3.61	+3.61	−3.61		0.00

If our assumption about the FEMs was correct then the end moments for the columns would be as shown in figure (b). The applied horizontal force R would be simply equal to the sum of the two horizontal support reactions, H_A and H_D, as shown in figure (a) or twice one support reaction as, by symmetry, H_A and H_D must be equal in value.

(a) Whole structure   (b) Free-body diagrams of columns only

Take moments about B to calculate H_A and hence calculate the value of R.

$$\boxed{H_A = 0.60 \text{ kN} \qquad R = 1.20 \text{ kN}}$$

Check your calculations. Taking moments about B for column AB:

$$\Sigma M_B \qquad\qquad -3.61 + (H_A \times 6) = 0$$
$$\therefore \quad H_A = 0.60 \text{ kN}$$

and

$$\Sigma H \qquad\qquad R - H_A - H_B = 0$$
$$\therefore \quad R = H_A + H_B = 2 \times 0.60 = 1.20 \text{ kN}$$

But we know that the actual value of R should be 16.26 kN so our assumption about the value of the fixed-end moments and the subsequent moment redistribution results were incorrect. However, if we multiply all our moments and the sway force by a correction factor of 13.55 ($= 16.26/1.20$) then we maintain equilibrium between our set of moments and the sway force whilst, at the same time, giving the set of moments which is correct for a sway force of 16.26 kN.

Hence the correct end moments are: $\pm 3.61 \times 13.55 = \pm 48.92$ kN m. This is the answer that would have been obtained no matter what value we originally assumed for the FEMs. If a different value had been assumed then a different set of end moments would have been obtained but the application of a different correction factor ($= 16.26/$calculated value of R) would produce the same answer for the moments. The final set of bending moments are obtained by combining the bending moment diagram for the non-sway case (Frame 45—figure (c)) and the sway case that we have just analysed as shown below:

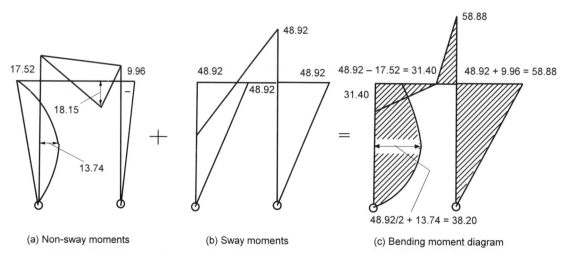

(a) Non-sway moments   (b) Sway moments   (c) Bending moment diagram

57

SWAY FRAMES WITH COLUMNS OF UNEQUAL LENGTH

To complete the programme we will consider briefly how to deal with sway frame problems having columns of unequal length, such as the example shown in figure (a) below. Figure (b) shows the joints prevented from rotation as the first stage in the moment distribution analysis of the sway case, identical to the example in Frame 52.

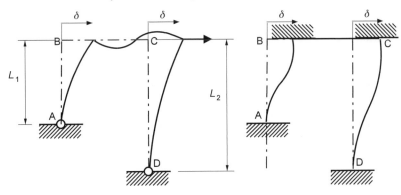

(a) Structure swaying under load   (b) Joints fixed against rotation

As the fixed-end moments are given by the standard formula, $-6EI\delta/L^2$ it follows that, when the columns are of different lengths, the fixed-end moments are in inverse proportion to the square of their lengths. If the EI values are not equal the fixed-end moments will also be in direct proportion to the EI values.

Assuming that the EI values are the same and that L_2 is, as an example, twice L_1 it follows that the fixed-end moments for AB will be four times those of CD. In this case, when setting up the distribution table for the arbitrary sway moments, as in frame 53, then arbitrary moments in the ratio of 1:4 should be chosen.

In the situation described, if the arbitrary FEMs for CD are taken as $-12\ kN\ m$ what will those be for AB?

58

$$\boxed{-48\ kN\ m}$$

Any other arbitrary set of moments in the ratio of 1:4 can be assumed. All the subsequent calculations will remain, in principle, the same as in the previous example.

59

60

FURTHER PROBLEMS

In the following problems try to sketch the deflected shape and hence the shape of the BM diagram before attempting any calculation of bending moments. The solutions to all the problems are given at the end of the programme.

1. For each of the beams shown calculate the end moments and the support reactions. Draw the deflected shape, shearing force and bending moment diagrams indicating key values on the diagrams. For figures Q1(a), Q1(c) and Q1(d) assume that the *EI* value is constant throughout.

Q1(a)   Q1(b)   Q1(c)   Q1(d)

2. For each of the frames shown in figure Q2(a) and Q2(b) calculate the joint moments and support reactions and draw the bending moment diagrams.

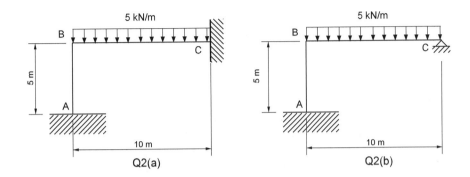

Q2(a)   Q2(b)

3. Calculate the joint moments and support reactions of the braced frame shown in figure Q3. Draw the bending moment diagram.

4. Repeat question 3 if the bracing at joint C in figure Q3 is removed making the structure into a sway frame.

5. Calculate the joint moments and support reactions of the unbraced frame shown in figure Q5. Draw the bending moment diagram.

Q3   Q5

SOLUTIONS TO PROBLEMS

61

Each of the following diagrams is drawn to an approximate scale but the same scale is not used throughout.

FRAME 33

FRAME 38

FRAME 51

Q1

Q2

FRAME 60

Q1(a)

Q1(b)

Q1(c)

Q1(d)

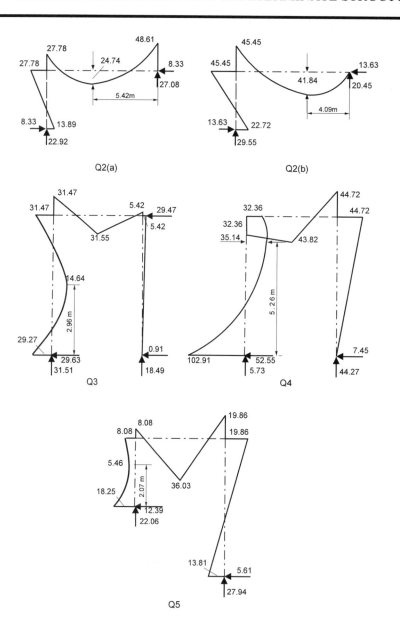

Q2(a)

Q2(b)

Q3

Q4

Q5

Programme 16

THE
SLOPE-DEFLECTION
METHOD

On completion of this programme you will be able to:

- Recall and state the slope-deflection equations
- Use the slope-deflection equations to calculate the joint rotations in continuous beam structures
- Calculate, directly, the end moments in the individual sections of continuous beam structures
- Modify the technique for the solution of beams with cantilevered ends
- Determine the effect of support settlement
- Apply the method to propped portal frame structures
- Apply the method to unpropped portal frame structures

1

In the previous programme we looked at the method of *moment distribution* and showed how we could use the method to analyse beam and frame type structures which are *statically indeterminate*. The method gave us a means of calculating, by a process of successive reduction, the end moments in the members of those structures and hence, using normal principles of statics, the support reactions and bending moments and shearing forces within the members.

In this programme we will introduce an alternative technique of solution known as the *Slope Deflection Method*. This alternative method will enable us to solve the same type of problems but has the added advantage that we will be able to calculate the deformations of (deflections and rotations) as well as the internal forces within the structure. This particular technique involves the solution of sets of simultaneous equations: the more sizeable and complex the structure the more equations will have to be solved. Hence it is only possible to manually solve structures of limited size using this method. However, the method can be readily adapted such that the equations can be solved by computer and, adapted in this way, is in fact the basis of modern computer techniques of structural analysis.

Before we start can you recall the expression that defines the 'stiffness' of a member in a rigidly jointed structure?

2

$$k = 4EI/L$$

You should recall from Programme 15 that the *stiffness* of a member is as given above and that the *carry-over factor* is 0.5. Let's make use of these facts by analysing again the continous beam problem that we first solved in Programme 15, Frames 15–18, reproduced in figure (a) below. The *EI* value is the same throughout. You will remember that to solve the problem we considered it to be equivalent to the superposition of the restrained and released systems shown in figures (b) and (c) where the moments of 40 kN m in figure (b) are the *fixed-end moments* for AB calculated using the data in Frame 20 of Programme 15.

(a) Original structure   (b) Restrained system – no joint rotation   (c) Released system – joint B free to rotate

Let's now assume that in figure (c) joint B rotates by an amount θ_B under the action of the 40 kN m joint moment. θ_B is at this stage an unknown value that we will eventually be able to calculate.

In terms of θ_B what is the value of the moment at the right-hand end of the left-hand span, M_{BA}?

$$M_{BA} = k\theta_B = \frac{4EI}{L} \times \theta_B = \frac{4EI}{8} \times \theta_B = \frac{EI}{2} \times \theta_B$$

Similarly the moment induced in the right-hand span, M_{BC} will be given by $4EI/4 \times \theta_B = EI \times \theta_B$ where both moments and rotation will be in the same (anti-clockwise) direction.

Again, referring to figure (c), what is the carry-over moment induced at A?

$$M_{AB} = \frac{1}{2}M_{BA} = \frac{1}{2} \times \frac{EI}{2} \times \theta_B = \frac{EI}{4} \times \theta_B$$

Where again the moment will be in the anti-clockwise direction. Similarly the moment carried over to end C (M_{CB}) will be one-half of M_{BC}. Now considering both figures (b) and (c) it can be seen that the actual end moments in the original structure of figure (a) are given by the *superposition* of the moments from the released and the restrained system. i.e.:

Final moments $(M') = Released\ system\ moments + Restrained\ system\ moments$

or:

$$M'_{AB} = \frac{EI}{4}\theta_B - 40 \qquad \text{(a)}$$

$$M'_{BA} = \frac{EI}{2}\theta_B + 40 \qquad \text{(b)}$$

$$M'_{BC} = EI\theta_B \qquad \text{(c)}$$

$$M'_{CB} = \frac{1}{2}M_{BC} = \frac{EI}{2}\theta_B \qquad \text{(d)}$$

where the clockwise fixed-end moments from figure (b) have been taken as positive values and anti-clockwise as negative. The expressions for the released system moments have been taken as positive such that for positive (clockwise) rotations they will give positive (clockwise) moments and, as in this example, will give negative (anti-clockwise) moments when the rotations are negative anti-clockwise. It can be seen from these equations that if θ_B were known then the end moments could all be calculated.

To solve these equations we need some additional information. This information comes from consideration of the *equilibrium* of joint B. We have expressions for the internal moments on either side of B and we know that, as there is no external moment at this joint, the two moments (M'_{BA} and M'_{BC}) must be equal and opposite or, in other words, algebraically they must sum to zero.

Can you make use of this fact to calculate the rotation at B from the equations given above?

5

$$\theta_B = -\frac{80}{3EI}$$

Check your calculations:

For equilibrium of joint B $\quad M'_{BA} + M'_{BC} = 0$

$$\therefore \quad \left(\frac{EI}{2} \times \theta_B + 40\right) + EI \times \theta_B = 0$$

which solves to give: $\qquad\qquad\qquad \theta_B = -\frac{80}{3EI}$

which, as expected, is a negative solution indicating anti-clockwise rotation.

Now use this expression for rotation and the moment equations (a) to (d) in Frame 4 to calculate the end moments in the beam.

6

$$M'_{AB} = -46.67 \quad M'_{BA} = +26.67 \quad M'_{BC} = -26.67 \quad M'_{CB} = -13.33 \text{ kN m}$$

Check your calculations:

$$M'_{AB} = \frac{EI}{4}\theta_B - 40 = \frac{EI}{4} \times \left(-\frac{80}{3EI}\right) - 40 = -6.67 - 40 \ = -46.67 \text{ kN m}$$

$$M'_{BA} = \frac{EI}{2}\theta_B + 40 = \frac{EI}{2} \times \left(-\frac{80}{3EI}\right) + 40 = -13.33 + 40 = +26.67 \text{ kN m}$$

$$M'_{BC} = EI\theta_B \qquad = EI \times \left(-\frac{80}{3EI}\right) \qquad\qquad\qquad = -26.67 \text{ kN m}$$

$$M'_{CB} = \frac{1}{2}M_{BC} \qquad = \frac{EI}{2}\theta_B = \frac{EI}{2}\left(-\frac{80}{3EI}\right) \qquad\qquad = -13.33 \text{ kN m}$$

If you compare these solutions to those obtained by moment distribution in Frame 18 of the previous programme you will see that we have obtained the same solutions but we have used direct calculation and have also determined the rotation of joint B as part of our calculations. As before, a positive solution indicates a clockwise moment and a negative solution an anti-clockwise moment.

Now that we have obtained the end moments the calculation of support reactions and the drawing of shear force and bending moment diagrams would now follow exactly the same principles as illustrated in Programme 15. We will not therefore repeat these stages of the calculation.

We will now consider the more general application of this solution method to continuous beam problems.

Consider the beam, AB, as shown which is one span of a continuous beam. Figure (a) shows the span initially 'fixed' by the application of fixed-end moments and figures (b) and (c) show each joint being 'released' in turn. The joints rotate by amounts θ_A and θ_B, respectively and the releasing moments are assumed to be positive, clockwise.

(a) Span restrained by fixed-end moments   (b) Joint A released   (c) Joint B released

From the superposition of the moments in the three figures it can be seen that the final moment at each joint will be given by

$$M'_{AB} = \frac{4EI}{L}\theta_A + \frac{1}{2}M_{BA} \pm FEM_A = \frac{4EI}{L}\theta_A + \frac{1}{2}\frac{4EI}{L}\theta_B \pm FEM_A$$

$$= \frac{2EI}{L}[2\theta_A + \theta_B] \pm FEM_A$$

and

$$M'_{BA} = \frac{4EI}{L}\theta_B + \frac{1}{2}M_{AB} \pm FEM_B = \frac{4EI}{L}\theta_B + \frac{1}{2}\frac{4EI}{L}\theta_A \pm FEM_B$$

$$= \frac{2EI}{L}[\theta_A + 2\theta_B] \pm FEM_B$$

The fixed-end moments are shown as either positive or negative as they can be clockwise or anti-clockwise depending on the nature and type of loading on the span. Usually they will be clockwise at the right-hand end and anti-clockwise at the left-hand end. Let's apply these equations to the solution of the continuous beam problem that you first saw in Programme 15, Frame 21.

Write down the equations for the end moments of each of the spans shown in the figure. There should be two equations for each span. Refer back to Programme 15, Frame 20 if you cannot remember how to calculate the FEMs.

STRUCTURAL MECHANICS

8

| Check your answers below |

$$M'_{AB} = \frac{2EI}{L}[2\theta_A + \theta_B] \pm FEM_A = \frac{2EI}{8}[2\theta_A + \theta_B] - 40$$

$$M'_{BA} = \frac{2EI}{L}[\theta_A + 2\theta_B] \pm FEM_B = \frac{2EI}{8}[\theta_A + 2\theta_B] + 40$$

$$M'_{BC} = \frac{2EI}{L}[2\theta_B + \theta_C] \pm FEM_B = \frac{2EI}{4}[2\theta_B + \theta_C] - 32$$

$$M'_{CB} = \frac{2EI}{L}[\theta_B + 2\theta_C] \pm FEM_C = \frac{2EI}{4}[\theta_B + 2\theta_C] + 32$$

$$M'_{CD} = \frac{2EI}{L}[2\theta_C + \theta_D] \pm FEM_C = \frac{2EI}{8}[2\theta_C + \theta_D] - 20$$

$$M'_{DC} = \frac{2EI}{L}[\theta_C + 2\theta_D] \pm FEM_D = \frac{2EI}{8}[\theta_C + 2\theta_D] + 20$$

Look at the diagram of the structure. What do you know about the rotation at A and the rotation at D? Hence, how can the above equations be simplified?

9

| Both joints are 'built-in' and therefore there is zero rotation |

The equations can hence be simplified by making both θ_A and θ_D equal to zero, i.e. deleting them from the equations as follows:

$$M'_{AB} = \frac{2EI}{8}[\theta_B] - 40 \qquad (a)$$

$$M'_{BA} = \frac{2EI}{8}[2\theta_B] + 40 \qquad (b)$$

$$M'_{BC} = \frac{2EI}{4}[2\theta_B + \theta_C] - 32 \qquad (c)$$

$$M'_{CB} = \frac{2EI}{4}[\theta_B + 2\theta_C] + 32 \qquad (d)$$

$$M'_{CD} = \frac{2EI}{8}[2\theta_C] - 20 \qquad (e)$$

$$M'_{DC} = \frac{2EI}{8}[\theta_C] + 20 \qquad (f)$$

What can you say about the moments M'_{BA} and M'_{BC}?

484

10

They must sum to zero

As in the last example, the two internal moments at joint B must be equal and opposite, that is sum to zero, as there is no external moment acting at the joint. This is again an expression of joint *equilibrium*. The same is true at joint C. Making use of this information equations (b) and(c) which apply to joint B together with (d) and (e) at joint C can be further modified as follows:

At joint B, equation (b) added to (c) gives:

$$M'_{BA} + M'_{BC} = \left(\frac{2EI}{8}[2\theta_B] + 40\right) + \left(\frac{2EI}{4}[2\theta_B + \theta_C] - 32\right) = 0$$

which can be simplified to:

$$1.5EI\theta_B + 0.5EI\theta_C + 8 = 0 \qquad\qquad (g)$$

and at joint C, equation (d) added to (e) gives:

$$M'_{CB} + M'_{CD} = \left(\frac{2EI}{4}[\theta_B + 2\theta_C] + 32\right) + \left(\frac{2EI}{8}[2\theta_C] - 20\right) = 0$$

which can be simplified to:

$$0.5EI\theta_B + 1.5EI\theta_C + 12 = 0 \qquad\qquad (h)$$

Equations (g) and (h) are a pair of simultaneous equations which can now be solved directly to give θ_B and θ_C.

Solve these equations to determine the values of the two joint rotations.

11

$$\theta_B = -3/EI \qquad \theta_C = -7/EI$$

Now substitute these values into equations (a) to (f) in Frame 9 to obtain the end moments in each span of the beam.

12

$M'_{AB} = -40.75$	$M'_{BA} = +38.50$	$M'_{BC} = -38.50$ kN m
$M'_{CB} = +23.50$	$M'_{CD} = -23.50$	$M'_{DC} = +18.25$ kN m

Compare these answers to those obtained by the moment-distribution method in Frame 22 of Programme 15. *Why do you think that they do not agree exactly?*

13

> Because the moment-distribution method is a technique of successive reduction with 'rounding-off' errors at each stage of the reduction. This method relies on the direct solution of sets of equations and will give slightly more 'exact' solutions. As in all calculations, there will still be some 'rounding-off' errors.

Again, once we have calculated the end moments the procedures for calculating the support reactions and drawing the shear force and bending moment diagrams are exactly the same as those given in Programme 15. You can look back at Programme 15 to see and revise the complete solution of this problem. Now let's summarise the steps that we have followed so far:

Step 1: For each span in the beam write down the two *slope-deflection equations* for the end-moments given in general terms as:

$$M'_{AB} = \frac{2EI}{L}[2\theta_A + \theta_B] \pm FEM_A \qquad M'_{BA} = \frac{2EI}{L}[\theta_A + 2\theta_B] \pm FEM_B$$

where the subscripts, A and B, refer to the left-hand end and the right-hand end of the span, respectively.

Step 2: At each joint, where joint rotation occurs, identify the *equilibrium condition* which requires that the sum of the two internal moments at the joint should sum to zero (or to the value of the external joint moment if there is such a moment applied). Summing the end-moment equations either side of each joint will give a series of simultaneous equations. Note that you must obtain one equation for each joint that is free to rotate; i.e. each joint rotation or *degree of freedom* of the structure will generate one additional equation to be solved.

Step 3: Solve these equations for the unknown joint rotations.

Step 4: Back-substitute the calculated joint rotations into the original set of *slope-deflection equations* to give the end moments in each span.

Step 5: Determine the support reactions, shear force and bending moment diagrams.

This solution approach is orderly and systematic and can be applied to any continuous beam or, as we will see later, framed-structure type problem. Its disadvantage, however, is that it generates a large number of equations and, as a manual solution method for anything other than small problems, it does therefore have its limitations. However, this orderly and systematic approach lends itself to adaptation for solution using computer methods whereby large numbers of simultaneous equations can readily be solved. This requires the equations to be formulated and presented in matrix form such that computer-based matrix solution techniques can be used for their solution. We will not be looking at these *matrix methods* in this book but, when working through the problems in this programme, it is worth remembering that this method is the basis of more advanced methods and is not in itself applicable, as a manual method, to other than problems of modest size. Now try some problems for yourself. These are identical to the problems in Frame 24 of Programme 15 so that you can compare and contrast the two solution methods of *moment distribution* and *slope deflection*.

14

PROBLEMS

For each of the continuous beams shown below calculate the joint rotations and the values of the end moments in each span. Assume that, in each case, the *EI* value is constant throughout.

Ans. (Moments in kN m—clockwise moments shown as positive)

	θ_B	θ_C	M'_{AB}	M'_{BA}	M'_{BC}	M'_{CB}	M'_{CD}	M'_{DC}
Q1:	0.83/EI		−5.83	+8.33	−8.33	+10.83		
Q2:	−1.50/EI		−16.00	+13.00	−13.00	+8.50		
Q3:	0	0	−13.33	+13.33	−13.33	+13.33	−13.33	+13.33
Q4:	4.44/EI	−4.44/EI	−8.89	+22.22	−22.22	+22.22	−22.22	+8.89

Q1   Q2

Q3   Q4

15

BEAMS WITH SIMPLY SUPPORTED ENDS

Let's look again at continuous beam problems where one or more ends is *simply supported* or *pinned* as illustrated in the example below.

Write down the slope-deflection equations for the two spans of the structure.

16

$$M'_{AB} = \frac{2EI}{L}[2\theta_A + \theta_B] \pm FEM_A = \frac{2EI}{8}[2\theta_A + \theta_B] - 40$$

$$M'_{BA} = \frac{2EI}{L}[\theta_A + 2\theta_B] \pm FEM_B = \frac{2EI}{8}[\theta_A + 2\theta_B] + 40$$

$$M'_{BC} = \frac{2EI}{L}[2\theta_B + \theta_C] \pm FEM_B = \frac{2EI}{4}[2\theta_B + \theta_C] = \frac{2EI}{4}[2\theta_B]$$

$$M'_{CB} = \frac{2EI}{L}[\theta_B + 2\theta_C] \pm FEM_C = \frac{2EI}{4}[\theta_B + 2\theta_C] = \frac{2EI}{4}[\theta_B]$$

Did you realise that there are no FEMs for span BC and, as C is a built-in support, the rotation θ_C is zero?

Consider the equilibrium conditions for joint A and joint B and, using the above equations, write down the expressions that we will use for calculating the joint rotations.

17

$$\frac{2EI}{8}[2\theta_A + \theta_B] - 40 = 0$$

$$\left[\frac{2EI}{8}[\theta_A + 2\theta_B] + 40\right] + \frac{2EI}{4}[2\theta_B] = 0$$

The first of the above equations comes from the fact that joint A is simply supported and, therefore the moment at A, M'_{AB}, must be zero. The second equation comes from consideration of the equilibrium of joint B and the fact that the sum of the two internal moments $(M'_{BA} + M'_{BC})$ at that joint must equate to zero. These equations may now be solved as simultaneous equations for θ_A and θ_B and these values then back-substituted into the *slope-deflection equations* given in Frame 16.

Complete the calculations to calculate the joint rotations and end moments.

18

θ_A	θ_B	M_{AB}	M_{BA}	M_{BC}	M_{CB}
$101.82/EI$	$-43.63/EI$	0	+43.64	−43.64	−21.82 kN m

19

BEAMS WITH CANTILEVERED ENDS

Let's look again at another example that we previously solved in Programme 15. In this case the left-hand support at A is not simply supported but continues as a cantilevered overhang.

What is the moment at A due to the load applied at the end of the cantilever?

20

$$\boxed{-5 \text{ kN m (anti-clockwise)}}$$

The solution to this problem is essentially no different to the almost identical problem that we considered in Frame 15. The only difference is that in the previous problem the moment at A was zero whilst in the above example the moment at A is -5 kN m. Hence, when considering the slope-deflection equation for the left-hand end of span AB we would write the expression:

$$M'_{AB} = \frac{2EI}{L}[2\theta_A + \theta_B] \pm FEM_A = \frac{2EI}{8}[2\theta_A + \theta_B] - 40$$

which, when considering the equilibrium of joint A, we would then express as:

$$\frac{2EI}{8}[2\theta_A + \theta_B] - 40 = -5$$

The solution to the problem then continues exactly as in the previous example.

Complete this example and calculate the joint rotations and the moments at the end of each span.

21

θ_A	θ_B	M'_{AB}	M'_{BA}	M'_{BC}	M'_{CB}
$90.91/EI$	$-41.82/EI$	-5	$+41.82$	-41.82	-20.91 kN m

22

$$\boxed{\text{PROBLEMS}}$$

For each of the continuous beams shown below calculate the joint rotations and the values of the end moments in each span. Assume that, in each case, the EI value is constant throughout.

Ans. (Moments in kN m—clockwise moments shown as positive)

	θ_A	θ_B	θ_C	M'_{AB}	M'_{BA}	M'_{BC}	M'_{CB}
Q1:	$3.33/EI$	0		0	$+10.00$	-10.00	$+10.00$
Q2:		0	$-5/EI$	-15.00	$+15.00$	-15.00	0
Q3:	$-19.52/EI$	$5.71/EI$		-40.00	-1.43	$+1.43$	$+15.71$
Q4:		$-1.76/EI$	$0.88/EI$	-16.18	$+12.65$	-12.65	$+10.00$

23

SETTLEMENT AT SUPPORTS

In all the problems that we have considered so far we have assumed that the structure has been supported on rigid supports which remain fixed in space. There are, however, situations where due to, for example, poor ground conditions one or more supports may 'settle' and this settlement will induce additional moments into the structure together with a redistribution of the total loads between the supports and a resulting change in the distribution of shearing forces.

This is illustrated in the two-span beam on the next page where the support at B is shown, in figure (a), to have settled by an amount δ. The solution procedure is very similar to that adopted to deal with the effect of normal loadings on the span in that each span is first 'fixed' by the application of fixed-end moments (figure (b)) and then 'released', as before, by the application of releasing moments as shown in figures (c), (d) and (e).

(a) Two span-beam with support settlement   (b) Spans 'fixed' by application of fixed-end moments

(c) Span AB restrained by fixed-end moments   (d) Joint A released   (e) Joint B released

The key to the analysis is in the calculation of the fixed-end moments and in this case the standard formulae given in the table in Frame 20 of Programme 15 can be used. In the left-hand span negative anti-clockwise fixed-end moments will be applied (see case 4 in the table) and in the right-hand span positive clockwise fixed-end moments will be applied (case 5 in the table).

If, for illustration purposes, we look at span AB only, as shown in figures (c) to (e) the superposition of the moments in these figures will give the final moments at each joint. That is:

$$M'_{AB} = \frac{4EI}{L}\theta_A + \frac{1}{2}M_{BA} - \frac{6EI\delta}{L^2} = \frac{4EI}{L}\theta_A + \frac{1}{2}\frac{4EI}{L}\theta_B - \frac{6EI\delta}{L^2}$$
$$= \frac{2EI}{L}\left[2\theta_A + \theta_B - \frac{3\delta}{L}\right]$$

and

$$M'_{BA} = \frac{4EI}{L}\theta_B + \frac{1}{2}M_{AB} - \frac{6EI\delta}{L^2} = \frac{4EI}{L}\theta_B + \frac{1}{2}\frac{4EI}{L}\theta_A - \frac{6EI\delta}{L^2}$$
$$= \frac{2EI}{L}\left[\theta_A + 2\theta_B - \frac{3\delta}{L}\right]$$

We have in this example shown only the effect of the sinking support but if there had been, as would be usual, actual loading on the span this would be accounted for, as before, by including the FEMs due to this loading as additional terms in the above equations, i.e.

$$M'_{AB} = \frac{2EI}{L}\left[2\theta_A + \theta_B - \frac{3\delta}{L}\right] \pm FEM$$
$$\text{and} \quad M'_{BA} = \frac{2EI}{L}\left[\theta_A + 2\theta_B - \frac{3\delta}{L}\right] \pm FEM$$

These two expressions represent the general form of the slope-deflection equations and *should be remembered*.

24

Let's look at an example of a problem with a sinking support. The two-span continuous beam shown below has a constant EI value throughout and the support C sinks by a fixed amount δ. We want to examine the effect of this sinking support on the bending moments induced in the structure.

How many unknown displacements are there to be solved in this problem?

25

Two

Joint A is fixed and hence there is zero rotation at this joint. However, both joint B and joint C will rotate giving two unknown joint displacements which must be determined. This implies that we must establish two simultaneous equations containing θ_B and θ_C which can then be solved.

Now let's write down the *slope-deflection* equations for each span of the beam as follows.

For span AB:

$$M'_{AB} = \frac{2EI}{L}\left[2\theta_A + \theta_B - \frac{3\delta}{L}\right] \pm FEM = \frac{2EI}{4}[\theta_B] \qquad (a)$$

$$M'_{BA} = \frac{2EI}{L}\left[\theta_A + 2\theta_B - \frac{3\delta}{L}\right] \pm FEM = \frac{2EI}{4}[2\theta_B] \qquad (b)$$

and for span BC:

$$M'_{BC} = \frac{2EI}{L}\left[2\theta_B + \theta_C - \frac{3\delta}{L}\right] \pm FEM = \frac{2EI}{8}\left[2\theta_B + \theta_C - \frac{3\delta}{8}\right] \qquad (c)$$

$$M'_{CB} = \frac{2EI}{L}\left[\theta_B + 2\theta_C - \frac{3\delta}{L}\right] \pm FEM = \frac{2EI}{8}\left[\theta_B + 2\theta_C - \frac{3\delta}{8}\right] \qquad (d)$$

Note that in the above equations θ_A has been set to zero as there is no joint rotation at A and the *FEM* terms have been set to zero as, in this example, there are no loads on the beam. Also, for span AB the deflection term δ is set to zero as the deflection only occurs in span BC at joint C.

Write down the equilibrium equations for joints B and C and hence calculate the joint rotations and end moments in terms of the displacement δ.

$$\theta_B = 3\delta/88 \quad \theta_C = 15\delta/88 \quad M'_{AB} = 3EI\delta/176$$
$$M'_{BA} = 6EI\delta/176 \quad M'_{BC} = -6EI\delta/176 \quad M'_{CB} = 0$$

Check your solution.

For equilibrium of joint B:

$$M'_{BA} + M'_{BC} = \frac{2EI}{4}[2\theta_B] + \frac{2EI}{8}\left[2\theta_B + \theta_C - \frac{3\delta}{8}\right] = \frac{2EI}{8}\left[6\theta_B + \theta_C - \frac{3\delta}{8}\right] = 0$$

and for equilibrium of the end support at C:

$$M'_{CB} = \frac{2EI}{8}\left[\theta_B + 2\theta_C - \frac{3\delta}{8}\right] = 0$$

These reduce to two simultaneous equations:

$$6\theta_B + \theta_C - \frac{3\delta}{8} = 0 \quad \text{and} \quad \theta_B + 2\theta_C - \frac{3\delta}{8} = 0$$

which can be solved directly to give $\theta_B = 3\delta/88$ and $\theta_C = 15\delta/88$. Back-substitution into equations (a) and (b) will give the end moments:

$$M'_{AB} = \frac{2EI}{4}[\theta_B] = \frac{2EI}{4} \times \frac{3\delta}{88} = \frac{3}{176}EI\delta$$
$$M'_{BA} = \frac{2EI}{4}[2\theta_B] = \frac{2EI}{4} \times 2 \times \frac{3\delta}{88} = \frac{6}{176}EI\delta$$

and we know that for equilibrium of joint B $M'_{BC} = -M'_{BA}$ and that, as C is simply supported, $M'_{CB} = 0$.

Can you sketch the bending moment diagram for the beam? As usual, think about the deflected shape of the beam to help you sketch the diagram.

The diagram illustrates the practical effect that a non-rigid support will have on the internal moments in the structure. If, as there would be in practice, there were actual loads on the structure the above diagram would be superimposed on to that for the loading to give the combined set of bending moments.

28

1. Calculate the end moments in the beam shown in figure Q1 if the support at C sinks by an amount δ. Take EI as constant throughout.

Ans. $(M'_{AB} = 0.176EI\delta \quad M'_{BA} = 0.353EI\delta \quad M'_{BC} = -0.353EI\delta \quad M'_{CB} = 0)$

2. Repeat question 1 if support A sinks by an amount δ. (Hint: when considering the *sign* of the deformation term δ in the slope-deflection equations for AB note that the left-hand end of the member has deflected relative to the right-hand end whereas in the previous example and in question 1 above it was the other way around.)

Ans. $(M'_{AB} = 0.510EI\delta \quad M'_{BA} = 0.353EI\delta \quad M'_{BC} = -0.353EI\delta \quad M'_{CB} = 0)$

3. Repeat question 1 if support B sinks by an amount δ. (Hint: Both members will now have fixed-end moments as a result of the sinking support – think carefully about the *signs* of the deformation term δ in each case.)

Ans. $(M'_{AB} = -0.686EI\delta \quad M'_{BA} = -0.706EI\delta \quad M'_{BC} = 0.706EI\delta \quad M'_{CB} = 0)$

Q1

29

PORTAL FRAME STRUCTURES: NON-SWAY

We will now illustrate the application of the slope-deflection method to portal frame problems. Consider the propped portal frame shown below:

How many simultaneous equations do we need to establish to solve this problem?

30

Joints A and D are fixed and do not, therefore, contribute to the development of the equations. Both joints B and C can rotate in space. If you examine joint B you might argue that it could move both vertically and horizontally due to the elastic shortening of the column AB and beam BC, respectively. However, we treat this type of deformation as second-order deformation which is negligibly small and therefore would consider that joint B (and C) can only rotate and remain fixed in position in space as shown in the deflected shape diagram. Joints B and C, therefore, each contribute one unknown joint rotation to the problem and hence we need to establish two simultaneous equations to determine these unknown rotations.

Can you calculate the fixed-end moments for each member of the frame?

31

$$\boxed{\begin{array}{ll} \text{Columns:} & FEM = 0 \\ \text{Beam:} & FEM = \pm 30 \text{ kN m} \end{array}}$$

Neither of the columns are subject to applied loads and hence the FEMs are zero. However, for the beam, the application of the standard FEM formula (case 2) from the table in Frame 20 of Programme 15 gives, for the left-hand 40 kN load:

$$FEM_B = -\frac{Wab^2}{L^2} = -\frac{40 \times 1 \times 3^2}{4^2} = -22.5 \text{ kN m}$$

$$FEM_C = +\frac{Wa^2b}{L^2} = +\frac{40 \times 1^2 \times 3}{4^2} = +7.5 \text{ kN m}$$

As the right-hand 40 kN load is symmetrically located on the beam the fixed-end moment calculations for this load will give identical results to those above with a FEM at B of −7.5 kN m and at C of +22.5 kN m. Superposition of these two sets of calculations will give the FEM at B as $(-22.5 - 7.5) = -30$ kN m and at C as $(+7.5 + 22.5) = +30$ kN m.

The slope-deflection equations can now be written down for each member and the appropriate modifications made to allow for the built-in supports ($\theta_A = 0$ and $\theta_D = 0$). Also, the fact that all the joints are fixed in position in space means that there is no sway term in any of the members ($\delta = 0$).

Write down and simplify the set of slope-deflection equations for the structure.

32

$$M'_{AB} = \frac{2EI}{L}\left[2\theta_A + \theta_B - \frac{3\delta}{L}\right] \pm FEM = \frac{2EI}{6}[\theta_B] \tag{a}$$

$$M'_{BA} = \frac{2EI}{L}\left[\theta_A + 2\theta_B - \frac{3\delta}{L}\right] \pm FEM = \frac{2EI}{6}[2\theta_B] \tag{b}$$

$$M'_{BC} = \frac{2EI}{L}\left[2\theta_B + \theta_C - \frac{3\delta}{L}\right] \pm FEM = \frac{2EI}{4}[2\theta_B + \theta_C] - 30 \tag{c}$$

$$M'_{CB} = \frac{2EI}{L}\left[\theta_B + 2\theta_C - \frac{3\delta}{L}\right] \pm FEM = \frac{2EI}{4}[\theta_B + 2\theta_C] + 30 \tag{d}$$

$$M'_{CD} = \frac{2EI}{L}\left[2\theta_C + \theta_D - \frac{3\delta}{L}\right] \pm FEM = \frac{2EI}{6}[2\theta_C] \tag{e}$$

$$M'_{DC} = \frac{2EI}{L}\left[\theta_C + 2\theta_D - \frac{3\delta}{L}\right] \pm FEM = \frac{2EI}{6}[\theta_C] \tag{f}$$

The solution method is now exactly as before. If the equilibrium of joint B is considered equations (b) and (c) can be combined. Likewise equations (d) and (e) can be combined by considering the equilibrium of joint C.

Calculate the joint rotations at B and C.

33

$$\boxed{\theta_B = 25.71/EI \qquad \theta_C = -25.71/EI}$$

Check your solution:

$$M'_{BA} + M'_{BC} = \frac{2EI}{6}[2\theta_B] + \frac{2EI}{4}[2\theta_B + \theta_C] - 30 = EI\left[\frac{5}{3}\theta_B + \frac{1}{2}\theta_C\right] - 30 = 0$$

$$M'_{CB} + M'_{CD} = \frac{2EI}{4}[\theta_B + 2\theta_C] + 30 + \frac{2EI}{6}[2\theta_C] = EI\left[\frac{1}{2}\theta_B + \frac{5}{3}\theta_C\right] + 30 = 0$$

This pair of simultaneous equations can readily be solved to give $\theta_B = 25.71/EI$ and $\theta_C = -25.71/EI$. The end moments for column AB can then be calculated by substituting these rotations back into equations (a) and (b):

$$M'_{AB} = \frac{2EI}{6}[\theta_B] = \frac{2EI}{6} \times \frac{25.71}{EI} = 8.57 \text{ kN m}$$

$$M'_{BA} = \frac{2EI}{6}[2\theta_B] = \frac{2EI}{6} \times \frac{51.42}{EI} = 17.14 \text{ kN m}$$

and for equilibrium of joint B we know that $M'_{BA} + M'_{BC} = 0$; therefore $M'_{BC} = -17.14$ kN m. To calculate the end moments in the rest of the structure we can substitute into the remaining equations (d) to (f).

Rather than carry out this substitution there is a a quicker way of establishing the end moments in the right-hand part of the structure. What is the principle that we can use to do this?

34

> The principle of symmetry

This structure is geometrically symmetrical and is loaded symmetrically and, therefore, the moments in the right-hand half of the structure will 'mirror' those in the left-hand half. Once we have established the end moments the calculation of support reactions and the drawing of the bending moment diagrams will, as in the previous continuous beam examples, follow the analytical principles established in Programme 15. We will not repeat this aspect of the analysis but ensure, by looking back at Programme 15, that you are thoroughly familiar with these stages of the analysis.

35

PORTAL FRAMES WITH SWAY

Portal frames where sway is permitted are tedious to solve manually as sway introduces additional degrees of freedom into the structure and additional equations then have to be solved. We will, however, illustrate the technique by further considering the portal frame that we have just analysed but we will now permit sway to take place as shown. Additionally, we will include a lateral wind force to act on the left-hand side of the structure.

How many simultaneous equations do we need to establish to solve this problem?

36

$$\boxed{\text{Three}}$$

As before, joints A and D are fixed and do not, therefore, contribute to the development of the equations. Both joints B and C can rotate in space but as sway is now permitted both joints deflect horizontally by an unknown amount δ.

Again we treat second-order deformations due to axial shortening of the columns as negligably small. We therefore consider that joint B (and C) do not displace vertically but they will rotate and now, as sway is permitted, will displace horizontally. If we neglect axial shortening in the beam then the side-sway of B and C can be considered to be of the same magnitude and of an unknown amount δ as shown in the figure in Frame 35.

Can you calculate the fixed-end moments for each member of the frame?

37

Column AB:	$FEM = \pm 15$ kN m
Column CD:	$FEM = 0$
Beam:	$FEM = \pm 30$ kN m

Column AB is subject to a distributed load of 5 kN/m and the FEMs can be calculated from the standard formula for a uniformly distributed load ($\pm wL^2/12 = \pm 5 \times 6^2/12 = \pm 15$ kN m). The right-hand column is not subject to load and hence the FEMs are zero. The FEMs for the beam were previously calculated as ± 30 kN m. The slope-deflection equations can now be written down for each member and the appropriate modifications made to allow for the built-in supports ($\theta_A = 0$ and $\theta_D = 0$). However, for both columns there will now be a sway term δ in each of the equations which introduces an additional 'unknown' into the problem. These can now be written as:

$$M'_{AB} = \frac{2EI}{L}\left[2\theta_A + \theta_B - \frac{3\delta}{L}\right] \pm FEM = \frac{2EI}{6}\left[\theta_B - \frac{3\delta}{6}\right] - 15 \tag{a}$$

$$M'_{BA} = \frac{2EI}{L}\left[\theta_A + 2\theta_B - \frac{3\delta}{L}\right] \pm FEM = \frac{2EI}{6}\left[2\theta_B - \frac{3\delta}{6}\right] + 15 \tag{b}$$

$$M'_{BC} = \frac{2EI}{L}\left[2\theta_B + \theta_C - \frac{3\delta}{L}\right] \pm FEM = \frac{2EI}{4}[2\theta_B + \theta_C] - 30 \tag{c}$$

$$M'_{CB} = \frac{2EI}{L}\left[\theta_B + 2\theta_C - \frac{3\delta}{L}\right] \pm FEM = \frac{2EI}{4}[\theta_B + 2\theta_C] + 30 \tag{d}$$

$$M'_{CD} = \frac{2EI}{L}\left[2\theta_C + \theta_D - \frac{3\delta}{L}\right] \pm FEM = \frac{2EI}{6}\left[2\theta_C - \frac{3\delta}{6}\right] \tag{e}$$

$$M'_{DC} = \frac{2EI}{L}\left[\theta_C + 2\theta_D - \frac{3\delta}{L}\right] \pm FEM = \frac{2EI}{6}\left[\theta_C - \frac{3\delta}{6}\right] \tag{f}$$

If you are unsure about the signs of the FEMs for column AB just rotate the diagram of the structure so that you are viewing the column as if it were a horizontal beam with the load acting vertically downwards. The signs of the FEMs (clockwise positive or anti-clockwise negative) should then be apparent. Likewise the sway-deflection term, δ, can be determined to be equivalent to a right-hand support settlement by viewing both columns in this way (see case 4 in the FEM table in Programme 15, Frame 20)

Using the slope-deflection equations write down the equilibrium equations for joint B and joint C.

38

$$M'_{BA} + M'_{BC} = \frac{2EI}{6}\left[2\theta_B - \frac{3\delta}{6}\right] + 15 + \frac{2EI}{4}[2\theta_B + \theta_C] - 30 = EI\left[\frac{5\theta_B}{3} + \frac{\theta_C}{2} - \frac{\delta}{6}\right] - 15 = 0$$

$$M'_{CB} + M'_{CD} = \frac{2EI}{4}[\theta_B + 2\theta_C] + 30 + \frac{2EI}{6}\left[2\theta_C - \frac{3\delta}{6}\right] = EI\left[\frac{\theta_B}{2} + \frac{5\theta_C}{3} - \frac{\delta}{6}\right] + 30 = 0$$

These two equations now contain the three variables, θ_B, θ_C and δ. To obtain a solution it is necessary to find a third equation which can be obtained by considering the horizontal equilibrium of the structure as shown in the figure below. The end moments are shown as clockwise positive to be consistent with our sign convention.

(a) Free-body diagram of whole structure

(b) Free-body diagrams of columns AB and CD

Now carry out the following exercise:
 (i) *From figure (a) write down the equation of horizontal equilibrium,*
 (ii) *From figure (b) take moments about the top of each column to express each of the horizontal reactions H_A and H_D in terms of the column end moments,*
 (iii) *Hence eliminate the horizontal reactions from the equation in (i) and produce a third equation of equilibrium for the structure.*

39

Check your solution against that given below

From figure (a) for the whole structure

$$\Sigma H = 0 \qquad\qquad -H_A - H_D + 30 = 0$$

From figure (b) for column AB:

$$\Sigma M_B = 0 \qquad H_A \times 6 + M'_{AB} + M'_{BA} - (5 \times 6) \times 3 = 0$$
$$\therefore \quad H_A = -(M'_{AB} + M'_{BA} - 90)/6$$

From figure (b) for column CD:

$$\Sigma M_C = 0 \qquad H_D \times 6 + M'_{CD} + M'_{DC} = 0$$
$$\therefore \quad H_D = -(M'_{CD} + M'_{DC})/6$$

Hence substituting for H_A and H_D from the last two equations into the first one gives:

$$-H_A - H_D + 30 = 0$$
$$\frac{(M'_{AB} + M'_{BA} - 90)}{6} + \frac{(M'_{CD} + M'_{DC})}{6} + 30 = 0$$
$$\therefore \quad M'_{AB} + M'_{BA} + M'_{CD} + M'_{DC} + 90 = 0$$

This final equation can now be combined with the slope-deflection equations in Frame 37 to give:

$$M'_{AB} + M'_{BA} + M'_{CD} + M'_{DC} + 90 = 0$$
$$\therefore \quad \frac{2EI}{6}\left[\theta_B - \frac{3\delta}{6}\right] - 15 + \frac{2EI}{6}\left[2\theta_B - \frac{3\delta}{6}\right] + 15 + \frac{2EI}{6}\left[2\theta_C - \frac{3\delta}{6}\right] + \frac{2EI}{6}\left[\theta_C - \frac{3\delta}{6}\right] + 90 = 0$$
$$\therefore \quad EI[3\theta_B + 3\theta_C - 2\delta] + 270 = 0$$

Hence, combining with the two joint equilibrium equations given at the top of Frame 38 the three equations can be simplified and expressed as:

$$EI[10\theta_B + 3\theta_C - \delta] - 90 = 0$$
$$EI[3\theta_B + 10\theta_C - \delta] + 180 = 0$$
$$EI[3\theta_B + 3\theta_C - 2\delta] + 270 = 0$$

These are three simultaneous equations which can now be solved.
Solve these equations to obtain the two joint rotations and the side-sway term, δ.

$$\theta_B = 28.28/EI \quad \theta_C = -10.29/EI \quad \delta = 162.00/EI$$

Now use the moment equations in Frame 37 to calculate the end moments in each member of the frame.

$$M'_{AB} = -32.57 \quad M'_{BA} = +6.85 \quad M'_{BC} = -6.85$$
$$M'_{CB} = +33.86 \quad M'_{CD} = -33.86 \quad M'_{DC} = -30.43 \text{ kN m}$$

Check your answers against the working below:

$$M'_{AB} = \frac{2EI}{6}\left[\theta_B - \frac{3\delta}{6}\right] - 15 = \frac{2EI}{6}\left[\frac{28.28}{EI} - \frac{3 \times 162}{6EI}\right] - 15 = -32.57 \text{ kN m}$$

$$M'_{BA} = \frac{2EI}{6}\left[2\theta_B - \frac{3\delta}{6}\right] + 15 = \frac{2EI}{6}\left[\frac{2 \times 28.28}{EI} - \frac{3 \times 162}{6EI}\right] + 15 = +6.85 \text{ kN m}$$

$$M'_{BC} = \frac{2EI}{4}[2\theta_B + \theta_C] - 30 = \frac{2EI}{4}\left[\frac{2 \times 28.28}{EI} - \frac{10.29}{EI}\right] - 30 = -6.85 \text{ kN m}$$

$$M'_{CB} = \frac{2EI}{4}[\theta_B + 2\theta_C] + 30 = \frac{2EI}{4}\left[\frac{28.28}{EI} - \frac{2 \times 10.29}{EI}\right] + 30 = +33.86 \text{ kN m}$$

$$M'_{CD} = \frac{2EI}{6}\left[2\theta_C - \frac{3\delta}{6}\right] = \frac{2EI}{6}\left[-\frac{2 \times 10.29}{EI} - \frac{3 \times 162}{6EI}\right] = -33.86 \text{ kN m}$$

$$M'_{DC} = \frac{2EI}{6}\left[\theta_C - \frac{3\delta}{6}\right] = \frac{2EI}{6}\left[-\frac{10.29}{EI} - \frac{3 \times 162}{6EI}\right] = -30.43 \text{ kN m}$$

Once the end moments have been calculated the bending moment diagram for the frame can be drawn by superimposing the 'end-moment' bending moment diagram on to the 'simply supported' bending moment diagram for each member of the frame. This was illustrated in the examples in Programme 15 and it is not necessary to repeat this stage of the calculations again.

If the horizontal support reactions at A and D are required then substitution for the end moments into the equilibrium equations given at the top of Frame 39 will give H_A and H_D directly. As in the portal frame example in Programme 15, the horizontal reaction at A, H_A, can then be used to determine the position of zero shear and, hence, maximum moment in the column AB. Consideration of the equilibrium of the free-body diagram of each element of the structure in turn will give the vertical components of reaction at the supports A and D.

42

TO REMEMBER

Standard fixed-end moment formula—see table in Frame 20 in Programme 15.

The standard form of the slope-deflection equations:

$$M'_{AB} = \frac{2EI}{L}\left[2\theta_A + \theta_B - \frac{3\delta}{L}\right] \pm FEM$$

and

$$M'_{BA} = \frac{2EI}{L}\left[\theta_A + 2\theta_B - \frac{3\delta}{L}\right] \pm FEM$$

and the sign convention:

(i) clockwise moments and rotations are taken as positive
(ii) right-hand (clockwise) support deformation is taken as positive and left-hand (anti-clockwise) support deformation is taken as negative.

Right-hand support deformation – positive

Left-hand support deformation – negative

43

FURTHER PROBLEMS

The majority of the problems given below are identical to those given in the 'Further Problem' section at the end of Programme 15. You should have already attempted them using the moment-distribution method; you can now attempt them using the slope-deflection method. In this case you are only required to calculate the joint deformations and end moments: it is assumed that you have already mastered drawing the bending moment diagrams from our work in the previous programme.

1. For each of the beams shown calculate the joint rotations and end moments. For figures Q1(a), Q1(b) and Q1(c) assume that the EI value is constant throughout.

Ans.	θ_A	θ_B	θ_C	θ_D	M'_{AB}	M'_{BA}	M'_{BC}	M'_{CB}	M'_{CD}	M'_{DC}
Q1(a)	11.90/EI	−7.14/EI	0		0	14.29	−14.29	−7.14		
Q1(b)	0	−2.80/EI	−3.6/EI		−30.40	19.20	−19.20	0		
Q1(c)	0	0.44/EI	1.56/EI	0	0.44	+0.89	−0.89	+6.89	−6.89	11.56
Q1(d)	0	−3.53/EI	3.43/EI		−17.53	10.29	−10.29	+10	−10	0

Q1(a)   Q1(b)   Q1(c)   Q1(d)

2. If the end support at A in the beam in figure Q1(a) settles by an amount δ calculate the *additional* moments that are induced in the beam due to the effect of settlement.

Ans	M'_{AB}	M'_{BA}	M'_{BC}	M'_{CB}
	0	0.429EIδ	−0.429EIδ	−0.214EIδ

3. If the support at B in the beam in figure Q1(b) settles by an amount δ calculate the *additional* moments that are induced in the beam due to the effect of settlement.

Ans.	M'_{AB}	M'_{BA}	M'_{BC}	M'_{CB}
	−1.147EIδ	−0.960EIδ	0.960EIδ	0

4. For each of the frames shown in figures Q4(a) and Q4(b), on the next page, calculate the joint rotations and end moments. Take the EI value as constant throughout.

Ans.	θ_A	θ_B	θ_C	M'_{AB}	M'_{BA}	M'_{BC}	M'_{CB}
Q4(a)	0	34.733/EI	0	13.88	27.77	−27.77	48.62
Q4(b)	0	56.84/EI	−132.61/EI	22.71	45.44	−45.44	0

Q4(a)   Q4(b)

5. Calculate the joint rotations and end-moments of the braced frame shown in figure Q5

Ans.

θ_A	θ_B	θ_C	θ_D	M'_{AB}	M'_{BA}	M'_{BC}	M'_{CB}	M'_{CD}	M'_{DC}
0	$2.199/EI$	$-10.903/EI$	$5.451/EI$	-29.29	31.45	-31.45	5.45	-5.45	0

6. Calculate the joint rotations and end-moments of the unbraced frame shown in figure Q6.

Ans.

θ_A	θ_B	θ_C	θ_D	δ	M'_{AB}	M'_{BA}	M'_{BC}	M'_{CB}	M'_{CD}	M'_{DC}
0	$26.01/EI$	$-18.12/EI$	0	$46.65/EI$	-18.26	8.09	-8.09	19.85	-19.85	-13.81

Q5

Q6

Programme 17

INFLUENCE LINES

On completion of this programme you will be able to:

- Draw influence-line diagrams for reactions, shearing forces and bending moments for simply supported and cantilevered beam structures
- Recognise the difference between influence-line diagrams for shearing force and shearing force diagrams
- Recognise the difference between influence-line diagrams for bending moment and bending moment diagrams
- Use influence-line diagrams for simple beam structures to determine the maximum effect(s) attributable to moving-point load systems
- Use influence-line diagrams for simple beam structures to determine the maximum effect(s) attributable to moving distributed load systems
- Draw influence-line diagrams for axial load in the members of pin-jointed frame structures
- Use influence-line diagrams for pin-jointed frame structures to determine the maximum axial loads attributable to moving load systems

1

In all the previous programmes we have studied structures, or structural components, where the loading system is considered to be fixed in position on the structure. There are, however, some types of engineering problems where the load system could be located at one of a number of possible positions and, indeed, may actually move across the structure.

Can you think of a situation where a structure may be subject to a moving load?

2

> A train crossing a rail bridge

You may equally have thought about a car or a number of cars simultaneously crossing a road bridge. If you were designing such a bridge constructed as, let's say, a simply supported beam you would need to determine the maximum bending moments and shear forces in order to subsequently carry out the design of the beam. In the case of a system of moving loads these maximum effects will depend not only on the number and magnitude of the loads but will also depend on their location and relative position on the span. In this programme we are going to examine a number of different types of structure and determine the effect of moving load systems. We will adopt graphical methods of presentation which are referred to as *influence lines*.

Let's start by looking at the simple case of a simply supported beam carrying a single point load of 1 kN located at a distance x from the left-hand support. The distance x defines the position of the load on the span as the load moves from left to right.

Calculate and tabulate the value of the left-hand reaction V_A for $x = 0, 2, 4, 6, 8$ and 10 m and plot, to scale, a graph showing how V_A varies in magnitude as the load moves across the span. Repeat for the right-hand reaction, V_B.

3

x (m)	0	2	4	6	8	10
W (kN)	1	1	1	1	1	1
V_A (kN)	1	0.8	0.6	0.4	0.2	0
V_B (kN)	0	0.2	0.4	0.6	0.8	1

You should have plotted graphs of support reaction as follows:

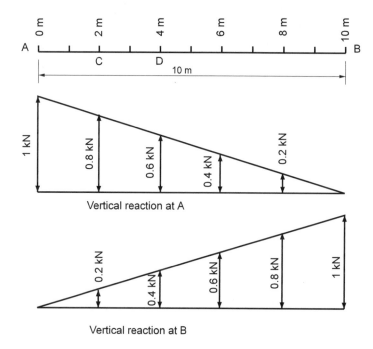

Vertical reaction at A

Vertical reaction at B

These graphs are the *influence-line diagrams* for the vertical reactions at A and B, respectively and indicate how these reactions vary as a *single load of unit value* of 1 kN moves across the span from left to right.

From the influence-line diagram for V_A it is apparent that a load of 1 kN at C, 2 m from A, will cause a reaction of 0.8 kN at A.

What would be the reaction at A if a load of 15 kN was placed on the beam at C?

4

12 kN

You should have realised that if a load of unit value (i.e. 1 kN) causes a reaction of 0.8 kN then a load of 15 kN will cause a reaction of $15 \times 0.8 = 12$ kN.

What would be the reaction at A if a load of 15 kN was placed on the beam at C together with a further load of 10 kN at D, 4 m from A? What would be the reaction at B when the loads are in this position?

5

$$V_A = 18 \text{ kN} \qquad V_B = 7 \text{ kN}$$

You should have used the *influence-line* diagrams to calculate the effect of the individual loads and then used the *Principle of Superposition* to add together the effect of the two separate loads, i.e.

$$V_A = (15 \times 0.8) + (10 \times 0.6) = 18 \text{ kN}$$

and

$$V_B = (15 \times 0.2) + (10 \times 0.4) = 7 \text{ kN}$$

Now let's extend the concepts that we have just introduced to look at the shearing force variation at a point on the beam when a unit load is located at a variable distance x along the span. Let's consider the shearing force at the mid-span position of the beam, marked as point M on the diagram shown below:

The shear force (*SF*) at M is taken as the algebraic sum of all the forces to the left of M.

Calculate and tabulate the value of the left-hand reaction V_A and hence the SF at M for $x = 0, 2, 4, 6, 8$ and 10 m and plot, to scale, a graph showing how SF_M varies in magnitude as the load moves across the span. (Hint: when the unit load is to the left of M [$x < 5$ m] $SF_M = V_A - 1$. When it is to the right of M [$x > 5$ m] $SF_M = V_A$.)

6

x (m)	0	2	4	6	8	10
W (kN)	1	1	1	1	1	1
V_A (kN)	1	0.8	0.6	0.4	0.2	0
SF_M (kN)	0	−0.2	−0.4	0.4	0.2	0

The *influence line for shearing force* at mid-span, drawn opposite, is very different from the *shearing force diagram* which would be drawn for the point load in a single static position (see Programme 4, Frame 16) It must be realised that the *influence-line diagram* shows the variation of shearing force at a *fixed point* on the beam for a unit load *at a variable position* whereas the *shearing force diagram* represents the variation of shear force at differing positions along the span for a fixed position of load. Note that there is an abrupt change in the shape of the diagram, from positive to negative shear as the load passes through the mid-span position.

Shearing force at mid-span

Now use the influence-line diagram to calculate the shear force at mid-span when a load of 15 kN is located at point C with an additional load of 10 kN at D.

7

$$-7 \text{ kN}$$

Did you calculate the shear force as follows?

$$SF_M = [15 \times (-0.2)] + [10 \times (-0.4)]$$
$$= -7 \text{ kN}$$

Now let's extend the concepts further to look at the bending moment variation at midspan when a unit load is at a variable distance x along the span. If you look back at the diagram in Frame 5 you should remember that the bending moment at M is given by the algebraic sum of the moments of all forces to the left *or* to the right of M. We will, as usual, take sagging moments as positive.

Calculate and tabulate the value of the left-hand reaction V_A, the right-hand reaction V_B and hence the BM at M for $x = 0, 2, 4, 6, 8$ and 10 m and plot, to scale, a graph showing how BM_M varies in magnitude as the load moves across the span. (Hint: when the unit load is to the left of M [$x < 5$ m] take moments of all forces to the right of M. When it is to the right of M [$x > 5$ m] take moments of all forces to the left of M.)

8

x (m)	0	2	4	6	8	10
W (kN)	1	1	1	1	1	1
V_A (kN)	1	0.8	0.6	0.4	0.2	0
V_B (kN)	0	0.2	0.4	0.6	0.8	1
BM_M (kN m)	0	1.0	2.0	2.0	1.0	0

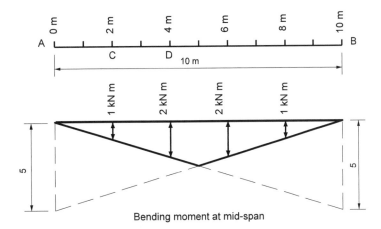

Bending moment at mid-span

Again you should note that the influence-line diagram represents the variation of bending moment at a fixed point at mid-span for a unit load acting at a variable position along the span.

Use the influence-line diagram to calculate the bending moment at mid-span when a load of 15 kN is located at point C with an additional load of 10 kN at D.

9

$$\boxed{35 \text{ kN m}}$$

Did you calculate the bending moment as follows?

$$BM_M = (15 \times 1) + (10 \times 2) = 35 \text{ kN m}$$

It should be apparent that the influence-line diagram for bending moment at mid-span consists of straight lines. Although, to draw the diagram, we calculated the bending moments for discrete positions of the load along the span, in practice it is only necessary to calculate and plot the bending moment values when the unit load is (i) at the supports at A and B and (ii) directly above the mid-span position. These plotted values can then be connected by straight lines to give the influence line for mid-span bending moment.

Similarly, for the influence line for mid-span shearing force, calculations are only necessary for the unit load placed (i) at the supports at A and B (ii) just to the left of the mid-span position and (iii) just to the right of the mid-span position. These values can be plotted and joined by straight lines to give the required diagram.

Hence to draw the influence lines for shearing force and bending moment at any other position along the span the same principles can be applied and a limited number of calculations will give the influence lines directly.

Draw the influence-line diagrams for shearing force and bending moment for point C which is at 2 m from the left-hand support.

10

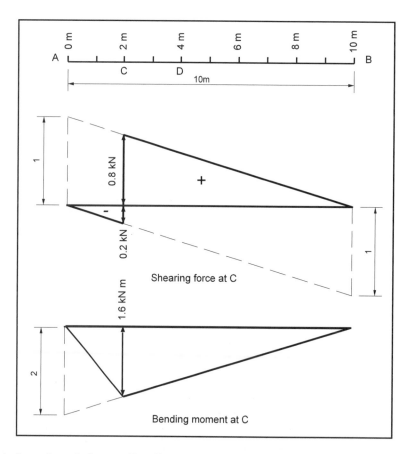

As before, these influence-line diagrams can be used to calculate the shearing force and bending moment at C, respectively for any combination of loads on the span. The ordinates corresponding to each load position can be scaled from the diagram, multiplied by the magnitude of the respective load and then the separate values calculated for each of the loads can be superimposed.

11

| PROBLEMS |

1. For the cantilever beam shown in figure Q1, draw the influence-line diagrams for (i) shearing force at B (ii) bending moment at B (iii) vertical reaction at C and (iv) the end-fixing moment at C.

Ans. (see below)

2. The load system shown in figure Q2 is applied to the cantilever beam. Use the influence-line diagrams drawn in question 1 to find (i) shearing force at B (ii) bending moment at B (iii) vertical reaction at C and (iv) the fixing moment at C.

Ans. ((i) −15 kN (ii) −12.5 kN m (iii) 15 kN (iv) −42.5 kN m)

Ans. to Q1:

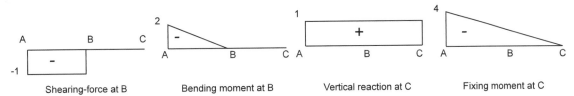

| Shearing-force at B | Bending moment at B | Vertical reaction at C | Fixing moment at C |

12

DISTRIBUTED LOAD SYSTEMS

The previous examples demonstrate how influence-line diagrams for a fixed point on a structure can be used to calculate reactions, shearing forces, bending moments and so on at that point for a system of one or more concentrated loads.

If we have a load system consisting of a moving distributed load, such as shown in the figure opposite, the influence-line diagrams can be used directly to determine the effect of the distributed load system. The figure shows the, previously determined, influence lines for shearing force and bending moment at the mid-span point, M, of a 10 m simply supported beam. In this case, a uniformly distributed load of 5 kN/m and of length 2 m crosses the beam from left to right and is instantaneously positioned as shown. If a small length of beam of length dx is considered then the shearing force due to the load on the length dx will be given by $SF = 5 \times dx$.

Hence the shear force at M due to the 2 m long distributed load is given by:

$$SF_M = \int_0^2 5\,dx = 5\int_0^2 dx$$

The integral expression is simply the area, shown shaded, of the influence-line diagram underneath the distributed load and can be simply calculated as:

$$SF_M = 5 \times \left[\left(\frac{-(0.2 + 0.4)}{2}\right) \times 2\right] = -3.0 \text{ kN}$$

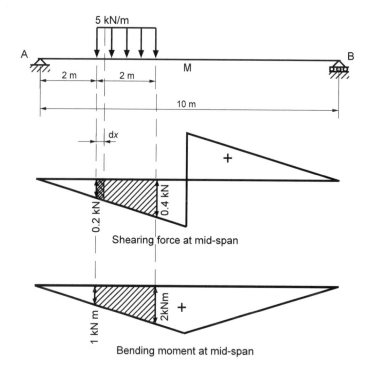

Shearing force at mid-span

Bending moment at mid-span

Similarly the bending moment at M can be calculated from the shaded area of the influence-line diagram as:

$$BM_M = 5 \times \left[\left(\frac{1.0 + 2.0}{2}\right) \times 2\right] = 15 \text{ kN m}$$

Where would this 2 m long load be positioned to give the maximum shearing force at mid-span. What would be the magnitude of this shearing force and what would be the corresponding mid-span bending moment?

13

$$SF_{\mathrm{M}} = -4.0 \text{ kN} \qquad BM_{\mathrm{M}} = +20 \text{ kN m}$$

Did you realise that the leading edge of the load system would have to reach the mid-span position in order to cover the maximum possible area under the influence-line diagram for shearing force as indicated below? The corresponding area under the influence-line diagram for bending moment would give the corresponding bending moment at mid-span.

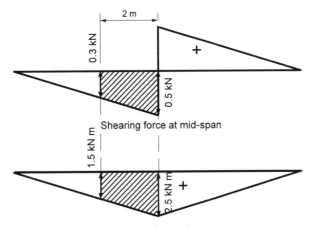

Shearing force at mid-span

Bending moment at mid-span

$$SF_{\mathrm{M}} = 5 \times \left[\left(\frac{-(0.3 + 0.5)}{2} \right) \times 2 \right] = -4.0 \text{ kN}$$

and

$$BM_{\mathrm{M}} = 5 \times \left[\left(\frac{2.5 + 1.5}{2} \right) \times 2 \right] = 20 \text{ kN m}$$

It should be apparent that the magnitude of the shearing force and bending moment, in this case at mid-span, depends on (i) the magnitude of the distributed load (ii) its length and (iii) its position on the span. If the load were longer than the span, entered the span at A and moved across from left to right the influence-line diagram for shear would indicate that the maximum negative shear at mid-span would occur when that part of the span to the left of M is fully loaded. When the load moves completely to the right of M and fully covers the part of the span to the right of M then maximum positive shear occurs at M. The influence-line diagram for bending moment shows, however, that the maximum bending moment occurs when the whole span is loaded. The same general observations would be true for other positions than mid-span.

Returning to the 2 m long distributed load shown in the figure in Frame 12, where would this be positioned to give the maximum bending moment at mid-span and what would be the value of this moment?

$$\boxed{BM_M = +22.5 \text{ kN m}}$$

The maximum area under the influence-line diagram occurs when the load is located symmetrically either side of the mid-span position.

Bending moment at mid-span

The maximum bending moment is then calculated as:

$$BM_M = 5 \times \left[2 \times \left(\frac{2.5 + 2.0}{2} \right) \times 1 \right] = 22.5 \text{ kN m}$$

For other points on the span it can be shown that the maximum bending moment occurs when the load is positioned such that the point divides the load in the same ratio as it divides the span. Thus the maximum moment at C, 2 m from the left-hand support in the figure below is given when $c/d = a/b$, or in other words when:

$$\frac{c}{(2-c)} = \frac{2}{8} \qquad \therefore \; 8c = 2(2-c) \quad \text{or} \; c = 0.4 \quad \text{and} \; d = (2 - 0.4) = 1.6 \text{ m}$$

Bending moment at C

The key values of bending moment shown on the influence-line diagram have been calculated by simple proportions. You should check these values.

Using the diagram above, what is the value of the maximum bending moment at C?

15

$$\boxed{14.40 \text{ kN m}}$$

The maximum bending moment is given by the shaded area under the influence-line diagram multiplied by the load per unit length. Check your calculations:

$$BM_{C(max)} = 5 \times \left[\left(\frac{(1.28 + 1.60)}{2} \right) \times 0.4 + \left(\frac{1.60 + 1.28}{2} \right) \times 1.6 \right] = 14.40 \text{ kN m}$$

16

$$\boxed{\text{PROBLEMS}}$$

1. The cantilever beam shown in figure Q1 is subject to a uniformly distributed load of 10 kN/m which can cover all or any part of the span. Calculate (i) the maximum shearing force at B (ii) the maximum bending moment at B (iii) the maximum vertical reaction at C and (iv) the end-fixing moment at C. (*Hint: refer back to the influence-line diagram solutions to Q1 in frame 11.*)

Ans. ((i) −20 kN (ii) −20 kN m (iii) 40 kN (iv) −80 kN m)

Q1   Q2

2. The beam shown in figure Q2 is simply supported at A and C and has a cantilever overhang between C and D. Draw the influence-line diagram for (i) the reaction at A (ii) the reaction at C (iii) the shear force at B and (iv) the bending moment at B.

Ans. (see opposite)

3. Use the influence-line diagrams drawn in Q2 to calculate the maximum value of each of the actions (i) to (iv) when a concentrated load of 10kN moves from left to right across the structure.

Ans. ((i)10 kN (ii) 14.29 kN (iii) ±5 kN (iv) 17.50 kN m)

4. Use the influence lines drawn in Q2 to calculate the maximum value of each of the actions (i) to (iv) when a distributed load of 10 kN/m, longer than the span, moves from left to right across the structure entering the structure at A and leaving at D.

Ans. ((i) 35 kN (ii) 71.43 kN (iii) −8.75 kN (iv) 61.25 kN m)

Ans. to Q2.

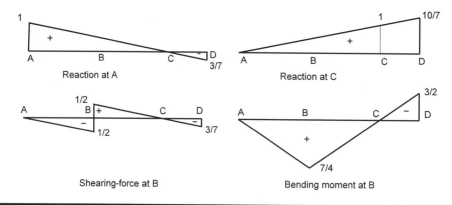

Reaction at A   Reaction at C

Shearing-force at B   Bending moment at B

17

FRAMED STRUCTURES

The principles that we have established for beams subject to moving loads can also be applied to bridge structures which are constructed as statically determinate pin-jointed trusses such as those studied in Programme 3. A system of loads moving across such a bridge will cause variation of force within each member of the structure. The application of influence-line theory will enable us to determine the greatest force developed in each member of the truss: the magnitude and sense (tension or compression) of these forces being required in the subsequent design of those members. Consider the truss shown below which we will use as an illustration of the application of influence lines to this type of problem.

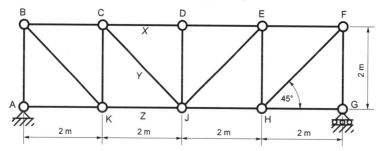

The truss is one of a pair of parallel trusses with a bridge deck suspended between them in such a way that the load from the deck is transmitted to the truss through the joints in either the top or bottom chord. We will assume that for the purposes of this illustration the load is transferred to the bottom chord, A–K–J–H–G, and moves from A to G. We will examine the influence-line diagrams for the members marked X, Y and Z.

If the truss was subject to a set of static loads and we were interested in determining the forces in the members marked X,Y and Z only what method of analysis would you use?

18

You should recall that we studied this method in Programme 3. If you do not recall the method you should revise Programme 3 before proceeding. Now let's take a section through the three members indicating the unknown member forces as if they are all in tension.

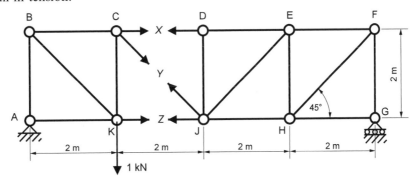

Assuming that a single unit load of 1 kN acts at joint K how would you calculate the member force marked X?

19

Take moments about joint J for the part
of the structure to the right of the section

You could also take moments about J for the part of the structure to the left of the section but the calculations are marginally simpler if you consider that part of the structure with the least number of forces acting on it.

What is the value of X when the unit load is at K, as shown?

20

-0.50 kN

You should have calculated that the reaction at G is 0.25 kN then

$$\Sigma M_J = 0: \qquad -(V_G \times 4) - (X \times 2) = 0 \quad \therefore X = -2 \times 0.25 = -0.5 \text{ kN}$$

Repeat this calculation when the unit load is located at each of the other joints on the bottom chord, tabulate your results and plot the influence-line diagram for X.

Location of load	A	K	J	H	G
V_A (kN)	1	0.75	0.50	0.25	0
V_G (kN)	0	0.25	0.50	0.75	1
X (kN)	$=-2V_G$ $=0$	$=-2V_G$ $=-0.50$	$=-2V_A$ $=-1.0$	$=-2V_A$ $=-0.50$	$=-2V_A$ $=0$

Force X in member CD

Note that the negative sign indicates that this member is in compression for any position of the unit load and hence for any load system that crosses the structure.

When the unit load was to the *left* of the section did you consider the equilibrium of the part of the frame to the *right* of the section? Likewise did you consider the left-hand side of the frame when the unit load was to the right? If you did you would realise that the force in member CD varies linearly with V_A when the load is to the right of the section and linearly with V_G when the load is to the left. If we apply this principle to the members marked Y and Z it should be apparent that it is only necessary to calculate the value of the member force when the unit load is located at either side of the section, i.e. at joints K and J, and to plot these two values. The two plotted points can be joined together and can be connected to the zero values when the unit load is at A and G to complete the diagram.

Referring back to the diagram in Frame 18 how would you now calculate the forces in the members marked Y and Z?

For Y: resolve forces vertically
For Z: take moments about C

By considering the unit load located firstly at J then at K calculate the forces in the members marked Y and Z. Use these values to plot the influence-line diagrams for these two members.

23

Check your calculations and diagrams against those given below

For force Y resolve vertically:

Location of load	K	J
Reaction V_A	0.75	0.50
Reaction V_G	0.25	0.50
Consider frame	to right of section	to left of section
$\Sigma V = 0$:	$Y\sin 45° + V_G = 0$	$-Y\sin 45° + V_A = 0$
	$\therefore\ Y = -V_G/\sin 45° = -0.25/\sin 45°$	$\therefore\ Y = +V_A/\sin 45° = +0.50/\sin 45°$
	$= -0.35$ kN	$= +0.71$ kN

Force Y in member CJ

Note that when the unit load, and hence the actual load system, is to the left of the section member CJ is in compression but when it is to the right of the section it is in tension. This means that when this member is being designed consideration must be given to the load case that causes maximum compression *and* to the load case that gives maximum tension. For force Z take moments about C:

Location of load	K	J
Reaction V_A	0.75	0.50
Reaction V_G	0.25	0.50
Consider frame	to right of section	to left of section
$\Sigma M_C = 0$:	$(Z \times 2) - (V_G \times 6) = 0$	$-(Z \times 2) + (V_A \times 2) = 0$
	$\therefore\ Z = +3 \times V_G = +3 \times 0.25 = +0.75$ kN	$\therefore\ Z = +V_A = +0.50$ kN

Hence the influence line diagram for Z:

Force Z in member JK

If a distributed load of magnitude 20 kN/m and of length 10 m crosses the structure from left to right what are the maximum forces in the members marked X, Y and Z?

$X = -80$ kN (compression)
$Y = +37.84$ kN (tension) and -9.35 kN (compression)
$Z = +60$kN (tension)

Did you realise that for maximum value of X and Z the load would cover the whole span? However, Y will have both a maximum tension and a maximum compression value when the load covers that part of the span corresponding to either the positive or negative section of the influence-line diagram.

TO REMEMBER

This programme has not introduced any new formulae to commit to memory and is concerned with techniques of drawing and interpreting influence-line diagrams. You should be absolutely confident that you:

- Can recognise the essential differences between influence-line diagrams for shearing-force and shearing force diagrams. Likewise for bending moments,
- Can draw influence-line diagrams for simple beam and framed structure problems,
- Can interpret the diagrams for moving load systems consisting of trains of point loads and/or distributed loads.

FURTHER PROBLEMS

The influence-line diagram solutions are given at the end of the programme.

1. A bridge consists of two cantilevered side-spans which support a simply-supported central span, as shown in figure Q1. Draw the influence-line diagram for (i) bending moment at the mid-span point C (ii) the vertical reaction at A and (iii) the end-fixing moment at A. What would be the maximum moment at support A for (i) a load of 10 kN/m over the whole structure and (ii) a load of 10 kN/m, 6 m long which traverses the bridge from left to right?

Ans. ((i) −160 kN m (ii) −150 kN m)

Q1

2. The bridge shown in figure Q2 supports a moving distributed load of magnitude 10 kN/m and of length 6 m which traverses the bridge from left to right. Evaluate the maximum value of (i) the vertical reaction at A (ii) the vertical reaction at B (iii) the vertical reaction at D and (iv) the hogging moment over the support at B.

Ans. ((i) 25 kN (ii) 52.50 kN (iii) 25 kN (iv) −30 kN m)

Q2   Q3

3. For the structure shown in figure Q3, draw influence lines for (i) the vertical reaction at A (ii) the vertical reaction at C and (iii) the end moment at the built-in support at A. If a pair of concentrated loads at a fixed distance 2 m apart, the leading load of magnitude 5 kN and the trailing load of magnitude 2 kN, cross the bridge from left to right which support, A or C, would be subject to the largest reaction and what would be the value of this reaction?

Ans. (Support C = 8.5 kN)

4. The deck loading on the road bridge shown in figure Q4 is transmitted to the *bottom* chord of the truss. Draw the influence-line diagrams for the forces in the members marked X and Y and calculate the maximum value of these forces if a distributed load of magnitude 20 kN/m and of length 10 m crosses the structure from left to right.

Ans. (X = −80 kN, Y = 56.8 kN)

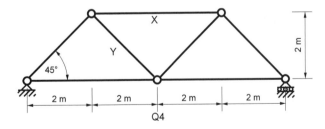

Q4

5. Figure Q5 shows a Warren girder which spans 15 m. Draw the influence-line diagrams for the members marked X and Y for loading applied to the *top* chord of the girder. Evaluate the maximum force in each member if a distributed load of 20 kN/m and 10 m long crosses the structure from left to right.

Ans. (X = −230 kN, Y = ±28.5 kN)

Q5

27

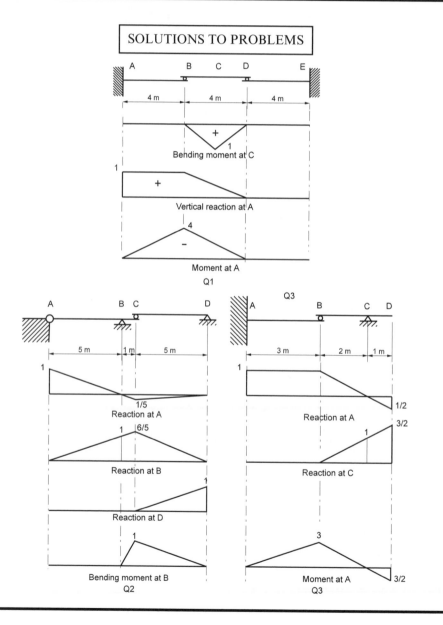

SOLUTIONS TO PROBLEMS

Bending moment at C

Vertical reaction at A

Moment at A

Q1

Q3

Reaction at A

Reaction at B

Reaction at D

Bending moment at B

Q2

Reaction at A

Reaction at C

Moment at A

Q3

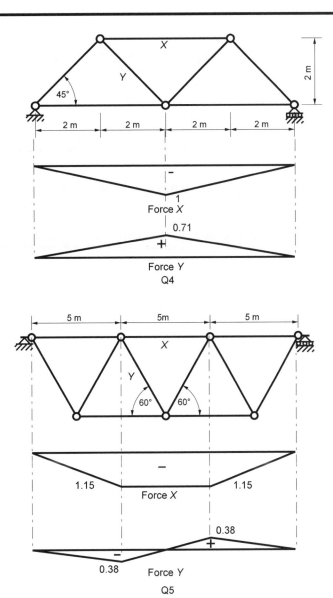

Q4

Q5

Programme 18

ELASTIC BUCKLING OF
AXIALLY LOADED
COMPRESSION MEMBERS

On completion of this programme you will be able to:

- Describe the different modes of failure of structural members loaded in axial compression
- Derive and quote the *Euler* buckling formula for a pin-ended member
- Define and calculate *radius of gyration*
- Define and calculate *slenderness ratio*
- State what is meant by a *short* member
- State what is meant by a *slender* member
- Describe the differences between the theoretical and actual behaviour of axially loaded members
- Define *effective length* (height)
- Describe the effect of different end conditions on the strength of members in axial compression
- Carry out simple theoretical calculations to determine the load carrying capacity of stanchions, columns and struts

1

In Programme 5 we examined the behaviour of axially loaded structural members including those subject to compressive loads; such as columns within buildings and struts within pin-jointed framed structures. We saw how any externally applied axial load would result in the development of an internal elastic stress system. Such structural members would continue to resist any increase in load by developing proportionately higher internal stresses providing that the material is not strained beyond the elastic range. If the loading is increased beyond this limit then some form of material failure will eventually occur such as crushing of the concrete in a concrete column or yielding of the steel in a steelwork stanchion.

However, such considerations neglect the possibility of other forms of structural failure. In this programme we will examine another type of failure that can occur in axially loaded members and is of particular concern when designing, for example, long slender columns within building and bridge structures. We will explain the principles within the context of steel construction although these principles are equally applicable to concrete, masonry and timber construction.

A hollow cylindrical steel tube is used as a column to support part of a bridge structure. It has an outer diameter of 200 mm and a wall thickness of 20 mm. Calculate the stress in the section if it is subject to an axial compressive load of 2500 kN.

2

$$\boxed{221.04 \text{ N/mm}^2}$$

You should have obtained the answer as follows:

$$\text{Area of section} = \frac{\pi}{4}\left(D_{\text{out}}^2 - D_{\text{in}}^2\right) = \frac{\pi}{4}\left(200^2 - 160^2\right) = 11.31 \times 10^3 \text{ mm}^2$$

$$\text{Stress, } \sigma = \frac{\text{Force}}{\text{Area}} = \frac{2500 \times 10^3}{11.31 \times 10^3} = 221.04 \text{ N/mm}^2$$

This is a relatively straightforward calculation and should not have caused you any problems. If we assume that the steel will yield at a typical value of yield strength of, say, 265 N/mm² the calculation implies that the column has a reserve of strength and will sustain a load somewhat greater than 2500 kN before failure by yielding of the steel occurs. The question did not give any information about the height of the column and, indeed, we did not have any need to know the height when calculating the axial compressive stress. This would imply that the column's response to loading is independent of its height. Let's, however, consider two columns (a) 1 m high and (b) 10 m high; both fabricated from the same hollow steel tubular section and both subject to an increasing axial load.

Would you expect the two columns to fail in the same way?

3

It's unlikely!

The first column is a *short* column and will probably fail by yielding of the steel when the load is of sufficient magnitude to develop a yield stress of 265 N/mm^2 within the section (figure (a)). The second column is *slender* and will fail by *buckling* as indicated in figure (b). If you want to demonstrate the effect of buckling for yourself take a long plastic ruler, stand it on a table and press downwards on the top as shown in figure (c) – it will be immediately apparent what we are talking about. Don't push too hard unless you want to break your ruler!

(a) Short column   (b) Tall slender column   (c) Simple test on a plastic ruler

Assuming that you haven't broken your ruler what happened to the ruler when you removed your hand?

4

The ruler straightened itself

In fact, if you repeat this simple experiment several times, each time the ruler will return to its original undeformed position. The ruler, and the slender column of figure (b), has become *unstable* and can carry much less load than if it were of a much shorter length.

What does the fact that the ruler fully recovers on unloading tell you about the stress–strain behaviour of the material of which the ruler is made?

5

> The material has remained elastic

The fact that the ruler fully recovers its shape means that the material of which it is made has not been strained beyond the elastic range and hence we can say that the ruler has become *elastically unstable*. In other words, it has buckled and failed whilst still strained within the elastic range of material behaviour. This is an important observation because it implies that axially loaded members that are slender, rather than short and stocky, may fail by buckling at stress levels well below those that would cause those members to fail by yielding (of steel) or crushing (of concrete, masonry or timber). Hence, in the design of such members, account must be taken of both likely modes of failure which will depend on whether or not the member is short or slender.

Now let's look at the classical theory of buckling which we can develop by considering the axially loaded, pin-ended, slender strut AB shown in figure (a). The strut is shown just on the point of buckling.

(a) Axially loaded strut   (b) Free-body diagram of A–C

If you look back at Programme 12, for a laterally loaded beam, of flexural stiffness *EI*, we developed the equation that describes the moment–curvature relationship at a section X–X, distance x from the left-hand origin as:

$$EI\frac{d^2v}{dx^2} = -M$$

This equation can now be applied to the above strut if the moment, M, at the section is defined. If we look at the free-body diagram of AC, by taking moments about C, we can see that the moment at C is given by $P \times v$. Hence we can write:

$$EI\frac{d^2v}{dx^2} = -Pv$$

which can readily be re-arranged and simplified to give:

$$\frac{d^2v}{dx^2} + k^2v = 0 \qquad \text{where } k = \sqrt{\frac{P}{EI}} \tag{18.1}$$

As in Programme 12, in order to determine the deflection, v, at the section we can carry out a double integration of equation (18.1) to determine an expression for v which can be applied at any point along the strut.

Can you carry out the double integration of this expression?

ELASTIC BUCKLING OF AXIALLY LOADED COMPRESSION MEMBERS

6

$$v = A\cos kx + B\sin kx \qquad (18.2)$$

The double integration of equation (18.1) is not as straightforward as some of the integral calculations that we carried out in Programme 12. If you are good at mathematics you may have obtained this solution. If you didn't, differentiate the solution, given in equation (18.2), twice and substitute back into equation (18.1) to show that the solution is valid.

A and *B* in equation (18.2) are constants of integration and can be determined by applying the equation at two known locations (two values of *x*) where the deformations (*v*) are known and hence *A* and *B* can be solved in terms of these known deformations. As in programme 12, these are known as the *boundary conditions*.

Can you identify appropriate boundary conditions?

7

$$\begin{array}{l} x = 0: \ v = 0 \\ x = L: \ v = 0 \end{array}$$

The boundary conditions come from the fact that the deflection at both ends of the strut must be zero. If we substitute these values into equation (18.2) we get:

when $x = 0$: $v = A\cos(k \times 0) + B\sin(k \times 0)$ $\therefore \ A = 0$

when $x = L$: $v = A\cos(k \times L) + B\sin(k \times L) = 0$ $\therefore \ B\sin kL = 0$ (18.3)

There are two possible solutions that can be derived from equation (18.3). The equation is satisfied if $B = 0$ but, if this were the case, then substitution of $A = 0$ and $B = 0$ into equation (18.2) implies that the deformation, *v*, is zero at all points along the strut. Thus the strut remains undeformed, which is not the case when it has buckled, and, hence, $B = 0$ is not a valid solution. The only other possible solution of equation (18.3) is that $\sin kL = 0$, in which case the constant *B* will have a real but not definable value. Thus equation (18.2) simply becomes:

$$v = B\sin kx \qquad (18.4)$$

where *B* is a constant of unknown value but, from equation (18.3), we know has non-zero values when $\sin kL = 0$. Hence when $\sin kL = 0$ equation (18.4) indicates that when the strut buckles it assumes a sinusoidal shape of unknown magnitude.

What values of k would ensure that $\sin kL = 0$?

8

$$k = 0 \quad k = \frac{\pi}{L} \quad k = \frac{2\pi}{L} \quad k = \frac{3\pi}{L} \quad \cdots \quad k = \frac{n\pi}{L}$$

As we are considering a sinusoidal equation there is an infinite number of solutions for k that will ensure that the equation equates to zero. In equation (18.1) k was defined as $k = \sqrt{P/EI}$ and, hence, $P = k^2 EI$. The above solutions for k therefore give the values of P shown in the table below together with the corresponding equation for the deformed shape of the strut as given by substitution into equation (18.4).

K	0	π/L	$2\pi/L$	$3\pi/L$	$\cdots$	$n\pi/L$
$P \, (= k^2 EI)$	0	$\dfrac{\pi^2 EI}{L^2}$	$\dfrac{4\pi^2 EI}{L^2}$	$\dfrac{9\pi^2 EI}{L^2}$	$\cdots$	$\dfrac{n^2\pi^2 EI}{L^2}$
$v \, (= B \sin kx)$	0	$B \sin \pi x$	$B \sin 2\pi x$	$B \sin 3\pi x$	$\cdots$	$B \sin n\pi x$

Each of the above values of P will result in instability in the strut and a different mode of buckling, each mode being defined by a sinusoidal curve. If we neglect the trivial solution of $P = 0$ then the critical (least) value of load at which buckling will occur is given by:

$$P_{\text{crit}} = \frac{\pi^2 EI}{L^2} \tag{18.5}$$

This critical value is often referred to as the *Euler buckling load* and is given the symbol P_e. The other values of P are mathematical solutions to the equations but have little practical relevance as the strut will buckle once it reaches the Euler buckling load. If we now apply the Euler formula to the problem of the hollow circular steel section that we first considered in Frames 1 and 2, we can calculate the buckling load of both the 1 m high column and the 10 m high column. The section had an outer diameter of 200 mm and a wall thickness of 20 mm and we can take E for steel as 200 kN/mm^2

I for the section $= \dfrac{\pi}{4}\left(R_{\text{out}}^4 - R_{\text{in}}^4\right) = \dfrac{\pi}{4}\left(100^4 - 80^4\right) = 46.37 \times 10^6 \text{ mm}^4$

For the 1m high column: $\quad P_{\text{crit}} = P_e = \dfrac{\pi^2 EI}{L^2} = \dfrac{\pi^2 \times 200 \times 46.37 \times 10^6}{(1 \times 1000)^2} = 91\,531 \text{ kN}$

For the 10m high column: $\quad P_{\text{crit}} = P_e = \dfrac{\pi^2 EI}{L^2} = \dfrac{\pi^2 \times 200 \times 46.37 \times 10^6}{(10 \times 1000)^2} = 915 \text{ kN}$

The original problem in Frame 1 indicated that, based on considerations of yield, this hollow section could safely carry a load of 2500 kN. What the above calculation shows is that, as the height of the column increases, buckling becomes a more important consideration and, when 10 m high, the column can only support a fraction of the original 2500 kN load.

PROBLEMS

1. A length of steel of rectangular section, 100×100 mm is used to support an axial compressive load. If the stress in the section should not exceed 265 N/mm^2 what is the maximum load that this section can sustain based on considerations of yield in the steel?

Ans. (2650 kN)

2. If the section in question 1 is to be used as a pin-ended strut, 5 m long, what is the maximum load that it can carry based on considerations of buckling? Take $E = 200$ kN/mm^2

Ans. (658 kN)

3. What would be the maximum length of strut such that yielding of the steel and buckling of the strut occur simultaneously?

Ans. (2.49 m)

10

What we have seen so far is that members loaded in axial compression can fail by yielding or crushing but, if they are very long and become *slender*, then they may fail by buckling. Question 3, above, indicates that there is a transition point when both modes of failure may occur simultaneously. Now, let's look in a little more detail at what we mean by a *slender* member. The Euler buckling formula tells us that the *load* at which buckling occurs is given by:

$$P_e = \frac{\pi^2 EI}{L^2}$$

If we divide the buckling load by the area of the section then we will get the *stress* at which buckling occurs:

$$\sigma_e = \frac{P_e}{A} = \frac{\pi^2 EI}{AL^2} = \frac{\pi^2 E(I/A)}{L^2} \qquad (18.6)$$

We can define the second moment of area, I, to be equal to $A \times r^2$ where A is the cross-sectional area of the section and r is the *radius of gyration* of the section. r is therefore equal to $\sqrt{I/A}$. Hence equation (18.6) becomes:

$$\sigma_e = \frac{P_e}{A} = \frac{\pi^2 EI}{AL^2} = \frac{\pi^2 E(I/A)}{L^2} = \frac{\pi^2 Er^2}{L^2} = \frac{\pi^2 E}{(L/r)^2} \qquad (18.7)$$

where L/r is a non-dimensional factor known as the *slenderness ratio* and is a simple way of expressing the length and geometrical shape of any axially loaded member.

Assuming that we are dealing with a steel strut, and taking $E = 200$ kN/mm^2, use equation (18.7) to calculate and plot a graph of σ_e for values of slenderness ratio varying from 100 to 350 in increments of 50.

11

L/r	100	150	200	250	300	350
σ_e (N/mm²)	197	88	49	32	22	16

The above values are obtained by simple substitution into equation 18.7. Plotted as a graph these will give:

from which we can see that as the *slenderness ratio* increases the stress at buckling reduces significantly indicating a rapid decline in the load carrying capacity of the strut due to buckling effects. Although stress values could be plotted for low values of slenderness ratio, in practice the stress values calculated from equation (18.7) are limited as the strut would fail by yielding in direct compression once it reaches the yield strength of steel. This transition from buckling to direct yielding is shown by the upper curtailment of the plotted diagram which you can draw on your graph at a typical value of 265 N/mm². Save your graph: you will need it to solve problems later.

The diagram clearly indicates that at a slenderness ratio somewhere between 50 and 100, depending on the yield strength of the steel, there is a transition in the mode of failure of the strut. For low slenderness ratios the strut can be considered to be *short* and failure will be by yielding in direct compression. For higher slenderness ratios the strut can be considered to be *slender* and failure will be by buckling at much reduced loads. The important fact is that this change of behaviour depends on the *slenderness ratio* which is a function of both the length and the geometry of the cross-section as expressed by its *radius of gyration*.

A steel stanchion, considered pinned at both ends, has a length of 6 m and, from standard steel tables, has a radius of gyration about the axis of buckling of 53 mm and a cross-sectional area of 11 000 mm². Use the graph that you have just plotted to determine the failure load of the stanchion. Is the stanchion short or slender?

> 1705 kN—slender

Check your calculations:

$$\frac{L}{r} = \frac{6000}{53} = 113 \text{ which is in the } slender \text{ zone of the graph}$$

$\therefore$ from the graph $\sigma_e \approx 155 \text{ N/mm}^2$

$\therefore P = \sigma_e \times A \approx 155 \times 11\,000 \times 10^{-3} \approx 1705 \text{ kN}$

The graph that we plotted and used to solve this last problem was based on the assumption of an idealised perfectly straight member with an idealised perfect axial load. In practice, columns and struts are rarely perfectly straight and loads are rarely perfectly axial and, often, are applied at an eccentricity to the longitudinal axis of the member. Hence, the real behaviour of a steel member in compression is more complex than indicated so far and is more likely to follow the typical design curve indicated below with, for a given slenderness ratio, lower stresses developing at failure than indicated by the Euler formula.

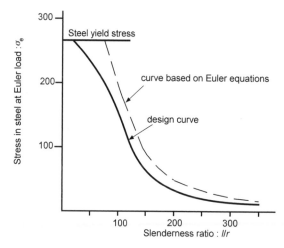

Such curves, which depend on the type of section (e.g. hollow circular or I shape) are based on the *Perry–Robertson* equation which take into account imperfections in geometry and loading. The derivation of the *Perry–Robertson* equation is beyond the scope of this book but you should recognise that modern codes of practice for structural steelwork are based on the principles that we have covered so far with, usually, tables of data, rather than graphs, given to determine maximum stress levels. These stress levels are, as indicated above, a function of *slenderness ratio* and depend on the type of structural section.

What is the factor which governs whether or not a steel compression member will fail by yielding of the steel or by buckling?

13

> Its slenderness ratio

When we introduced the idea of slenderness in Frame 3 we looked at a short steel column and a tall steel column made from the same section. Conceptually we could see that the tall column was likely to buckle whereas the short one was more likely to fail by yielding. What is important to remember is that it is not the height that is important but rather the *slenderness ratio*. Hence a very high column can still be classified as *short* if its section has a sufficiently high radius of gyration.

So far we have discussed the *radius of gyration* as if any shape of section has just one radius. In fact sections, such as I shape column sections, will have two radii of gyration, each one corresponding to one of the two axes of bending. Consider the standard column section shown. $I_{XX} = 9000$ cm^4, $I_{YY} = 3000$ cm^4 and its cross-sectional area is 100 cm^2.

What are the radii of gyration for both the X–X axis and the Y–Y axis respectively?

14

> $r_{XX} = 94.9$ mm $r_{YY} = 54.8$ mm

Check your calculations:

$$r_{XX} = \sqrt{\frac{I_{XX}}{A}} = \sqrt{\frac{9000}{100}} \times 10 = 94.9 \text{ mm} \qquad r_{YY} = \sqrt{\frac{I_{YY}}{A}} = \sqrt{\frac{3000}{100}} \times 10 = 54.8 \text{ mm}$$

Hence we can see that whereas square or circular sections have the same second moment of area about both axes of bending, and hence just one radius of gyration, sections such as I shapes or T shapes will have two radii of gyration corresponding to the different I values of the two axes of bending.

The I-shape section shown above is used as a pin-ended column of 4 m height. Calculate the slenderness ratio for each axis and state which is the most critical axis as far as buckling is concerned.

$$\frac{l}{r_{XX}} = \frac{4000}{94.9} = 42.2 \qquad \frac{l}{r_{YY}} = \frac{4000}{54.8} = 73.0$$

The Y–Y axis is critical

The calculation shows that the column is most slender in relation to the Y–Y axis and, hence this axis will be the most critical in the design of the column which should be based on the slenderness ratio of 73.0.

EFFECTIVE LENGTH OR EFFECTIVE HEIGHT

The theory that we have looked at so far has been based on members that are considered to be pinned at both ends. In practice, one or both ends of a steel strut may be welded or heavily bolted at the end connections and the ends may, more realistically, be considered to be rigidly fixed rather than pinned. Such fixity implies that the ends of the member cannot rotate freely. If full fixity is not certain then the actual fixity may be considered to be somewhere between pinned and fully fixed. Similarly, a stanchion in a building at foundation level, may be constrained by a heavily bolted and stiffened base plate that prevents end rotation and a certain degree of fixity may be assumed.

Such types of connections will have the effect of stiffening the member and increasing its buckling load. If you take your plastic ruler, grasp both ends firmly in your fists and push hard to compress the ruler, you will see that it will buckle at a greater load than when you simply pressed down on it as indicated in Frame 3. The effect of providing rigid or partially rigid end joints is to increase the buckling load and hence the load carrying capacity of the member. Rather than develop different buckling theories for each different type of end condition most codes of practice use the concept of *effective lengths* or *effective heights* where the theory and data developed for pin-ended members is used but calculations are based on *effective* rather than actual lengths. The *effective length* is taken as the actual length multiplied by a coefficient which is dependent on the type of connection at each end of the member. In such cases the *slenderness ratio* is defined as:

$$\text{Slenderness ratio} = \frac{\text{Effective length (height)}}{\text{radius of gyration}}$$

This modified slenderness ratio is then used in the calculation of the buckling stress and, hence, the buckling load. For example, a strut that is rigidly fixed at both ends has a theoretical effective length of one-half of its actual length. We are not going to derive the theoretical values of effective length for different end conditions but we will examine, briefly, the practical implications of this concept.

A slender pin-ended steel strut will fail, theoretically, at its Euler buckling load. If its ends are rigidly restrained such that its effective length can be considered to be one-half of the actual length, what is the increase in its load carrying capacity? Assume that the steel does not yield.

16

> It will increase by a factor of 4

We know from the Euler equation that $P_{crit} = \pi^2 EI / L_e^2$ where L is now taken as the *effective length*. If the effective length (L_e) is taken as one-half of the actual length (L) then:

$$P_{crit} = \frac{\pi^2 EI}{L_e^2} = \frac{\pi^2 EI}{(0.5L)^2} = \frac{4\pi^2 EI}{L^2} = 4P_e$$

Hence we can see that restraining the ends of the strut leads to a dramatic increase in its theoretical load carrying capacity. This concept of effective length is shown diagrammatically below. The effective length is, in effect, that length of the member that buckles between contraflexure points and can be determined theoretically using similar integration techniques to those used in Frame 5 when we first examined the case of a pin-ended member.

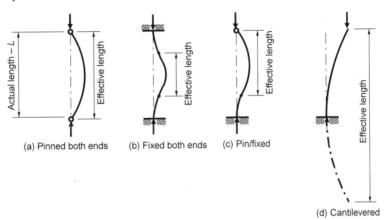

(a) Pinned both ends   (b) Fixed both ends   (c) Pin/fixed   (d) Cantilevered

The table below gives the theoretical values of effective length and, for comparison and interest, typical, more conservative values used in modern steel design codes for the four cases shown above:

Case	(a)	(b)	(c)	(d)
Theoretical effective length	L	$0.5L$	$0.7L$	$2.0L$
Design values of effective length	L	$0.7L$	$0.85L$	$2.0L$

A steel stanchion has a length of 6 m, a radius of gyration about the axis of buckling of 53 mm and a cross-sectional area of 11000 mm². Use the graph that you plotted in Frame 10 to determine the theoretical failure load of the stanchion for each of the cases (a) to (d) above. In each case state whether or not the stanchion is short or slender?

(a) 1705 kN   (b) 2915 kN   (c) 2915 kN   (d) 440 kN

Check your calculations:

Case (a)

$\dfrac{l}{r} = \dfrac{6000}{53} = 113$ which is in the *slender* zone of the graph

$\therefore$ from the graph $\sigma_e \approx 155$ N/mm^2

$\therefore P = \sigma_e \times A \approx 155 \times 11000 \times 10^{-3} \approx 1705$ kN

Case (b)

$\dfrac{l}{r} = \dfrac{0.5 \times 6000}{53} = 57$ which is in the *short* zone of the graph

$\therefore$ from the graph $\sigma_e \approx 265$ N/mm^2

$\therefore P = \sigma_e \times A \approx 265 \times 11000 \times 10^{-3} \approx 2915$ kN

Case (c)

$\dfrac{l}{r} = \dfrac{0.7 \times 6000}{53} = 79$ which is in the *short* zone of the graph

$\therefore$ from the graph $\sigma_e \approx 265$ N/mm^2

$\therefore P = \sigma_e \times A \approx 265 \times 11000 \times 10^{-3} \approx 2915$ kN

Case (d)

$\dfrac{l}{r} = \dfrac{2.0 \times 6000}{53} = 226$ which is in the *slender* zone of the graph

$\therefore$ from the graph $\sigma_e \approx 40$ N/mm^2

$\therefore P = \sigma_e \times A \approx 40 \times 11000 \times 10^{-3} \approx 440$ kN

These calculations clearly show the effect, and importance, of different end conditions on the load carrying capacity of stanchions and other axially loaded members and are important considerations in the design of such members.

18

TO REMEMBER

The Euler formula: $\quad P_e = \dfrac{\pi^2 EI}{L^2}$

Radius of gyration: $\quad r = \sqrt{\dfrac{\text{second moment of area}}{\text{cross-sectional area}}} = \sqrt{\dfrac{I}{A}}$

Slenderness ratio: $\quad SR = \dfrac{\text{effective length (height)}}{\text{radius of gyration}} = \dfrac{l}{r}$

Effective length $\quad = \text{coefficient} \times \text{actual length}$

19

To answer these questions you may use the graph that you plotted in Frame 11. Assume that $\sigma_y = 265$ N/mm^2 and $E_s = 200$ kN/mm^2.

1. A hollow steel scaffolding tube of outer diameter 50 mm and wall thickness 5 mm is to be used as a strut. Determine the critical length of the tube such that it will reach its full yield strength before it buckles for (a) both ends pinned (b) both ends fully fixed

Ans. (1.38 m, 2.76 m)

2. A steel column, of circular section, is 3 m long. It is fully fixed at its base but is completely unrestrained at the top end such that the column is effectively a cantilever. It carries an axial load of 500 kN. Determine the minimum required diameter of the column.

Ans. (117 mm)

3. A stanchion within a steel-framed building has an I_{XX} value of 36×10^6 mm^4 and I_{YY} value of 15×10^6 mm^4. For bending about the Y–Y axis it may be considered to be fully fixed at both ends and for bending about the X–X axis it may be considered to be pinned at both ends. Its height is 6 m. Determine the buckling load for both axes of bending and, hence, state the maximum load that the column can carry.

Ans. (1974 kN [X–X], 3290 kN [Y–Y], max = 1974 kN)

4. A stanchion is fabricated from two 203 × 133 × 25 universal beam sections which are connected by two steel plates as shown in figure Q4. Calculate the spacing, s, between the two beams such that the radius of gyration about both the X–X axis and the Y–Y axis are the same. $I_{XX} = 2360$ cm^4, $I_{YY} = 310$ cm^4 and $A = 32.3$ cm^2. Ignore the contribution of the steel plates.

Ans. (159.33 mm)

Q4

5. The stanchion in question 4 is 8 m high. It is bolted to a very rigid foundation at its base and at the top supports a pair of steel beams. The connection between the stanchion and the beams can be considered to be a pinned connection. What is the maximum axial load that this stanchion can carry?

Ans. (2970 kN)

Programme 19

PLASTIC ANALYSIS

On completion of this programme you will be able to:

- Draw the plastic stress distribution diagram across a beam's cross-section
- Calculate the fully plastic moment of standard shapes of section
- Calculate the plastic modulus of standard shapes of section
- State the number of plastic hinges required for the development of a full collapse mechanism
- Calculate the collapse load factor of one-span beams, continuous beams and portal frame structures under the action of point and distributed load systems
- Carry out simple design procedures to determine the required value of a fully plastic moment of a steel structure for a given collapse load system
- Apply the plastic collapse theorems to the analysis of beam and simple frame structures

1

In this programme we will develop the principles of plastic theory which are widely used in the analysis and design of steel structures including portal frames, multi-bay frames and continuous beams. All the techniques of analysis and design that we have developed so far have been based on the principles of linear elastic theory, i.e. that stress is directly proportional to strain and that stress and strain are directly related by *Young's Modulus of Elasticity*. However, if we consider the behaviour of steel as a structural material we know that once the material is strained beyond the *limit of proportionality* its stress–strain behaviour changes and stress is no longer related to strain by *Young's Modulus of Elasticity*.

If a sample of mild steel is strained beyond the limit of proportionality what term do we use to describe the next stage of its stress–strain behaviour?

2

The plastic stage

The figure below shows the idealised stress–strain behaviour of mild steel. We can see that initially it behaves in a linear elastic manner but once the strain exceeds the limit of proportionality the material *yields* and it continues to strain without any corresponding increase in stress. We assume that the stress–strain behaviour is identical in both tension and compression.

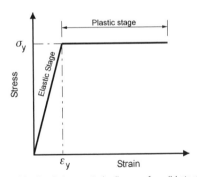

Idealised stress–strain diagram for mild steel

Other types and grades of steel will exhibit similar characteristics, the essential feature of which is that once the material is strained beyond the limiting strain, ε_y, it will exhibit plastic behaviour with a constant stress, σ_y, as the strains increase.

Plastic theory is concerned with the behaviour of structures, parts of which have been strained by the load system beyond the yield point such that the structure reaches a state of collapse. Application of the theory allows us to analyse a given structure to determine its collapse load or, alternatively, to design a structure to just collapse under the action of a given set of collapse loads. The collapse loads are obtained by applying suitable factors of safety to the actual loads supported by the structure. Safety in design therefore comes from an appropriate choice of safety factor applied to the load system.

Let's start by examining the behaviour of a typical simply supported steel I beam with an applied central point load, as shown. Let's assume that the load is such that the strains across the section X–X at mid-span are initially well within the elastic range of the material's stress–strain behaviour.

Simply supported steel beam   Section X–X

Can you sketch the shape of both the strain and stress distribution diagrams at the mid-span cross-section marked X–X?

3

Cross-section   Strain distribution   Stress distribution
                       diagram   diagram

The above triangular strain and stress diagrams should be quite familiar to you: if not revise Programme 6 before proceeding. As the load on the beam increases the strain and stress diagrams will increase proportionately but will remain triangular in shape as long as the strains remain within the elastic range of the material behaviour.

If the beam is loaded further such that the strains in the top and bottom of the section exceed the yield strain, ε_y, can you sketch the likely shape of the stress distribution diagram across section X–X?

4

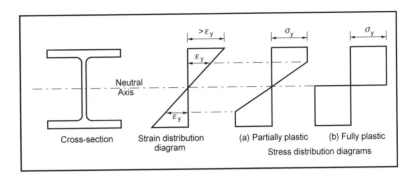

You may have drawn one or the other of the above stress distribution diagrams. Once the strains in the top and bottom fibres of the beam reach the yield strain then the material in these outer layers will yield and become plastic at a constant stress value of σ_y. As the load on the beam increases further then the strains across the section will increase and the zone of plasticity will spread from the outer fibres towards the centre of the section as indicated in figure (a). For even more increase in load the zone of plasticity will spread even further into the section until a stage of full plasticity is reached as shown in figure (b). At this stage the section has reached its full moment capacity as there is no more stress capacity available to develop any additional moment of resistance within the section.

What do you think will happen to the beam when the mid-span cross-section becomes fully plastic as indicated above?

5

> The beam will collapse

The figure below indicates that once the mid-span section becomes fully plastic a *plastic hinge* forms at mid-span and a *collapse mechanism* develops. The load at this stage has increased beyond the normal working load which the beam is intended to carry but with a margin of safety indicated by the *load factor* λ which is the ratio of the collapse load W_c to the normal working load W_w.

Plastic methods are therefore concerned with the identification of the most likely collapse mechanisms of a steel structure and the analysis of those mechanisms to determine the collapse load W_c or the collapse load factor λ. Conversely, we may be interested in designing a steel structure to carry a set of loads which will be the working loads multiplied by an appropriate value of load factor, λ. The safety of such structures against collapse will be assured by the inclusion of the load factor, λ.

Before we consider these methods of analysis and design it is first necessary to establish how to calculate the *moment of resistance* of a steel section under fully plastic stress conditions when a plastic hinge forms in the section. This moment of resistance is usually referred to as the section's *fully plastic moment* and is given the symbol, M_P. Consider the section shown below with its associated plastic stress diagram and with compression above and tension below the neutral axis.

Cross-section   Plastic stress diagram

The compressive stress, σ_y, acting on the area, A_1, above the neutral axis will give rise to a compressive force, C. Likewise the stress, σ_y, acting on the area, A_2, below the neutral axis will give rise to a tensile force, T. C and T, acting together, result in an internal couple system which creates the internal moment of resistance, or the *fully plastic moment*, of the section.

Hence: $C = A_1 \times \sigma_y$ and $T = A_2 \times \sigma_y$

But for equilibrium of the cross-section $C = T$

$$\therefore \quad A_1 \times \sigma_y = A_2 \times \sigma_y$$

What does this last equation tell you about the relationship between A_1 and A_2 and, hence, where does the position of the neutral axis lie?

6

A_1 equals A_2: the neutral axis is the axis of equal area

In other words the neutral axis bisects the section such that the area above equals the area below the neutral axis.

How does this differ from the neutral axis position when we used elastic theory?

7

In elastic theory the neutral axis passes through the centroid of the section

In the case of the symmetrical I beam shown in the last figure the centroidal axis and the axis of equal area will be the same. However, in the case of non-symmetrical sections they will not be the same which means that as the section moves from an elastic to a plastic state the neutral axis moves from the centroidal axis to the axis of equal area. Ensure that you are fully conversant with this fact as it is important! Consider the steel channel section shown below. All dimensions are in mm.

Can you identify the position of the plastic neutral axis?

8

20 mm above the base

By inspection, it can be seen that the plastic neutral axis lies along the top surface of the flange where the combined area of the two web sections equals the area of the flange section. If this were a more complex problem then the neutral axis position would have to be determined by geometric calculation. Now let's look at the plastic stress diagram for this section:

Can you calculate C and T and hence, by taking moments, calculate the fully plastic moment of the section? Take $\sigma_y = 275\ N/mm^2$.

9

$$\boxed{C = T = 1100 \text{ kN} \qquad M_\mathrm{p} = 66 \text{ kN m}}$$

Check your calculations:

$$C = T = A_1 \times \sigma_\mathrm{y} = (2 \times 100 \times 20) \times 275 \times 10^{-3} = 1100 \text{ kN}$$

and, noting that both C and T pass through the centre of each of the respective rectangular stress blocks, the fully plastic moment is obtained by taking moments about either the line of action of C or the line of action of T to give:

$$M_\mathrm{p} = C \times z = T \times z = 1100 \times (60 \times 10^{-3}) = 66 \text{ kN m}$$

PLASTIC MODULUS

If you refer back to Programme 6 you will recall that we used the *elastic section modulus* $(Z_\mathrm{e} = I/y_\mathrm{max})$ to calculate the moment of resistance of a beam section as:

$$M = Z_\mathrm{e} \times \sigma_\mathrm{max}$$

This formula was based on elastic theory and assumed elastic behaviour. In an analogous way the fully plastic moment, M_p can be expressed as:

$$M_\mathrm{p} = Z_\mathrm{p} \times \sigma_\mathrm{y}$$

where Z_p is the *plastic section modulus*. Whereas the elastic modulus is calculated from the geometry of the section, the plastic modulus can be calculated by re-arranging the above expression to give:

$$Z_\mathrm{p} = M_\mathrm{p}/\sigma_\mathrm{y}$$

Calculate the plastic section modulus of the steel channel section shown in Frame 8.

10

$$\boxed{Z_\mathrm{p} = M_\mathrm{p}/\sigma_\mathrm{y} = 66 \times 10^6/275 = 240 \times 10^3 \text{ mm}^3}$$

Note that the units of plastic modulus are identical to those of elastic modulus. Readily available tables of section properties give the value of plastic modulus for each of the wide range of standard steel sections that are used in structural steel design. These tabulated values of Z_p can be multiplied by the value of σ_y, appropriate to the grade of steel being used, to calculate the value of M_p for each standard section. Thus, in practice, it is not necessary to calculate Z_p from basic principles as in the previous example.

11

PROBLEMS

1. Determine the expression for the *plastic section modulus* of the solid rectangular section shown in figure Q1. Compare your answer to the standard formula for the *elastic section modulus* which we developed in Programme 6.

Ans. $(Z_p = bd^2/4; \; Z_e = bd^2/6)$

2. Calculate the plastic section modulus of the channel section shown in figure Q2.

Ans. $(560 \times 10^3 \; mm^3)$

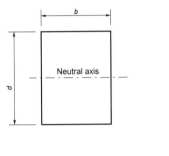

Q1   Q2

12

SHAPE FACTOR

In Frame 9 we showed that the fully plastic moment of a beam section is given by:

$$M_p = Z_p \times \sigma_y$$

and the corresponding equation for elastic behaviour is:

$$M = Z_e \times \sigma_{max}$$

This last equation applies within the elastic range including the point of first yield in the outer fibres of the section. Hence, the moment that causes yield to first occur can be expressed as:

$$M_y = Z_e \times \sigma_y$$

The *shape factor*, α, of a section is defined as:

$$\alpha = \frac{M_p}{M_y} = \frac{Z_p \times \sigma_y}{Z_e \times \sigma_y} = \frac{Z_p}{Z_e}$$

The shape factor of a section is a convenient way of relating the fully plastic moment of the section to the moment at first yield and indicates the reserve of moment capacity once yield has first developed.

Using the solution to Q1 above, what is the shape factor of a solid rectangular section?

13

$$\alpha = \frac{Z_p}{Z_e} = \frac{bd^2/4}{bd^2/6} = 1.5$$

This calculation demonstrates that the fully plastic moment of a solid rectangular section is 50% greater than the moment when yield first develops. For I section shapes, commonly used in structural steelwork, α is typically of the order of 1.15 indicating a 15% reserve of moment capacity beyond the point of first yield.

Now that we have seen how to determine the values of M_p, Z_p and α for typical structural sections, let's move on to see how to use this information in the analysis and design of steel structures. Let's start by examining the case of a built-in beam supporting a centrally placed point load W_w, as shown; the load factor at collapse being λ. The beam has a fully plastic moment of M_p throughout and we want to determine the value of λ at collapse.

Sketch the shape of the bending moment diagram and identify the likely positions where plastic hinges might form in the structure.

14

You should have drawn a bending moment diagram as indicated above with local maximum hogging moments at the supports A and C, together with a maximum sagging moment at the mid-span position B. A, B and C are therefore the locations where plastic hinges are likely to form as the beam is loaded towards collapse.

How many of these three possible hinges need to develop before the beam reaches a state of collapse?

15

We haven't actually identified the values of the local maximum bending moments but let's assume that the first hinge forms at the mid-span position, B. The formation of this single hinge would not give rise to a collapse mechanism as both AB and BC would then act as separate cantilevered beams effectively hinged together at B. It is only when additional plastic hinges form at A and C will a collapse mechanism develop. The figure below shows this collapse mechanism where it is assumed that the parts of the beam AB and BC deflect as separate rigid bodies and rotation takes place at each plastic hinge position.

The figure is exaggerated for clarity and the angle of rotation of A and, by symmetry, C is taken as a small angle θ. The rotation at the mid-span hinge is obtained by simple geometry. For *small angles* the deflection at mid-span is obtained from the geometry of the triangle ABB' as $\theta \times L/2$.

To calculate the relationship between the external load system and the internal moment system we are going to apply the principle of *virtual work* which we studied in Programme 14. This principle can be expressed as:

$$\sum W \times \delta = \sum M \times \theta$$

where W and M are the set of external forces and internal moments respectively which must be in equilibrium. δ and θ are *any* compatible set of virtual displacements which must be geometrically possible. If we take the displacements represented by the actual collapse mechanism of the structure as the set of virtual displacements then the above expression can be used to find the relationship between the external loads and internal moment system. The left-hand side of the equation represents the *external work* done by the load system which can be calculated by multiplying the external load(s) by their corresponding displacement(s) measured in the direction of the load(s). The right-hand side of the equation represents the *internal work* done by the plastic hinges which can be determined by multiplying the moment, M_p, at each of the plastic hinges by the angle through which that hinge rotates.

Can you write down the equations for external and internal work done?

16

$$EWD = \lambda W_{\mathrm{w}} \times \theta L/2$$
$$IWD = 4M_{\mathrm{p}} \times \theta$$

Check your calculations:

External work done $= \text{load} \times \text{displacement} = \lambda W_{\mathrm{w}} \times \theta L/2$

Internal work done $= \text{work done by hinge } (A + B + C)$

$$= (M_{\mathrm{p}} \times \theta) + (M_{\mathrm{p}} \times [\theta + \theta]) + (M_{\mathrm{p}} \times \theta) = 4M_{\mathrm{p}} \times \theta$$

Equate internal to external work and hence determine the expression for the load factor, λ.

17

$$\lambda = \frac{8M_{\mathrm{p}}}{W_{\mathrm{w}}L}$$

Note that the rotation term θ is eliminated from the equation and does not appear in the final answer. Now let's look at the same beam which is now carrying a uniformly distributed load w_{w} per unit length. Again, we wish to calculate the load factor λ, at collapse.

Sketch the shape of the bending moment diagram and identify the likely positions where plastic hinges might form in the structure.

18

Plastic hinge

Now sketch the collapse mechanism of the structure, calculate the internal work and external work done and hence determine the collapse load factor, λ.

19

$$\lambda = \frac{16M_{\mathrm{p}}}{w_{\mathrm{w}}L^2}$$

Did you obtain the solution as follows:

$\lambda\, w_{\mathrm{w}}$ /unit length

A θ $\theta L/2$ B' θ M_{p} C

$L/2$ $L/2$

Again, you should have used the general shape of the bending moment diagram to indicate where the hinges are most likely to develop, i.e. at the points of maximum moment. The collapse mechanism will be identical to that of the previous example with a plastic hinge at mid-span and at each support and the expression for internal work done will be, as before, given by:

Internal work done   = work done by hinge $(A + B + C)$

$$= (M_{\mathrm{p}} \times \theta) + (M_{\mathrm{p}} \times [\theta + \theta]) + (M_{\mathrm{p}} \times \theta) = 4M_{\mathrm{p}} \times \theta$$

However, in the case of the work done by the external load system you should have realised that it is necessary to consider the work done by the load on the span AB separate to that on BC which, by symmetry, will be the same. In the case of span AB, A doesn't deflect at all whilst B deflects by an amount $\theta L/2$. Hence the *average* deflection over AB, which is the deflection at the centroid of the load on this span, will be $\theta L/4$. Hence:

External work done   = work done by load on AB + work done by load on BC

$$= 2 \times \text{work done by load on AB}$$

$$= 2\left[\left(\lambda \times w_{\mathrm{w}} \times \frac{L}{2}\right) \times \theta\frac{L}{4}\right] = \lambda w_{\mathrm{w}} \times \frac{\theta L^2}{4}$$

Note that the load per unit length has been multiplied by the span length to give the total load. Therefore equating external to internal work:

$$\lambda w_{\mathrm{w}} \times \frac{\theta L^2}{4} = 4M_{\mathrm{p}}\theta$$

$$\therefore \quad \lambda = \frac{16M_{\mathrm{p}}}{w_{\mathrm{w}}L^2}$$

In deciding on the location of the plastic hinges it is worth noting that we sketched the *shape* of the bending moment diagram: we didn't attempt to calculate any values of moment on the diagram. This is because the bending moment diagram under *elastic conditions*, which you are familiar with, has the same shape but not the same proportions as the diagram under *plastic collapse conditions*.

If, in the last example, we had calculated the elastic moments under the working load, w_w, the support moments would be shown to be twice those at mid-span (figure (a)). Hence, as the load increases, plastic hinges will first develop at A and C when the moments at these positions reach the fully plastic moment, M_p (figure (b)). Once these hinges have formed the beam is able to rotate freely at the supports. At this stage the beam hasn't collapsed but any additional loading is now carried by the beam behaving effectively as if it were simply supported. The support moments remain fixed at M_p but the mid-span moment will increase until it reaches a value of M_p equal to the end-span moments. At this stage a third hinge forms and the structure collapses (figure (c)). This process is often referred to as *moment redistribution* but the essential fact is that the bending moment diagram at collapse has the same intrinsic shape as the elastic moment diagram but has different relative proportions.

Sketching the shape of the bending moment diagram, using your knowledge developed through consideration of elastic analysis, therefore helps in determining the location of likely hinge positions. It is, however, of no relevance to attempt to calculate the value of elastic moments when carrying out a plastic collapse analysis.

(a) B.M diagram – working loads   (b) B.M diagram – first hinge formation   (c) BM diagram – at collapse

Now let's consider an example of a propped cantilever supporting a uniformly distributed load:

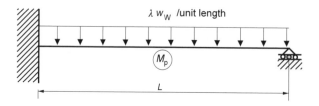

Sketch the shape of the bending moment diagram and identify the likely positions where plastic hinges might form in the structure.

20

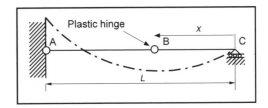

Did you realise that a plastic hinge will form at the support at A, where the hogging moment, is a maximum and somewhere along the span where the sagging moment is a maximum. Although we can sketch the shape of the diagram, we do not know the position of maximum sagging moment. Bearing in mind the discussion in Frame 19, we know that the position of maximum moment on the *plastic* bending moment diagram is not necessarily in the same location as that on the *elastic* bending moment diagram. Hence we can only say that the position of maximum moment is at some location, say distance x from the right-hand support and express our calculations in terms of this variable x.

Now sketch the collapse mechanism of the structure and calculate the expressions for internal work and external work done giving your answer in terms of x. Assume that the right-hand support, C, rotates through an angle, θ.

21

$$EWD = \lambda w_w L \times x\theta/2$$
$$IWD = M_p \left[\frac{x\theta}{(L-x)} + \left(\theta + \frac{x\theta}{(L-x)} \right) \right]$$

Check your calculations:

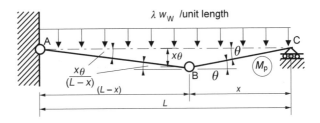

The geometry of the collapse mechanism is as shown with all rotations and displacements related to the rotation θ at the support, C. The expression for internal work done will be given by:

Internal work done $\quad\quad$ = work done by hinge $(A + B + C)$

$$= M_{\mathrm{p}}\left[\frac{x\theta}{(L - x)} + \left(\theta + \frac{x\theta}{(L - x)}\right) + 0\right]$$

Why is the work done at hinge C given as zero in the above expression?

> Because C is simply supported and is not a
> plastic hinge. Although C rotates no work is
> done as the moment at C is zero

Ensure that you remember this fact. If we consider the external work done, the *average* deflection for both spans AB and BC is $\frac{1}{2}x\theta$ which represents the deflection at the centroid of the load system. Hence:

External work done $= \left[(\lambda \times w_{\mathrm{w}} \times L) \times \dfrac{x\theta}{2}\right]$

Therefore if external work is equated to internal work:

$$\left[(\lambda \times w_{\mathrm{w}} \times L) \times \frac{x\theta}{2}\right] = M_{\mathrm{p}}\left[\frac{x\theta}{(L - x)} + \left(\theta + \frac{x\theta}{(L - x)}\right) + 0\right]$$

which re-arranges to give:

$$\lambda = \frac{2(L + x)M_{\mathrm{p}}}{w_{\mathrm{w}}Lx(L - x)}$$

This equation defines λ as a function of the variable distance x and λ will therefore vary depending on the value of x used in the equation. However, what we are looking for is the *least* possible value of λ representing the lowest possible collapse load of the structure. This can be obtained by finding the minimum value of the above function which is given when $\mathrm{d}\lambda/\mathrm{d}x$ is equal to zero, i.e. when:

$$\frac{\mathrm{d}\lambda}{\mathrm{d}x} = \frac{2M_{\mathrm{p}}}{w_{\mathrm{w}}L}\left[\frac{x(L - x) - (L + x)(L - 2x)}{x^2(L - x)^2}\right] = 0$$

or simply when:

$$x(L - x) - (L + x)(L - 2x) = 0$$

Solve this quadratic equation to determine x and back-substitute into the equation for λ *to determine the optimum solution for* λ.

23

$$x = 0.414L \qquad \lambda = \frac{11.66M_\text{p}}{w_\text{w}L^2}$$

This last example illustrates the procedure to be followed if the location of a hinge is not exactly defined from the shape of the bending moment diagram.

CONTINUOUS BEAMS

The principles that we have looked at so far can be also be applied to continuous multi-span beams. The essential difference is that a multi-span beam may potentially have more than one possible collapse mechanism and each must be investigated to determine which is the critical one. Consider the two-span beam shown below where we wish to determine the load factor, λ, at collapse.

Sketch the shape of the bending moment diagram and identify the likely positions where plastic hinges might form in the structure.

24

If you consider the likely position of the plastic hinges you will see that if hinges were to form at A, B and C then a collapse mechanism will form in span AB. Likewise if hinges form at C and D then a collapse mechanism will form in span CE. It is not possible to determine which possible mechanism is the most critical without examining each one in turn.

Calculate the load factor, λ, for each of the two possible mechanisms and state which of the two answers is most critical.

$$\boxed{1.46M_\text{p}; \; 0.47M_\text{p} \quad \text{Critical solution} = 0.47M_\text{p}}$$

Check your solution:

Mechanism 1:

This mechanism is identical to that of the built-in beam that we considered in our second example in Frame 17. We can therefore use the previous solution to obtain:

$$\lambda = \frac{16M_\text{p}}{w_\text{w}L^2} = \frac{16 \times M_\text{p}}{2 \times 2^2} = 2M_\text{p}$$

Mechanism 2:

Did you get the geometry of the collapse mechanism correct? You may have taken the rotation at the left-hand hinge as θ, in which case the deflection under the point load would be $1 \times \theta$ and the rotation at the right-hand support will be $\theta/3$. This would be equally correct and will give the same final answer.

External work done by the point load $= \lambda 5 \times 3\theta$

Internal work done by the two plastic hinges $= M_\text{p}[3\theta + (3\theta + \theta)]$

Equating external and internal work: $\lambda 5 \times 3\theta = M_\text{p}[3\theta + (3\theta + \theta)] \quad \therefore \; \lambda = 0.47M_\text{p}$

The *lowest* of the two calculated values of λ is $0.47M_\text{p}$ corresponding to the second collapse mechanism which is, hence, the most critical one.

26

1. A beam of length L carries a concentrated collapse load, λW_w, at a point $L/3$ from the left-hand end. If the beam has a fully plastic moment of M_p, determine the value of λ for (i) both ends simply supported (ii) the left-hand end built-in and the right-hand end simply supported and (iii) both ends built-in.

Ans. $(4.5M_p/W_wL, \ 7.5M_p/W_wL, \ 9.0M_p/W_wL)$

2. A simply supported beam spans 10 m and carries a working load of 100 kN at 2 m from the left-hand support and 100 kN at 1 m from the right-hand support. If the load factor at collapse is to be 1.75 what is the minimum value of M_p required? (*Hint: (i) There are two possible collapse mechanisms. When considering each one you should recognise that each external load will displace and do work and not just the load at the location where a hinge develops (ii) the minimum value of M_p will be the largest of the two solutions obtained in order to ensure collapse at the required load factor of 1.75.)*

Ans. *(315 kN m)*

3. The continuous steel I beam in figure Q3 supports the set of loads shown. Taking the yield stress of steel as 275 N/mm^2 and the load factor at collapse as 2.0, determine the minimum value of M_p and, hence, the required plastic modulus of the section. *(Hint: look back at hint (ii) in the last problem.)*

Ans. *(200 kN m, 0.73×10^6 mm^3)*

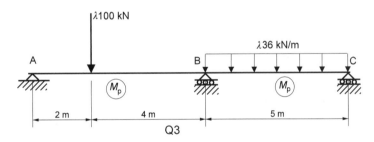

Q3

4. The continuous beam in figure Q3 is to be redesigned such that AB has a fully plastic moment, $2M_p$, and BC has a fully plastic moment of M_p. Determine the minimum value of M_p and the required plastic modulus of both AB and BC. *(Hint: When considering the hinge formation at B in which of the two joining members will the hinge form: the weaker or the stronger?)*

Ans. *(154.37 kN m, 1.12×10^6 mm^3, 0.56×10^6 mm^3)*

27

PORTAL FRAME STRUCTURES

Let's now apply the principles that we have developed to the analysis of portal frame structures such as shown below. This type of structure is more complex than the simple beam structures that we have considered so far and greater consideration needs to be given to the number and types of possible failure mechanisms.

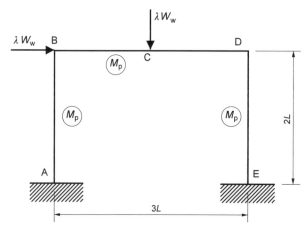

Can you sketch the bending moment diagram for the portal frame and indicate the locations where plastic hinges might form.

28

In thinking about the shape of the diagram did you consider the deflected shape of the structure and, hence, which side of each member is in tension? You should have drawn the bending moment diagram on the tension side of each member. The points of local maximum bending are where hinges *might* form.

Can you visualise what will happen to the structure if hinges form at the top and bottom of each column i.e. at A, B, D and E?

29

The frame will collapse in a 'side-sway' mode of failure

This is illustrated in the figure below which shows the displaced shape of the structure at collapse. The columns, AB and DE, and the beam, BD, displace as rigid bodies and rotation takes place at each plastic hinge position. Again, the diagram is exaggerated and the angles of rotation shown are, in reality, small angles. The angular rotations and the side-sway of the structure are related from the geometry of the structure.

Calculate the internal work and external work done and hence determine the expression for the collapse load factor, λ.

30

$$\lambda = \frac{2M_p}{W_w L}$$

Check your calculations:

Internal work done   = work done by hinge $(A + B + D + E)$

$$= (M_p \times \theta) + (M_p \times \theta) + (M_p \times \theta) + (M_p \times \theta) = 4M_p \times \theta$$

External work done   = work done by side-load at B +

                     work done by vertical load at C

$$= \lambda W_w \times 2L\theta + 0$$

Note that the vertical load at C moves horizontally but does not move vertically. It therefore does not contribute to the work equation. Therefore equating external to internal work:

$$\lambda W_w \times 2L\theta = 4M_p\theta$$

$$\therefore \quad \lambda = \frac{2M_p}{W_w L}$$

Does this represent the correct collapse load factor of the structure?

31

$$\boxed{\text{Not necessarily}}$$

It might be the correct solution but in a complex structure such as this there may be other more critical failure mechanisms that should be considered. The mechanism that gives the lowest value of the collapse load factor, λ, will be the correct one. Let's now consider the possibility of plastic hinges forming in the beam at B, C and D.

Draw the collapse mechanism; calculate the internal work and external work done and hence determine the expression for the collapse load factor, λ.

32

$$\boxed{\lambda = \frac{8M_\text{p}}{3W_\text{w}L}}$$

This is an identical problem to the built-in beam that we considered earlier on in this programme. Check your calculations:

Internal work done	$=$ work done by hinge $(B + C + D)$
	$= (M_\text{p} \times \theta) + (M_\text{p} \times 2\theta) + (M_\text{p} \times \theta) = 4M_\text{p} \times \theta$
External work done	$=$ work done by side-load at B $+$
	work done by vertical load at C

$$= 0 + \lambda W_\text{w} \left[\frac{3L}{2} \times \theta\right]$$

Note that in this case it is only the vertical load at C which contributes to the work equation. Therefore equating external to internal work:

$$\lambda W_\text{w} \left[\frac{3L}{2} \times \theta\right] = 4M_\text{p}\theta$$

$$\therefore \quad \lambda = \frac{8M_\text{p}}{3W_\text{w}L}$$

Which of the two values of λ that we have calculated so far is the most critical solution?

33

$$\lambda = \frac{2M_\mathrm{p}}{W_\mathrm{w}L}$$

This is the least of the two calculated values and, hence, is the most critical of the solutions so far. However, there is another possible mode of failure as indicated below, where hinges form at A, C, D and E. This is referred to as a *combined mechanism* as effectively it is a combination of the beam and side-sway mechanism with a 'cancelling out' of the joint rotations at B such that B remains a rigid, right-angled joint without the formation of a plastic hinge.

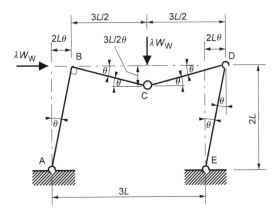

The geometry of this mechanism is a little more complex. If we start by taking the rotation of joint A as θ then the horizontal movement of B is $(2L \times \theta)$. As joint B remains rigid, it follows that the whole of the part of the structure ABC rotates through an angle θ. Hence the vertical deflection at C will be $3L/2 \times \theta$. From the deflection at C we can deduce that the rotation of CD is $(3L/2 \times \theta)/(3L/2) = \theta$. Also, for small deformations the horizontal movement of D must be the same as B and hence the rotation of DE is $(2L \times \theta)/2L = \theta$ These angular rotations and deformations are shown on the diagram of the collapse mechanism.

Calculate the internal work and external work done and hence determine the expression for the collapse load factor, λ.

34

$$\lambda = \frac{12M_\mathrm{p}}{7W_\mathrm{w}L}$$

Check your calculations:

Internal work done $=$ work done by hinge $(A + C + D + E)$

$$= (M_p \times \theta) + (M_p \times 2\theta) + (M_p \times 2\theta) + (M_p \times \theta) = 6M_p \times \theta$$

External work done $=$ work done by side-load at B $+$

work done by vertical load at C

$$= (\lambda W_w \times 2L\theta) + \left(\lambda W_w \times \frac{3L}{2}\theta\right) = \lambda W_w \times \frac{7L}{2}\theta$$

Equating external to internal work:

$$\lambda W_w \frac{7L}{2}\theta = 6M_p\theta$$

$$\lambda = \frac{12M_p}{7W_w L}$$

What do you now deduce is the correct answer for the collapse load factor λ?

35

$$\boxed{\lambda = \frac{12M_p}{7W_w L}}$$

This is the lowest of all the three calculated values of λ and indicates that this frame will collapse as a combined mechanism.

NUMBER OF HINGES FOR A MECHANISM

If you look at both the *side-sway* and *combined* mechanisms for the last problem you will see that they both required four hinges to develop for the complete structure to collapse. The structure is externally statically indeterminate and has three external redundancies (see Programme 2; Frames 30–31). This follows a general rule that for complete plastic collapse of a structure the number of hinges (N) required is one more than the degree of redundancy (R) of the structure. That is:

$$N = R + 1$$

This is a general rule that can be applied when attempting to determine the number and type of mechanisms that might develop in more complex structures. However, the rule only applies to *complete* collapse and, as we saw in the last example, the formation of three hinges ($< R + 1$) in the beam led to collapse of the beam or *partial* collapse of the structure. Hence, when attempting to evaluate the possible mechanisms, this simple rule can be used as a guide but the possibility of partial collapse mechanisms should not be overlooked.

36

PROBLEMS

1. Calculate the collapse load factor, λ, of the portal frame shown in figure Q1. Take $M_p = 100$ kN m. *(Hint: (i) when considering the hinge formation at the junction of the beam and either column in which of the two joining members will the hinge form: the weaker or the stronger? (ii) don't forget that the right-hand column has a pinned base.)*
Ans. (6.0)

2. Calculate the collapse load factor, λ, of the portal frame shown in figure Q2 if all the members have a fully plastic moment of 40 kN m. *(Hint: the rotation of the right-hand column will not be the same as the left-hand column – use the geometry of the failure mechanism to determine the relative values of the rotations.)*
Ans. (2.37)

37

PLASTIC COLLAPSE THEOREMS

The analysis techniques that we have considered so far require the identification and analysis of all possible collapse mechanisms in order to determine the critical solution. This is an acceptable approach when considering structures of modest size and where it is possible and not too arduous to carry out a comprehensive analysis. However, when structures are more complex such an approach may not be possible and, indeed, there is always the danger of overlooking what may be the most critical mechanism.

562

We will now consider an alternative approach to analysis which, whilst not necessarily giving the exact answer for the collapse load factor, λ, will allow us to define *upper* and *lower bound solutions* to λ. These upper and lower bound solutions will be answers which are greater than and less than the exact value of λ respectively. Providing they are sufficiently close together these answers can, for all practical purposes, give an acceptable approximation to the true solution which we will now refer to as λ_{true}.

The approach is based on three theorems which are themselves based on three necessary, and almost self-evident, conditions which must be satisfied by any structure at the point of plastic collapse. We are not going to prove these theorems but will see how to apply them.

The three necessary conditions:

- **Equilibrium condition:** the bending moments within the structure must be in equilibrium with the applied external load system.
- **Mechanism condition:** for a mechanism to develop sufficient plastic hinges must have formed to ensure either a *complete* or *partial* collapse mechanism.
- **Yield condition:** the bending moments in any section of a structure cannot exceed the fully plastic moment of that section.

The three plastic collapse theorems:

- **The uniqueness theorem:** if a set of bending moments can be found that satisfy the equilibrium condition *and* the mechanism and yield conditions then the calculated load factor must be the true collapse load factor for the structure ($\lambda = \lambda_{\text{true}}$).
- **The upper bound theorem:** if a set of bending moments can be found that satisfy the equilibrium condition *and* the mechanism condition but not necessarily the yield condition then the calculated load factor must be an upper bound to the true collapse load factor ($\lambda > \lambda_{\text{true}}$).
- **The lower bound theorem:** if a set of bending moments can be found that satisfy the equilibrium condition *and* the yield condition but not necessarily the mechanism condition then the calculated load factor must be a lower bound to the true collapse load factor ($\lambda < \lambda_{\text{true}}$).

It is true to say that *any* valid mechanism will give an upper bound solution to λ_{true}. If you refer back to the portal frame problem in Frame 27 you will see that we examined three valid mechanisms only one of which gave the correct (lowest) answer for λ_{true}. All the others gave answers which were greater than λ_{true} and, hence, can all be considered as *upper bound solutions*. Only one (the lowest) could be considered to be the *unique solution*. Move on to the next frame to see how to apply these theorems.

38

We will demonstrate the application of the theorems by considering the built-in beam shown below. The beam has an M_p value of 150 kN m and supports a combination of uniformly distributed and point loads.

How many hinges are necessary for a mechanism and where are they likely to occur?

39

Three: at A, C and the point of maximum sagging moment

We do not know the exact position of the point of maximum sagging moment but it would be reasonable to *assume* that it is directly underneath the point load. If we make this assumption and a hinge forms at this position then, together with hinges at the supports, we would have a valid mechanism. In other words we will have satisfied the *mechanism condition*. Writing down and equating the expressions for internal and external work will give the relationship between external loads and internal moments and is, effectively, a means of enforcing equilibrium onto the problem. In other words our application of virtual work methods will ensure that the *equilibrium condition* is satisfied. From the *upper bound theorem*, the fact that we have satisfied both the equilibrium and mechanism conditions will guarantee that the answer we obtain will automatically be an *upper bound* to the true collapse load factor, λ_{true}.

Now sketch the collapse mechanism of the structure, calculate the internal work and external work done and hence determine the upper-bound solution to the collapse load factor, λ_{true}.

40

$$\lambda = 2.25$$

Check your calculations on the page opposite:

Internal work done $= \text{work done by hinge } (A + B + C)$

$$= (M_\mathrm{p} \times \theta) + (M_\mathrm{p} \times 3\theta) + (M_\mathrm{p} \times 2\theta) = 6M_\mathrm{p} \times \theta$$

$$= 6 \times 150 \times \theta = 900\theta$$

External work done $= \text{work done by point-load at } B + \text{work done by UD load}$

$$= \lambda 25 \times 4\theta + \left((\lambda 25 \times 6) \times \frac{4\theta}{2} \right) = \lambda 400\theta$$

Equating external to internal work:

$$\lambda 400\theta = 900\theta$$

$$\therefore \quad \lambda = 2.25$$

This answer is automatically an upper-bound solution. Hence we can say that $\lambda_\text{true} \leq 2.25$. In fact, if we had located the hinge, B, at any other position the same process of calculation would give us an equally valid, although different, solution as *any* valid mechanism, upon which we impose equilibrium, will, according to the upper bound theorem, give an upper bound to λ_true.

This solution tells us that λ_true is less than 2.25 but doesn't tell us by how much the true answer is less than this value. Hence, to be of any use, we need to relate this upper-bound solution to a lower-bound solution so that λ_true can be sandwiched between finite and sensible limits.

Now let's examine the assumptions that we have made. We have assumed that the moment at B is equal to M_p but this tells us nothing about the moments at any other location along the span. If, indeed, B is the point of maximum moment, then we would have found the unique solution as, in addition to satisfying the equilibrium and mechanism conditions, we would have also satisfied the yield condition. The yield condition, you will recall, requires that nowhere can the moments exceed the fully plastic moment of the section. If we can show that the maximum sagging moment occurs at B and is, as assumed, equal to M_P then it is not possible that anywhere else along the span can the moment exceed M_p. Hence the yield condition will be satisfied.

What do you know about the value of the shearing force at any position of local maximum bending moment? How can this be used to identify the position of maximum moment?

41

> The shearing force has a value of zero

You will recall from Programme 4 that the bending moment is a local maximum where the shear force is zero. We can locate the position of zero shear and hence determine the value of maximum bending moment at this position. Let's assume that the point of maximum moment, if not at B, lies somewhere between A and B. The figure below shows the free-body diagram of AB. The moments at A and B are of value $M_p = 150$ kN m, as assumed in our original calculations.

Take moments about B to calculate the vertical reaction at A.

42

$$V_A = 187.5 \text{ kN m}$$

Check:

ΣM_B: $(V_A \times 4) - 150 - 150 - [(2.25 \times 25 \times 4) \times 2] = 0$

$$\therefore \quad V_A = 187.5 \text{ kN}$$

Now, assuming that the point of zero shear is at point X, x m from A, find this position and hence the corresponding value of maximum bending moment.

43

$$x = 3.33 \text{ m} \quad M_X = 162.5 \text{ kN m}$$

Check:

Looking to the left of X, the location of the point of zero shear is given by:

$$187.5 - (2.25 \times 25) \times x = 0$$

$$\therefore \quad x = 3.33 \text{ m}$$

And taking moments of all forces and moments to the left of X:

$$\Sigma M_X: \quad (187.5 \times 3.33) - 150 - (2.25 \times 25) \times \frac{3.33^2}{2} - M_X = 0$$

$$\therefore \quad M_X = 162.5 \text{ kN m}$$

Hence, we can see that the yield condition has been violated as we have found a moment which exceeds the fully plastic moment of the section (150 kN m). From the upper bound theorem we can say that the value of λ, calculated assuming a plastic hinge at B, can only be an upper bound and is not the unique answer, Our original assumption has been found to be invalid but nevertheless still gives a valid upper-bound solution.

Our original calculations gave us the upper-bound value of λ as 2.25, based on an assumed bending moment distribution with values of 150 kN m at A, B and C resulting in plastic hinges at these positions and a possible, valid mechanism. We have now calculated that this assumed distribution has a maximum moment at X of 162.5 kN m. The ratio of M_p to M_x is given by 150/162.5 = 0.92. *If* we now multiply *all* the loads and *all* the bending moments by this factor of 0.92 we will:

• Maintain equilibrium, as loads and moments are reduced in the same proportion (*equilibrium condition* still satisfied).
• Reduce the maximum bending moment at X to a value of 150 kN m and to less than this figure everywhere else on the bending moment diagram (*yield condition* now satisfied).
• Have only one plastic hinge at X and hence we will not have a mechanism (*mechanism condition* not *satisfied*).

The lower bound theorem thus gives a lower-bound value of λ as $\lambda = 2.25 \times 0.92 = 2.07$. Combining the upper-bound and lower-bound solutions together we can now say:

$$2.07 < \lambda_{\text{true}} < 2.25$$

Although we haven't calculated the true value of the collapse load factor or the correct position of the internal hinge the answers we have obtained are, however, sufficiently close together to say, with confidence, that the collapse load factor of this structure is approximately 2.

If the solution had not yielded a pair of results that were sufficiently close together to be of value how could we obtain better estimates of the upper- and lower-bound solutions?

44

Repeat the calculations with a hinge at A, C and at X

The example that we have considered was a fairly trivial illustration of the technique. Let's now look at a slightly more complex example where we will show that we can obtain upper- and lower-bound solutions even though we may not be sure that we have even identified the correct form of the collapse mechanism.

Figure (a) shows a simple steel structure subject to vertical and horizontal loading. Both members AB and BD have M_p values of 200 kN m. Figure (b) shows the general shape of the bending moment diagram, from which it can be seen that plastic hinges may form at A, B and at the point of maximum sagging moment between B and D. The structure is externally statically indeterminate with a single redundancy ($R = 1$). Hence the number of hinges (N) needed for a full collapse mechanism is ($R + 1$) = 2.

There are a number of possible mechanisms that could be considered but we are going to consider the 'combined' mechanism shown in figure (c). We do not know the correct position of the hinge between B and D but we will *assume* it to be at the mid-span point of the beam. By virtue of this being a possible valid mechanism the solution will automatically give us an upper-bound solution to λ_{true}, from which we will be able to deduce a lower-bound solution. It should be recognised that, although the mechanism may be valid it may not be the correct mechanism: this doesn't matter as the upper-bound theorem only requires us to consider any valid possible mechanism.

In figure (c) the hinge at A is assumed to rotate through a small angle θ and the geometry of the rest of the structure is related to this angle.

(a)   (b)   (c)

Study the geometry in figure (c). Write down the internal and external work equations from which you should then calculate the upper bound to λ_{true}

$$\boxed{\lambda_{\text{true}} \leq 2.4}$$

Check your calculations:

Internal work done $= $ work done by hinge $(A + C + D)$
$$= (M_p \times \theta) + (M_p \times 2\theta) + 0 = 3M_p \times \theta = 3 \times 200 \times \theta = 600\theta$$

External work done $=$ work done by point-load at B $+$ work done by UD load

$$= \lambda 25 \times 6\theta + \left((\lambda 25 \times 4) \times \frac{2\theta}{2} \right) = \lambda 250\theta$$

Equating external to internal work:

$$\lambda 250\theta = 600\theta$$
$$\therefore \quad \lambda = 2.4$$

As we have satisfied the *mechanism condition* and *the equilibrium condition* the upper bound theorem tells us that this solution is an upper bound to λ_{true}.

We now need to examine the validity of this solution to see if the position of the hinge that we chose to be at C is, in fact, in the correct location of maximum sagging bending moment. If we can show that this is the case then our solution, from the uniqueness theorem, will be the exact solution; if not then it will simply be an upper-bound solution. As in the last example, we will determine the position of maximum moment as the point of zero shearing force.

Sketch the free-body diagram of CD and use this diagram to calculate the vertical reaction at D. Hence determine the position of zero shearing force, measured from D and the value of the maximum sagging moment at this point.

46

(a)

$$\boxed{V_D = 160.0 \text{ kN} \quad x = 2.67 \text{ m from D} \quad M_X = 213.33 \text{ kN m}}$$

Check your calculations overleaf:

47

(a)   (b)

From figure (a), which shows the free-body diagram of CD:

$$\Sigma M_C: \quad -(V_D \times 2) + 200 + [(2.4 \times 25 \times 2 \times 1] = 0$$
$$\therefore \quad V_D = 160.0 \text{ kN}$$

From figure (b): assuming that the point of zero shear is at point X, x m from D. Looking to the right of X, the location of the point of zero shear is given by:

$$160.0 - (2.4 \times 25) \times x = 0$$
$$\therefore \quad x = 2.67 \text{ m}$$

And taking moments of all forces to the right of X:

$$\Sigma M_X: \quad -(160.0 \times 2.67) + (2.4 \times 25) \times \frac{2.67^2}{2} + M_X = 0$$
$$\therefore \quad M_X = 213.33 \text{ kN m}$$

Again, we can see that the yield condition has been violated as we have found a moment which exceeds the fully plastic moment of the section (200 kN m). From the upper bound theorem we can say that the value of λ, calculated assuming a plastic hinge at C, can only be an upper bound and is not the unique answer. Our original assumption has been found to be invalid but nevertheless still gives a valid upper-bound solution.

Can you now calculate a lower-bound solution to λ_{true}?

48

$$\boxed{2.25 \leq \lambda_{\text{true}}}$$

The ratio of M_P to M_X is given by $200/213.33 = 0.94$. If we now multiply *all* the loads and *all* the bending moments by this factor of 0.94 we will (i) maintain equilibrium, (ii) reduce the maximum bending moment at X to a value of 200 kN m and to less than this figure everywhere else on the bending moment diagram (*yield condition* satisfied) and (iii) have only one plastic hinge at X and hence will not have a mechanism (*mechanism condition* not satisfied) The lower bound theorem thus gives a lower-bound value of λ as $\lambda = \overline{2.4} \times 0.94 = 2.25$. Combining the upper-bound and lower-bound solutions together we can now say:

$$2.25 < \lambda_{\text{true}} < 2.40$$

This example demonstrates the application of the plastic collapse theorems to the analysis of steel structures. The theorems can also be applied to the design of such structures. The lower bound theorem, whilst placing a lower bound on λ for a given value of M_p, effectively establishes an upper bound on M_p for a given value of λ (e.g. if $\lambda \geq 2M_p$ then $M_p \leq 0.5\lambda$). Any upper bound on M_p would be a safe value to use in the design. Hence, the designer can establish *any* bending moment distribution which is in equilibrium with the collapse loads (equilibrium condition satisfied) and can then choose a value of M_p which is equal to, or greater than, the maximum moment on that bending moment diagram which will ensure the yield condition is satisfied and that a mechanism will not develop. This satisfies the requirements of the lower-bound theorem thus ensuring a safe, although not necessarily economical, design. The application of the theorems in this way is outside the scope of this book and you should consult more advanced texts on this topic if study to greater depth is required.

You should also consult such texts to look in more detail at the assumptions that we have implicitly made in the application of plastic theory. For example, large axial loads in the columns of steel structures can significantly reduce the fully plastic moment of the column section and we have not accounted for this in the simplified analytical techniques that we have considered.

49

TO REMEMBER

The shape of the plastic stress distribution diagram and how it differs from the elastic stress distribution diagram.

The neutral axis under plastic conditions is the axis of equal area.

The ultimate moment of resistance of a steel section under plastic conditions is known as its *fully plastic moment*, M_p.

The plastic section modulus of a steel section $Z_p = M_p/\sigma_y$.

The plastic section modulus of a solid rectangular section $= bd^2/4$.

The shape factor of a section is given by $\alpha = \dfrac{Z_p}{Z_e}$

The number of hinges (N) required for a full collapse mechanism $= R + 1$.

The statement of each of the three plastic collapse theorems.

50

1. Determine the plastic section modulus and the fully plastic moment of the section shown in figure Q1. Take the yield stress of the steel as 275 N/mm^2.

Ans. (2500 mm^3, 0.69 kN m)

40 mm

5 mm

20 mm

10 mm

Q1

2. The propped cantilever in figure Q2 has the cross-section shown in figure Q1. If the load can be applied anywhere along the span, determine the minimum value of W_c that will cause collapse and the distance, x, where the load is applied.

Ans. (2.0 kN, 0.83 m)

3. The steel section shown in figure Q1 is used to construct the propped cantilever shown in figure Q3. Determine the exact position of the plastic hinge that will form in the span and hence the collapse load factor, λ, of the beam.

Ans. (0.65 m, 2.21)

$x = ?$ W_c

2 m

Q2

$\lambda 1.5 \text{ kN/m}$

1 m   1 m

Q3

4. For the beam considered in question 3 assume that a plastic hinge forms at mid-span and use the plastic collapse theorems to determine upper- and lower-bound solutions to the collapse load factor, λ.

Ans. ($2.07 < \lambda_{\text{true}} < 2.76$)

5. Figure Q5 shows a continuous beam carrying a combination of distributed and point loads. It varies in section with fully plastic moments of each span as shown in the figure. Determine the collapse load factor, λ, of this structure.

Ans. (2.67)

6 m   4 m   4 m   2 m   2 m   2 m

M_p= 200 kNm M_p= 150 kNm M_p= 100kNm

Q5

6. The portal frame shown in figure Q6 is fixed at A and pinned at E. The fully plastic moment of AB and BD is 50 kN m and that of DE is 75 kN m. Determine the value of λ at collapse.

Ans. (2.08)

7. The portal frame shown in figure Q7 is subject to a side wind load. The fully plastic moment of the beam and stanchions are both 50 kN m. By considering all possible mechanisms calculate the load factor, λ, at collapse. *(Hint: sketch the bending moment diagram to identify possible hinge positions. There are three mechanisms; one of which will have a hinge at an unknown position up the height of the left-hand column: the optimum position of this hinge must be determined.)*

Ans. (4.67)

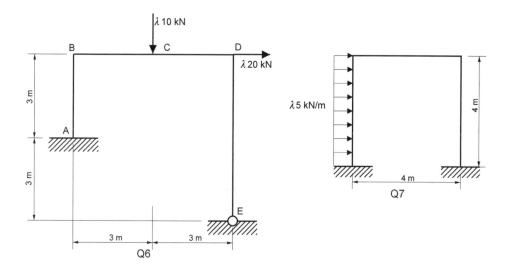

8. For the frame shown in figure Q7, assume a 'side-sway' collapse mechanism with hinges at the top and bottom of each column. Use the plastic collapse theorems to determine an upper-bound solution and, hence, a lower-bound solution to λ.

Ans. ($4 < \lambda_{true} < 5$)

9. Figure Q9 shows the structure that we considered in Frame 44. Both members AB and BD have M_p values of 200 kN m. Consider a possible 'side-sway' mechanism with plastic hinges forming at the top and bottom of the column and determine an upper bound to the load factor λ. Hence calculate a lower-bound solution to λ.

Ans. (2.12 < λ_{true} < 2.67)

10. All members of the frame shown in figure Q10 have a fully plastic moment of 150 kN m. Assume a 'combined' mechanism with a hinge at the mid-span of the beam and determine an upper bound to the load factor λ. Hence calculate a lower-bound solution to λ.

Ans. (3.20 < λ_{true} < 3.75)

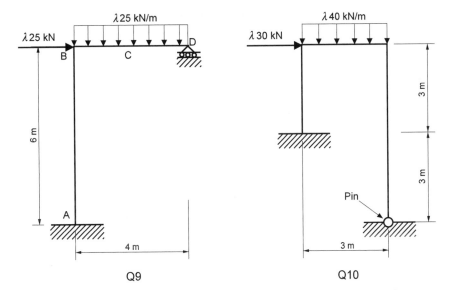

Q9   Q10

Appendix

SOME USEFUL
MATHEMATICAL
REVISION NOTES

You may find the following a useful summary of some of the key mathematical principles that we have used in this book:

(i) Geometry
In a right-angled triangle:

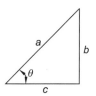

$$\sin\theta = \frac{\text{opposite}}{\text{hypotenuse}} = \frac{b}{a} \quad \cos\theta = \frac{\text{adjacent}}{\text{hypotenuse}} = \frac{c}{a} \quad \tan\theta = \frac{\text{opposite}}{\text{adjacent}} = \frac{b}{c}$$

And from Pythagoras' theorem:
$$a^2 = b^2 + c^2 \quad \text{or} \quad a = \sqrt{b^2 + c^2}$$

Area of a triangle: $A = \frac{1}{2} \times \text{base} \times \text{height} = \frac{1}{2} \times c \times b$

Area of a circle: $A = \pi r^2 = \pi \frac{D^2}{4}$ where r=radius; D=diameter of circle

(ii) Arithmetic
When operating on arithmetic expressions you should do so with the following precedence:

(a) terms in brackets   followed by
(b) powers and roots   followed by
(c) multiplication and division   and finally
(d) addition and subtraction

(iii) Algebraic identities

$$(a+b)^2 = a^2 + b^2 + 2ab$$
$$(a-b)^2 = a^2 + b^2 - 2ab$$
$$a^2 - b^2 = (a-b)(a+b)$$

(iv) Indices

Indices are a convenient mathematical shorthand used to express a repetitive multiplication in a simple way: e.g $y \times y \times y \times y \times y = y^5$, the index in this case being 5. When manipulating expressions written in index form the following rules apply:

$$y^a \times y^b = y^{(a+b)}$$
$$\frac{y^a}{y^b} = y^{(a-b)}$$
$$(y^a)^b = y^{(a \times b)}$$
$$y^0 = 1$$

(v) 10 in index form

When multiples of 10 are written in index form it is preferable to express them in *standard form* as multiples of 3, e.g.

$$10^3 \qquad 10^6 \qquad 10^9 \qquad 10^{-3} \qquad 10^{-6} \qquad 10^{-9}$$

(vi) Simultaneous linear equations

Simultaneous linear equations, relating two or more variables, are of the form:

$$3x + 2y = 19 \qquad (1)$$
$$2x + 3y = 16 \qquad (2)$$

To solve such equations the coefficient of one of the variables must be reduced to the same value in both equations. If we consider the variable x then equation (1) is multiplied by 2 (the coefficient of x in equation (2)) and the equation (2) is multiplied by 3 (the coefficient of x in equation (1)) to give:

$$6x + 4y = 38 \qquad (3)$$
$$6x + 9y = 48 \qquad (4)$$

Hence equation (3) may now be subtracted from equation (4) to give $5y = 10$; $\therefore\ y = 2$ and substituting back into equation (1) $3x + (2 \times 2) = 19$; $\therefore\ x = 5$. Other simultaneous equations will be solved in a similar way.

(vii) Solution of quadratic equations

The roots of quadratic equations of the form $ax^2 + bx + c = 0$ can be determined from the expression:

$$x = \frac{-b \pm \sqrt{b^2 - 4ac}}{2a}$$

or, if possible, by factorising the equation. For example, the equation $x^2 + 5x + 6 = 0$ can be factorised to give $(x+3)(x+2) = 0$ from which we can say that $x = -3$ or $x = -2$.

(viii) Trigonometric relationships

$$\sin^2 \theta + \cos^2 \theta = 1 \qquad \sin 2\theta = 2 \sin \theta \cos \theta$$

$$\cos 2\theta = \cos^2 \theta - \sin^2 \theta = 1 - 2\sin^2 \theta = 2\cos^2 \theta - 1$$

(ix) Standard differential equations

$y =$	1	x	x^2	x^3	x^4	x^n
$\frac{dy}{dx}$	0	1	$2x$	$3x^2$	$4x^3$	nx^{n-1}

$y =$	$\sin x$	$\cos x$	$\tan x$
$\frac{dy}{dx}$	$\cos x$	$-\sin x$	$\sec^2 x$

Differentiation of a quotient of two functions of the form $y = \frac{u}{v}$ is given by:

$$\frac{dy}{dx} = \frac{v\frac{du}{dx} - u\frac{dv}{dx}}{v^2}$$

(viii) Standard integral expressions

$y =$	1	x	x^2	x^3	x^4	x^n
$\int y\,dx$	$x + C$	$\frac{x^2}{2} + C$	$\frac{x^3}{3} + C$	$\frac{x^4}{4} + C$	$\frac{x^5}{5} + C$	$\frac{x^{n+1}}{n+1} + C$

$y =$	$\sin x$	$\cos x$
$\int y\,dx$	$-\cos x + C$	$\sin x + C$

UNITS

Units are important! You may find the following a useful summary of the units used in this book:

	Usual units	Alternative units	
Area (A)	m^2	mm^2	$1\ m^2 = 10^6\ mm^2$
Density (ρ)	kg/m^3		
Elastic modulus (E)	kN/mm^2	N/mm^2	$1\ kN/mm^2 = 10^3\ N/mm^2$
Energy (U)	joule		$1\ joule = 1\ N\ m$
Force (P, F)	N	kN	$1\ kN = 10^3\ N$
Length (L)	m	mm	$1\ m = 10^3\ mm$
Mass	kg		
Moment (M)	kNm	Nmm	$1\ kN\ m = 10^6\ N\ mm$
Polar second moment of area (J)	mm^4	cm^4	$1\ cm^4 = 10^4\ mm^4$
Power	watts		$1\ watt = 1\ Joule/second$
Second moment of area (I)	mm^4	cm^4	$1\ cm^4 = 10^4\ mm^4$
Section Modulus (Z) (elastic and plastic)	mm^3	cm^3	$1\ cm^3 = 10^3\ mm^3$
Shear modulus (G)	kN/mm^2	N/mm^2	$1\ kN/mm^2 = 10^3\ N/mm^2$
Stiffness	kN/m	N/mm	$1\ kN/m = 1\ N/mm$
Stress (σ, τ)	N/mm^2	kN/m^2	$1\ N/mm^2 = 10^3\ kN/m^2$
Torsional moment or Torque (T)	kNm	Nmm	$1\ kN\ m = 10^6\ N\ mm$
Volume (V)	m^3	mm^3	$1\ m^3 = 10^9\ mm^3$

INDEX